# Geographic Citizen Science Design

# Geographic Citizen Science Design

*No one left behind*

Edited by Artemis Skarlatidou
and Muki Haklay

First published in 2021 by
UCL Press
University College London
Gower Street
London WC1E 6BT

Available to download free: www.uclpress.co.uk

Collection © Editors, 2021
Text © Contributors, 2021
Images © Contributors and copyright holders named in captions, 2021

The authors have asserted their rights under the Copyright, Designs and Patents Act 1988 to be identified as the authors of this work.

A CIP catalogue record for this book is available from The British Library.

This book is published under a Creative Commons 4.0 International licence (CC BY 4.0). This licence allows you to share, copy, distribute and transmit the work; to adapt the work and to make commercial use of the work providing attribution is made to the authors (but not in any way that suggests that they endorse you or your use of the work). Attribution should include the following information:

Skarlatidou, A. and Haklay, M. (eds.). 2021. *Geographic Citizen Science Design: No one left behind*. London: UCL Press. https://doi.org/10.14324/111.9781787356122

Further details about Creative Commons licences are available at
http://creativecommons.org/licenses/

Any third-party material in this book is published under the book's Creative Commons licence unless indicated otherwise in the credit line to the material. If you would like to reuse any third-party material not covered by the book's Creative Commons licence, you will need to obtain permission directly from the copyright holder.

ISBN: 978-1-78735-614-6 (Hbk.)
ISBN: 978-1-78735-613-9 (Pbk.)
ISBN: 978-1-78735-612-2 (PDF)
ISBN: 978-1-78735-615-3 (epub)
ISBN: 978-1-78735-616-0 (mobi)
DOI: https://doi.org/10.14324/111.9781787356122

Printed and bound by CPI Group (UK) Ltd, Croydon, CR0 4YY

## Dedicated to the memory of Gill Conquest

This book is a marker in time, cataloguing the combined efforts of many people in seeking to make the tools of citizen science available to anybody who might wish to use them. One of the pioneers in this process was Gill Conquest, a polymath who beamed with intelligence and friendliness in equal measure, who brought laughter and joy to those lucky enough to spend time with her, but who tragically left us far too soon.

Arriving as a master's student in anthropology in 2011, Gill immediately applied her considerable talent to examine the potential of new technologies to support environmental justice movements led by indigenous peoples. The success of this research led to her recruitment into the Extreme Citizen Science research group (ExCiteS) for a PhD in 2013 supervised by Jerome Lewis and Haidy Geismar. Characteristically, her research project crossed many disciplinary and international boundaries as she undertook fieldwork with groups of indigenous peoples in the Congo Basin and the computer scientists/anthropologists working with them to develop mobile applications to address pressing issues that they identified. Working in Congo Brazzaville, Central African Republic and the Democratic Republic of Congo, she examined how different ways for expressing environmental knowledge by disparate groups such as Pygmy hunter-gatherers, forest farmers, commercial loggers and international conservation non-governmental organisations (NGOs) could be organised so as to interact more equally to reduce discrimination and biases in representation. Her fieldwork, and the anthropological perspectives she was developing, were groundbreaking, interrogating the idea of a pluriverse and how facilitating and supporting it might translate in anthropological practice and as digital technologies and tools. She contributed to the development of new ways of presenting this knowledge side by side so that more just and environmentally sound management decisions are made concerning the exploitation of forest people's land and resources.

Gill spent four years dedicated to supporting many local communities' struggles for environmental justice across the Congo Basin. While this practical work was vital to her methodology, her theoretical work focused on how to articulate the process of 'futuring' – of turning dreams, or myths as she considered them, into reality. By understanding and demystifying the processes involved in invention and innovation, she articulated the vital productive importance of mess and challenge in achieving the dream. She emphasised that the dream is a myth because we articulate it as success. Her words are particularly apt at the outset of a book that seeks to share the outcomes of our work:

Success stories are a powerful trope of international development and conservation work. They structure the way NGOs, governments and companies engage with powerful donors and public opinion. They also smooth over complexities, efface failures, and ignore contradictions. They present ongoing situations as if they are done, dusted, and thus can legitimately stand as a lesson to others who seek to achieve similar goals. They clean up mess. But conservation and development work is all mess.

As my ethnographic work proceeds here in Congo, it's hard to keep a lid on all the mess I encounter. As I sit in Brazzaville trying to make yet another workplan for the coming months that I know is unlikely to stick, I fear the mess may overflow and overwhelm me. It's been there from the start, from the very first day I started working on Extreme Citizen Science, but it's here – in the field, at the point of implementation – where mess makes itself most apparent. Success stories are powerful (particularly those that *begin* with failure) and certainly I've told my fair share of them. But I'm beginning to understand that without accounting for mess it's impossible to make sense of what is actually happening in the complex, multi-layered, power-laden realities of development and conservation work, or the global industrial processes this work simultaneously denigrates and supports. If we are to be faithful to mess, and we *urgently* need to be faithful to mess, then we need to unlearn old narrative tropes and engage with something quite different. We need to learn how to tell, and we need to learn how to listen to, unsuccess stories.

This is a tall order in an industry dominated by one principle agenda – securing the next round of project funding. It's a tall order in a world obsessed with the inspirational force of TED Talks, Disney movies and motivational posters. It's a tall order when the storyteller risks upsetting or offending friends and colleagues, and the listeners risk treating unsuccess as *failure*, which is quite a different beast. Unsuccess is not intended to be judgemental. It's just a story of what is, on the ground, sur le terrain. A story of mess.

Extreme citizen science is a mess as well, and we've also told our share of success stories; we've kept with the trends of the circles in which we move. But at the end of the day I'm a social anthropologist and I'm starting to realise that it's just as important – no, scrap that – I'm starting to realise that it's *more* important to tell our unsuccess stories. They don't (yet) have endings, and they probably won't make you feel all that warm and fuzzy inside. They're not what you'll get from TED Talks or from BBC documentaries or from Disney. They're just what is, on the ground, sur le terrain. They're just mess.

Gill would have contributed a chapter on unsuccesses, and it would certainly have been the most informative. But tragically, during fieldwork in the Central African Republic in 2016, Gill fell ill and had to be repatriated. Diagnosed with late-stage cancer, Gill approached her illness with dignity, courage and positivity, bringing out the best in the community of friends and family that surrounded her until the end. Gill passed away on 5 May 2017.

To work with Gill was a pleasure shared by all lucky enough to do so. She was so attentive to what needed doing and proactively doing it, whether collecting firewood, providing support with a thoughtful comment or simply cheering everyone up with a loud joke and big smile. She was the emotional and intellectual glue of the ExCiteS team, ensuring that despite our different disciplinary biases, we still managed to understand each other. Gill is terribly missed.

This book honours her memory. With this dedication, we acknowledge the extraordinary contribution that she made to the development of what is now called extreme citizen science, and also the huge importance of all the mess that we have collectively struggled with to arrive at some of the outcomes you will read about in what follows.

Jerome Lewis and Haidy Geismar
UCL Anthropology

# Contents

*List of figures* xiii
*List of tables* xxi
*List of contributors* xxiii
*Foreword by Jennifer Preece* xxix
*Acknowledgements* xxxiii

**Introduction**

    Geographic citizen science design: no one left behind    3
    *Artemis Skarlatidou and Muki Haklay*

**PART I    Theoretical and methodological principles**

1  Geographic citizen science: an overview    15
   *Muki Haklay*

2  Design and development of geographic citizen science: technological perspectives and considerations    38
   *Vyron Antoniou and Chryssy Potsiou*

3  Design approaches and human–computer interaction methods to support user involvement in citizen science    55
   *Artemis Skarlatidou and Carol Iglesias Otero*

4  Methods in anthropology to support the design and implementation of geographic citizen science    87
   *Raffaella Fryer-Moreira and Jerome Lewis*

## PART II  Interacting with geographic citizen science in the Global North

5  Geographic expertise and citizen science: planning and co-design implications     107
   *Robert Feick and Colin Robertson*

6  Citizen science mobile apps for soundscape research and public spaces studies: lessons from the Hush City project     130
   *Antonella Radicchi*

7  Using mixed methods to enhance user experience: developing Global Forest Watch     149
   *Jamie Gibson*

8  Path of least resistance: using geo-games and crowdsourced data to map cycling frictions     165
   *Diego Pajarito Grajales, Suzanne Maas, Mara Attard and Michael Gould*

9  Geographic citizen science in citizen–government communication and collaboration: lessons learned from the ImproveMyCity application     186
   *Ioannis Tsampoulatidis, Spiros Nikolopoulos, Ioannis Kompatsiaris and Nicos Komninos*

## PART III  Geographic citizen science with indigenous communities

10  Developing a referrals management tool with First Nations in northern Canada: an iterative programming approach     209
    *Jon Corbett and Aaron Derrickson*

11  Lessons from recording Traditional Ecological Knowledge in the Congo Basin     228
    *Michalis Vitos*

12  Co-designing extreme citizen science projects in Cameroon: biodiversity conservation led by local values and indigenous knowledge     247
    *Simon Hoyte*

13  Community monitoring of illegal logging and forest resources using smartphones and the Prey Lang application in Cambodia     266
    *Ida Theilade, Søren Brofeldt, Nerea Turreira-García and Dimitris Argyriou*

14  Representing a fish for fishers: geographic citizen science
    in the Pantanal wetland, Brazil                              282
    *Rafael Morais Chiaravalloti*

15  Digital technology in the jungle: a case study from the
    Brazilian Amazon                                             302
    *Carolina Comandulli*

16  Community mapping as a means and an end: how mapping
    helped Peruvian students to explore gender equality          317
    *Peter Ward and Rebecca Firth*

## Synthesis and epilogue

Geographic citizen science design: no one left behind –
an overview and synthesis of methodological, technological
and interaction design recommendations                           339
*Artemis Skarlatidou and Muki Haklay*

*Index*                                                          355

# List of figures

1.1 Different levels of engagement in citizen science (after Haklay 2013). Reprinted by permission from Springer: 'Citizen science and volunteered geographic information: Overview and typology of participation' by Muki Haklay, 2013. 22
1.2 The Five Cs model of participation (after Cooper et al. 2007). 23
1.3 Conceptual overlap between volunteered geographic information and citizen science – the boundaries of geographic citizen science. DIY: do-it-yourself. Source: author. 30
3.1 Overview of design approaches and methods used to engage users in the design, development and evaluation of citizen science applications. This conceptual model is rather generalised, and it captures methods as they are currently being implemented in similar initiatives. Methods can be modified and designed to address different aims and purposes, and therefore they can be used in different contexts supported by all or few of the proposed design approaches. Source: authors. 57
3.2 Persona example used for an evaluation of the Gender and Tech Magazines citizen science application (created with Xtensio.com). Source: authors. 77
4.1 Example of decision tree flow diagram. Source: Lewis, J. 2012. Technological leap-frogging in the Congo basin, pygmies and global positioning systems in central Africa: What has happened and where is it going?, *African Study Monographs*, Suppl. 43: 15–44. 95

| | | |
|---|---|---|
| 5.1 | Expertise Space Diagram (ESD; after Collins 2013). Reprinted by permission from Springer Nature: 'Three dimensions of expertise' by Harry Collins 2013. | 112 |
| 5.2 | Geographic ESD (GESD). Source: author. | 114 |
| 5.3 | RinkWatch: rink conditions data entry form. Source: RinkWatch.org. | 118 |
| 5.4 | RinkWatch: visualisation of rink 'skateability' across North America. Source: RinkWatch.org. Basemap © Esri. | 119 |
| 5.5 | GrassLander: sample farm and field boundaries, and bird observations. Source: Grasslander.org. Basemap © Esri. | 120 |
| 5.6 | GrassLander: adding a bird observation (desktop computer view). Source: Grasslander.org. Basemap © Esri. | 121 |
| 5.7 | Wildlife Health Tracker. Source: Wildlifehealthtracker.com. Basemap © OpenStreetMap. | 122 |
| 5.8 | Dimensions of expertise for reviewed geographical citizen science projects (after Collins 2013). | 125 |
| 6.1 | Interface of the Hush City mobile app. © Antonella Radicchi 2017. Basemap © Google Maps. | 137 |
| 6.2 | Concept of the Hush City mobile app data-collection process. © Antonella Radicchi 2019. Basemap © Google Maps. | 137 |
| 6.3 | Hush City Map as displayed on the web-based version of the Hush City app. © Antonella Radicchi 2020. Basemap © Google Maps. | 138 |
| 7.1 | Screenshot of the Global Forest Watch map, showing protected areas in Brazil in relation to forest cover and tree-cover loss and gain. Basemap © Mapbox, © OpenStreetMap contributors. Source: globalforestwatch.org. | 155 |
| 8.1 | Maps of the distribution of bicycle trips recorded by volunteers using the Cyclist Geo-C app. Basemap © OpenStreetMap contributors. Bicycle trip data: research data from Geotec research group – Universitat Jaume I. Approximated scale 1:50,000. | 173 |
| 8.2 | Examples of frictions and their cartographic representation. Source: Cyclist Geo-C app. Basemap © OpenStreetMap. Orthophoto: Castelló from Infraestructura Valenciana de datos espaciales (IDEV), Münster from Geoportal Münsterland, Valletta pictures by Diego Pajarito Grajales. | |

|  | Bicycle trip data: research data from Geotec research group–Universitat Jaume I. Approximated scale 1:10,000. | 174 |
| --- | --- | --- |
| 8.3 | Cyclist Geo-C app interface. Left: participant's trips as a percentage of the city's total contributions. Right: participant's position on city leader board based on the number of trips. Source: Cyclist Geo-C app. | 176 |
| 8.4 | Maps representing the iterative analysis from bicycle trips recorded using the app to the overlay grid and identified frictions. Basemap © OpenStreetMap. Bicycle trip data: research data from Geotec research group – Universitat Jaume I. Approximated scale 1:20,000. | 177 |
| 9.1 | Interactive map-based visualisations from ImproveMyCity (IMC). Clockwise from top left: heat map, scaled circle markers, shaded markers and geohash grid. Exported by authors using the IMC analytics. Basemap © OpenStreetMap contributors. Visualisations created with Kibana from Elasticsearch BV ('Elastic'). | 192 |
| 9.2 | The web-based IMC application front end. Source: Powered by improve-my-city.com. Basemap © Google Maps. | 194 |
| 9.3 | A set of screenshots from the IMC mobile app depicting the differences between iOS and Android. The map, issue details and timeline features are shown as well. Source: Powered by improve-my-city.com. Basemap © Google Maps. | 196 |
| 9.4 | IMC analytics dashboard with an indicative set of various visualisations also combining external data sources which are interactively interconnected. Dashboards can be embedded to existing websites or included in reports. Basemap © OpenStreetMap contributors, Elastic Maps Service. | 199 |
| 10.1 | Location of Gather project partners. Source: author. | 214 |
| 10.2 | Hypothetical example of the extent of an industry-proposed project, represented by the grey polygon, overlaying community-contributed geographic information. Source: Gather app. Credit: Spatial Information for Community Engagement Lab (SpICE). Basemap © Google Maps. | 217 |
| 10.3 | Referrals management tool – initial map-centric information management interface. Source: Gather app. Credit: SpICE. Basemap © Google Maps. | 218 |

| | | |
|---|---|---|
| 10.4 | Redesigned referrals management tool interface. Source: Gather app. Credit: SpICE. | 220 |
| 11.1 | Sapelli platform. (a) Participant using the application. © (2014) IEEE. Reprinted with permission from: M. Stevens, M. Vitos, J. Altenbuchner, G. Conquest, J. Lewis and M. Haklay, 'Taking participatory citizen science to extremes'. Credit: Michalis Vitos using Sapelli app, UCL Extreme Citizen Science research group (ExCiteS). (b) Decision tree designed in collaboration with Forests Monitor, CAGDF and local communities. Credit: Sapelli platform UCL Extreme Citizen Science research group (ExCiteS). | 234 |
| 11.2 | Prototype version of Tap&Map. (a) Printed prototype cards. (b) Picking the appropriate card. (c) Mapping a medicinal tree. Credit: Sapelli platform UCL Extreme Citizen Science research group (ExCiteS). | 236 |
| 12.1 | Community members lead icon design in whichever form they see fit (left). Credit: Photograph taken by Simon Hoyte 2018. The digitised icon (right). Credit: 2018 Bemba II village; Sapelli platform UCL Extreme Citizen Science research group (ExCiteS). | 257 |
| 12.2 | A red indicator to show when audio recording is ongoing helped to reduce confusion and mistakes by participants. Credit: Photograph taken by Simon Hoyte 2017 (right); Sapelli platform UCL Extreme Citizen Science research group (ExCiteS; left). | 258 |
| 12.3 | Example of the participant interaction required to take a data point. The participant begins by selecting a top-level category (1; 'killed animal'), followed by an intermediary-level icon (2; 'trap'). The choice is then presented to take a photo or not (3; the latter represented with a red cross). The participant can take the photo using the 'take photo' icon at the bottom of the screen (4), and verify it on the succeeding screen (5), followed by the option to take an audio recording (6). Audio is captured by pressing the 'take audio' icon (7), and can be subsequently verified (8). Sapelli then takes a Global Positioning System point automatically (9), after which the participant confirms the report (10). Credit: 2018 Bemba II village; Sapelli icons – Sapelli platform UCL Extreme Citizen Science research group (ExCiteS). | 260 |

| | | |
|---|---|---|
| 12.4 | An example of the Community Maps online geographic information system used for the project. The geolocated data points are displayed on the satellite map by their icons (right). On selecting a point (yellow circle), the relevant data are presented (left). Credit: Bemba II village. Community Maps platform © 2019 Mapping for Change. Basemap © Mapbox © OSM contributors. | 261 |
| 13.1 | Map showing Cambodia and Prey Lang Wildlife Sanctuary in red (left), and satellite image showing forest loss 2000–2016 (red), forest cover 2016 (green), forest gain 2000–2016 (blue), both gain and loss 2000–2016 (purple) and other land uses (black). An asterisk shows the location of the capital city, Phnom Penh. Source: Brofeldt et al. 2018. | 270 |
| 13.2 | Interface of the Prey Lang app, showing the opening page, interface for taking photos, interface for recording an audio and the visual representations of the three main categories documented by the app (upper row). The red squares indicate selection of a category (here illegal activities). The lower row of interfaces shows a sequence of the next four points in the decision tree: illegal logging, logging of single tree/stump, offender was a foreigner and patrol member interacted with offender. Credit: Prey Lang data-collection app by Prey Lang Community Network (PLCN). | 273 |
| 13.3 | PLCN patrol member documenting transport of illegal timber in Prey Lang Wildlife Sanctuary. Credit: Photograph taken by Ida Theilade, December 2018. | 275 |
| 14.1 | Current location and extent of protected areas and human settlements. Inset: location of the Pantanal in South America (top right). Credit: Map created by Rafael Chiaravalloti 2019. | 285 |
| 14.2 | Part of the first decision tree built for the project. It zooms into specific decisions made regarding resources from 'water'. Another part of the decision tree (not shown) dealt with resources related to 'land', such as wood and honey. Source: author. | 289 |
| 14.3 | Screenshot of the first version of the mobile app. The example shows the branch focused on recording the presence of *dequada* and its extent (large or small). | |

|  |  |  |
|---|---|---|
|  | Credit: Sapelli platform UCL Extreme Citizen Science research group (ExCiteS). | 290 |
| 14.4 | Photo of a fisher showing locations and types of natural resource use in the region. Credit: Photograph taken by Rafael Chiaravalloti. | 291 |
| 14.5 | Final decision tree used for the Pantanal version of the software. Source: author. | 293 |
| 14.6 | Figure of a crab initially used to represent the 'gathering crab' in the software (left). Final figure used to represent 'gathering crab' (right). The image is a scientific illustration of *Dilocarcinus pagei* – exactly the same species that local people collect as bait. Source: Pixabay.com | 294 |
| 14.7 | Sequence of photos showing the same fisherman recording his natural resource use in the Western Border of the Pantanal. In the first two frames, he is fishing, and in the second sequence, he is gathering bait. Credit: Photographs taken by Rafael Chiaravalloti. | 295 |
| 14.8 | Territory defined by local people from Settlement 1. Inset: location of the Pantanal in South America (right) and location of the study area in the Pantanal (left). Credit: Map created by Rafael Chiaravalloti 2019. | 297 |
| 15.1 | Group discussion about their understanding of the pictograms. Credit: Photograph taken by Carolina Comandulli 2017. | 308 |
| 15.2 | Testing the hot-pot phone charger. Credit: Photograph taken by Carolina Comandulli 2017. | 309 |
| 15.3 | Example of a sequence of data collection. Credit: Sapelli platform UCL Extreme Citizen Science research group (ExCiteS). | 311 |
| 16.1 | The conservative parents' group 'Con Mis Hijos No Te Metas', or 'Don't Mess With My Children', protesting in Lima, Peru. Credit: Mayimbú. CC BY-SA. | 321 |
| 16.2 | HOT Tasking Manager screen once a project is selected. Source: Humanitarian OpenStreetMap team (tasks.hotosm .org/project/5807?task=35). Basemap © Humanitarian OpenStreetMap. CC BY-SA. | 325 |
| 16.3 | OpenStreetMap iD Editor, with a building mapped. Basemap © OpenStreetMap contributors. CC BY-SA. | 326 |
| 16.4 | KoboToolbox data-analysis screen. In this example, responses have been disaggregated by responses to the question 'Is there gender in the use of public space in |  |

|  | your area?' Credit: Work developed by a group of secondary school students from Jose Maria Arguedas, District of Accha, Province of Paruro. Source: KoboToolbox.org. | 328 |
|---|---|---|
| 16.5 | An example of poor-quality community mapping. Note the misalignment of buildings and duplication of mapping. This work would be checked and corrected at the validation stage. Basemap © OpenStreetMap contributors. CC BY-SA. | 330 |
| 16.6 | KoboToolbox screen for starting a questionnaire from scratch. Source: KoboToolbox.org. | 331 |

# List of tables

| | | |
|---|---|---|
| 1.1 | Types of projects, typologies and some implications for design. | 33 |
| 5.1 | Case-study characteristics. | 116 |
| 5.2 | Geographic expertise and participant involvement in design. | 123 |
| 6.1 | Reporting the questionnaire in English embedded in the Hush City mobile app. | 135 |
| 6.2 | Framework of 15 people-centred recommendations for the design, build and use of citizen science mobile apps for soundscape and public spaces studies. | 144 |
| 8.1 | General context of the cities where the app was tested. | 169 |
| 9.1 | Annual number of reports submitted through ImproveMyCity by registered citizens in the municipality of Thessaloniki, Greece. Source: ImproveMyCity analytics reports. | 192 |
| 16.1 | Technologies used as part of the project. Source: KoboToolbox.org GNU AfferoGeneral Public License v3.0. | 324 |

# List of contributors

**Vyron Antoniou** is an army officer in the Hellenic Army geographic directorate. His research interests include geographic information science, spatial databases, web mapping, applications of volunteered geographic information and implementation of machine learning/deep learning in the geospatial domain.

**Dimitris Argyriou** is a data consultant and has been working with participatory monitoring for the protection and conservation of tropical forests. His research interests include community monitoring of natural resources through innovative technologies; indigenous knowledge; citizen science and near real-time data; and grass-roots forest governance.

**Maria Attard** is an associate professor in geography at the University of Malta. Her research interests focus on sustainable mobility, active travel and mobility justice in policy and planning. She co-edits the *Journal Research in Transportation Business and Management* and is associate editor of *Case Studies on Transport Policy*.

**Søren Brofeldt** is an ethnobotanist who finished his PhD with the University of Copenhagen, Denmark, in 2018. He currently works with LIFE Learning Center on the development of citizen science–based educational materials for the Danish school system. This includes research on the potential of citizen science to support students' self-efficacy in STEM subjects.

**Rafael Morais Chiaravalloti** is a postdoctoral fellow at Smithsonian Conservation Biology Institute in the United States, and senior researcher at IPE – Institute of Ecological Research. He has been working in the Pantanal for more than 10 years, supporting scientific research and public policies in the region. His research focuses mainly on socio-ecological systems.

**Carolina Comandulli** has a PhD from the Anthropology Department at UCL. She has been working and doing research with indigenous peoples since 2004, especially in the Amazon and Atlantic Forest regions. She participates in many multidisciplinary research groups concerned with issues of sustainability, social change and indigenous action.

**Jon Corbett** is an associate professor at University of British Columbia Okanagan and the director of the Spatial Information for Community Engagement Lab (SpICE). His research explores how maps can be used by communities to document, store and communicate their unique spatial knowledge, and how communicating this knowledge can impact a community. All aspects of his research include a core community element. This means that the research is of tangible benefit for the communities with whom he works, and that those communities feel a strong sense of ownership over the research process.

**Aaron Derrickson** is a PhD candidate at University of British Columbia Okanagan. His research centres on leadership and governance originating within traditional Syilx oraliture. These teachings are paralleled and compared with mainstream and academic literature as a means of seeking best cultural practices that are situated in Syilx knowledge.

**Robert Feick** is an associate professor in the School of Planning at the University of Waterloo. His research interests include volunteered geographic information and citizen science applications and data quality, public participation geographic information systems, representation of place-based knowledge and spatial multi-criteria analysis.

**Rebecca Firth** leads partnerships and community programmes for the Humanitarian OpenStreetMap Team (HOT), engaging mapping volunteers around the world to contribute to HOT's goals of mapping an area that is home to one billion people by 2025.

**Raffaella Fryer-Moreira** is a PhD candidate at the Department of Anthropology, UCL. Her doctoral research examines indigenous experiences and understandings of climate change among Guarani Kaiowá communities in Brazil, and incorporates diverse multimedia formats in the collection and presentation of research data.

**Jamie Gibson** is a user researcher and geographer whose research interest is how people influence the creation of places and things, and how those places and things influence people's behaviour.

**Michael Gould** is professor of information systems at University Jaume I, Castellón, Spain, and since 2009 is global education manager at Esri, Inc. His research interests include smart cities, mobility and international development.

**Muki Haklay** is a professor of geographic information science at UCL. His research interests include public access, use and creation of environmental information; interaction with geographical technologies; participatory mapping; and citizen science. He is the co-director of UCL's Extreme Citizen Science research group (ExCiteS) and co-founder of the social enterprise Mapping for Change.

**Simon Hoyte**'s research at UCL spans environmental anthropology, political ecology, participative technology and biodiversity conservation. Working in southern Cameroon alongside indigenous Baka communities, he is focused on utilising extreme citizen science and other tools to address issues of environmental injustice and to strengthen biocultural diversity.

**Carol Iglesias Otero** holds an MA from the Center for Research Architecture at Goldsmiths, and is a doctoral student in anthropology. Her work engages issues of epistemic politics in environmental modelling, infrastructures of climate-change adaptation and enduring histories of colonial toxicity in the Caribbean and Central and North America.

**Nicos Komninos** is professor emeritus at the School of Engineering, Aristotle University of Thessaloniki, Greece, and founder of URENIO Research. His research focuses on intelligent cities and systems of innovation. He is the author of 170 publications, including 13 books, on intelligent and smart cities, urban planning, innovation territories and strategies.

**Ioannis Kompatsiaris** is a research director with CERTH/ITI, leading the Multimedia, Knowledge and Social Media Analytics Laboratory. His research interests include image and video analysis, big data and social media analytics, semantics, human–computer interfaces (AR and BCI), eHealth, security and culture applications.

**Jerome Lewis** is associate professor of anthropology, director of the Centre for the Anthropology of Sustainability and co-director of the Extreme Citizen Science research group (ExCiteS) at UCL. Jerome's research focuses on hunter-gatherer social organisation, ritual and environmental relations, the political ecology of Congo Basin forests and participative research approaches to promote environmental justice and biocultural diversity.

**Suzanne Maas** has a BSc in Environmental Planning (2009) and an MSc in Environmental Sciences (2012). She is a PhD candidate at the Institute for Climate Change and Sustainable Development at the University of Malta. Her research interests are sustainable mobility, cycling and public participation in policy and planning.

**Spiros Nikolopoulos** has a BSc in computer engineering (2002) and an MSc in computer science (2004) from the University of Patras, Greece. He also has a PhD in semantic multimedia analysis using knowledge and context from the Queen Mary University of London (2012). His research interests include multimedia analysis, eGovernment, brain–computer interfaces, eHealth and virtual and augmented reality.

**Diego Pajarito Grajales** is a research fellow at the Institute for Advanced Architecture of Catalonia. After being a Marie Curie Fellow at Universitat Jaume I, Castellón, Spain, his research has combined citizen science, artificial intelligence for geoinformatics and the use of open and crowd-sourced geodata for decision making.

**Chryssy Potsiou** is professor of the School of Rural and Surveying Engineering, National Technical University of Athens, Greece, in Cadastre, property valuation and spatial information management, Honorary FIG President and UNECEWPLA Vice Chair. She was Hellenic Cadastre Board Member (2009–2012); FIG Commission 3 Chair (2007–2010); FIG Vice President (2011–2014); FIG President (2015–2018); and bureau member of UNECEWPLA (2001–2021).

**Antonella Radicchi** is a senior research associate at TU Berlin, Institute of Urban and Regional Planning. Her research interests include public spaces, sustainability, healthy cities, sensory urbanism, citizen science and citizen-generated data. She is the inventor of Hush City, a citizen science app to map and assess quiet areas.

**Colin Robertson** is an associate professor in the Department of Geography and Environmental Studies at Wilfrid Laurier University. Broadly trained in geographic information science and spatial analysis, Colin frequently works across disciplinary domains to solve problems at critical interfaces using spatial analysis, geographic information systems and citizen and community-based monitoring.

**Artemis Skarlatidou** is a senior research associate in the Extreme Citizen Science research group (ExCiteS) at UCL. Her research interests include risk communication, human–computer interaction (HCI) and user experience aspects (e.g. usability, aesthetics and trust) of citizen science applications, geospatial technologies and their spatial representations for expert and public use.

**Ida Theilade** is a professor at the Faculty of Science, University of Copenhagen. Her research interests include the role of local and indigenous knowledge in environmental monitoring and protection, near real-time forest monitoring, the use of information and communication technologies and citizen science. Ida and partners have received several innovation awards for the Prey Lang app, which is powered by Sapelli and used to monitor forest crime in Cambodia.

**Ioannis Tsampoulatidis** holds a BSc in computer science, and he is currently a PhD candidate in the Aristotle University of Thessaloniki, Greece, in the field of smart cities and collaborative eGovernment. He is a co-founder and the managing director of INFALIA PC, a spin-off company of CERTH/ITI.

**Nerea Turreira-García** has a PhD in environmental management from the University of Copenhagen, Denmark, where she continues as a postdoc. Her research interests include the role of local and indigenous knowledge in environmental monitoring and protection, public participation in research and development and the use of information and communications technologies.

**Michalis Vitos** is a software engineering consultant with a BSc (2008) and MSc (2010) in computer science. He also has a PhD (2017) in developing innovative ICT and GIS tools that could be used by semi-nomadic and non-literate indigenous communities to monitor environmental changes and resource extraction in scientifically validated ways. His research interests include human–computer interaction and software engineering in combination with the emerging area of participatory geographic information systems and citizen science.

**Peter Ward** works at the intersection of education, health and technology. With an MA in Education, Health Promotion and International Development, he currently focuses on using technology to enable Latin American youth to have an active voice in their education and in their societies more broadly.

# Foreword

*Geographic Citizen Science Design: No one left behind* by Artemis Skarlatidou and Muki Haklay and their colleagues is terrific! I couldn't stop reading it! This is the first book that I have read that brings together citizen science (CS) with technology and project design. The storytelling style infused with technical descriptions underpinned by a strong community-driven philosophy make this book a joy to read. It provides a deep learning experience through the 12 case studies that illustrate how different groups and projects succeeded in different geographical situations.

*Geographic Citizen Science Design: No one left behind* has the potential to change CS by enabling technology designers and project leaders to develop more innovative, useful and satisfying experiences for projects that harness technology. Specifically targeted at geographic citizen science (GCS), this book will inspire CS and community-driven projects in general. It speaks to the design and deployment of technology and project procedures. Drawing on the experience of a wide range of projects and a team of authors with interdisciplinary skills, this book emphasises the importance of interweaving community participation in project and technology design. It presents the philosophy and process of participatory design as the way to engage communities so that no one is left behind.

Readers are treated to a definition of GCS that situates the perspective taken by the authors within the broader sphere of CS. According to Muki Haklay, GCS is defined as scientific work undertaken by members of the general public where the data generated have a deliberate and explicit geographic aspect. Typically, GCS is a place-based activity, though not always (Haklay, Chapter 1). CS shares these values without explicitly emphasising the role of geography, though the significance of place is accepted in place-based projects.

The multidisciplinary team of authors includes anthropologists and linguists, as well as geographers, human–computer interaction (HCI) specialists, scientists and citizen participants. This team shines a much-needed light on the role of culture, community norms and linguistic differences in the projects featured in the book. In addition, the authors highlight how project procedures and technology design must be tailored to communities' needs and differences, and the environmental and geographical contexts in which the projects are situated.

In recent years, GCS and CS have looked to technology to leverage data collection in diverse environments across the world, as well as to scale up the volume of data collected. The goal of using technology is to broaden the scope and depth of GCS. However, this will only happen if the technology is well designed, which means it must go beyond simply being usable. It must engage and offer new opportunities to the citizens that use it. Such technology doesn't just happen by clever designers in Western labs coming up with design ideas. Project participants must have a stake in the design process so that they embrace and take ownership of the projects in which the technology is embedded.

Participatory design, as the name suggests, provides the techniques and philosophy for achieving participant involvement and buy-in. It was originally created in Scandinavia in the 1960s and 1970s in response to new laws that gave workers the right to have a say in how their working environment would be shaped by technology and other innovations. The field of HCI soon adopted this human-centred system development approach. Participatory design has evolved greatly over the years, and it now includes a broad collection of techniques to ensure that those who will use the technology also participate in its design. Done well, participatory design ensures that participants' culture, language and skills shape both the technology and the project in which it is embedded. In essence, participatory design is a democratic process, and democracy can also be slow, but when done well, it can produce amazing results.

The role of participatory design is described in the 12 cases studies, some of which include technology, while others focus more on project design. These cases range from stories about how to design technology with pygmies in the Congo, who are not literate according to a Western interpretation of literacy, to First Nations in British Columbia, Canada, to projects in Peru, Australia and Europe. The authors describe how they worked to gain the acceptance and the trust of the communities with whom they worked. This involved understanding each community's culture, language, politics and policies and other geographical features such as those related to climate and the land where they are located.

The participatory design principles articulated in the early chapters coupled with the 12 case studies plus the summary of lessons learned at the end of the book provide a practical and inspiring foundation for readers to develop their own projects. These principles apply to the entire project and not just to the technology component. In an ideal world, the technology is seamlessly integrated with the rest of the project. There may also be projects that don't need or want technology. Some communities may prefer to use existing technologies such as feature phones, smartphones, digital cameras, email, sensors, Instagram and Facebook. Sometimes, these can be made more convenient by integrating them into a website. Involving HCI specialists in a project can be helpful when it comes to deciding which technologies to use and how to present them, especially if the HCI specialist is also passionate about the topic that the project focuses on. While funding agencies sometimes strive to push technical development, mature technologies may be more appropriate for some projects.

Whether your project involves easily available technology or aims to leverage advances in machine learning, artificial intelligence and new ways to compensate for limited battery life and satellite coverage in remote areas, this book speaks to the fundamental issues of importance in CS projects. *Geographic Citizen Science Design: No one left behind* is a must-read book for students, researchers and everyone involved in GCS or any kind of CS and community-driven project.

Jennifer Preece, July 2020
Professor and Dean Emerita, University of Maryland Information School
Editor-in-Chief, *Citizen Science: Theory and Practice*

# Acknowledgements

We would like to thank everyone who contributed to the preparations and publishing of this volume. Special thanks to every member of the ExCiteS team at UCL for their support in the development of this volume and especially to Mrs Judy Barrett for her diligent work and input at all stages of the publication process. Special thanks to our reviewer, Dr. Susanne Hecker. Her scientific expertise and critical insight helped improve this book and bring it into its current form. We are grateful to our publisher, UCL Press, and the teams at Yellowback and Westchester Publishing for their attention to detail and their guidance to make this volume ready for publication.

The idea for *Geographic Citizen Science Design: No one left behind* was born during the interdisciplinary workshop *'Lessons learned from Volunteers Interactions with Geographical citizen science'* which took place 27 April 2018 at University College London, where more than 60 scholars from Europe, Africa, South America, Asia and Canada joined to share their experiences in developing and using geographic citizen science applications in various contexts. We would like to thank everyone who joined the workshop; their enthusiasm and contributions are invaluable for the geographic citizen science community.

We would like to thank the funding bodies which supported the development of this book. These are the European Research Council (ERC) project Extreme Citizen Science: Analysis and Visualisation (Grant Agreement No 694767) and the *COST Action CA15212, supported by COST (European Cooperation in Science and Technology)*.

Last but not least, we owe a debt of gratitude and admiration to the communities and individuals who contributed to the research that is described in these pages. Unfortunately, it is not possible to list the names of all the people who participated in the activities that are described here, but it is their effort that makes geographical citizen science research and work possible – this investment of time, effort, and working with researchers and practitioners is recognised here.

Artemis Skarlatidou and Muki Haklay
UCL, UK

# Introduction

# Geographic Citizen Science Design: *No one left behind*

Artemis Skarlatidou and Muki Haklay

## 1. Overview and definitions

Little did Isaac Newton, Charles Darwin and other 'gentlemen scientists' know, when they were making their scientific discoveries, that centuries later they would inspire a new field of scientific practice, research and innovation called citizen science. Citizen science can be defined in lay terms as 'scientific work undertaken by members of the general public, often in collaboration with or under the direction of professional scientists and scientific institutions' (Oxford English Dictionary 2014), foregrounding the role of the professional scientist, which has become established as a profession in the past two centuries. The current growth and availability of citizen science projects and relevant web-based applications and mobile apps to support citizen involvement in scientific discovery cannot be overstated. In principle, almost everyone has the opportunity to become a citizen scientist and to contribute to a scientific discipline or topic of interest – often without having any relevant professional qualifications. Instead of 'gentlemen scientists', we now have a much larger group of usually well-to-do and highly educated citizens contributing to scientific discovery (and, in some cases, the over-representation of men persists). To turn this true potential of citizen science into reality, however, there is a need to overcome the challenges of literacy in general and scientific literacy in particular (especially access to technology), as well as supporting citizens to find the time to engage with such activities and effectively interact with the digital technologies which enable them.

Geographic interfaces are now commonly used, and they are constantly evolving, to support the collection, analysis and dissemination of geographic data contributed by volunteers – from OpenStreetMap, which involves hundreds of thousands of volunteers in the systematic collection of geographic objects and data to create an open-source map of the world, to countless mobile apps that use mobile device sensors and the power of maps to collect and analyse information about noise pollution, environmental resources, accessibility barriers and so on. Not all citizen science initiatives use geographic interfaces or any digital technologies for that matter. Also, not all cases of utilising geographic information technology and interfaces to engage members of the public require the collection of data in a systematic and objective way. For example, there are numerous cases of volunteered geographic information (VGI) – defined in Chapter 1 as 'digital geographic information generated and shared by individuals' – which do not always fit the geographic citizen science context which is discussed in this volume.

These types of geographic activities are thoroughly analysed in Chapter 1 of this volume, where Haklay explains that geographic citizen science lies at the intersection of the fields of VGI and citizen science. Geographic citizen science therefore entails the utilisation of geographic information technology to collect, analyse and disseminate data collected by non-professional participants in a systematic and objective way. Geographic citizen science covers a wide breadth of initiatives which serve different purposes and have different characteristics. Like citizen science activities in general, these fall under different typologies due to their characteristics (e.g. the degree of participants' involvement and collaboration with scientists, the stage at which they are involved in the scientific process, etc.) and priorities (e.g. contributing to scientific research, increasing awareness, reaching out to and educating new audiences, etc.). Examples in geographic citizen science include participants collecting or analysing geographic data to assist scientists in answering research questions (as is the case with the Cyclist Geo-C mobile app described in Chapter 8 of this volume); to participate in problem-solving practices and decision making in local government (as is the case with the ImproveMyCity application described in Chapter 9); or for advocacy purposes, including in volunteer-initiated participatory action research projects which can be used to uncover and address issues of local and global concern (as in several of the case studies presented in this volume).

Geographic citizen science is approached from different angles, and it has the potential to have a massive impact on science, society, social innovation, public awareness and even participants' well-being. As the next section argues, a fundamental requirement for achieving this is that the interfaces, which support volunteers to collect, analyse or disseminate their contributions, are user-friendly and consider end-user needs as well as the local cultural and environmental conditions of the contexts where they are being implemented.

## 2. Background and the scope of this volume

A range of social and technical possibilities has resulted in the realisation and growth of geographic citizen science over the past 15 years or so. First, there was a series of technological developments that created new possibilities in the way geographic information science and its relevant technologies are currently being utilised to support citizen science. The two developments perhaps most crucial for geographic citizen science are: (1) the emergence of the Internet in the 1990s, which was a milestone for people's interactions with maps and in terms of how content and geographic information is created and disseminated online; and (2) the removal of the selected availability of the Global Positioning System (GPS) signal in the early 2000s, which enabled anyone to capture the accurate digital location of any geographic object and led to the proliferation of GPS-enabled sensors in many everyday devices (e.g. phones, cameras, car navigation systems, etc.).

Second, there were several social changes that contributed to the current state of geographic citizen science. As Haklay discusses extensively in Chapter 1, one of the most important is an increase in literacy levels and the continuously growing numbers of people completing secondary and even tertiary education. This undoubtedly had a massive impact on the way citizen science evolved over the centuries. Nevertheless, it should not be forgotten that there is still a significant proportion of the population with no access to education, people who have no access to technology (e.g. mobile devices, portable computers, etc.) and technological infrastructure (e.g. electricity, the Internet, etc.), and those who lack the financial resources to own the equipment which would enable them to participate in a citizen science project. Also, there are still a significant number of people who are completely unaware that these opportunities exist and how they can benefit from them because they are marginalised or completely excluded from existing scientific conservations or other types of projects. Despite the existence of these digital and the other socio-economic divides, everyone's knowledge, skills, efforts and, most importantly, everyone's voice is equally important, not only in advancing science but also in the context of environmental governance and sustainable development, which are major themes within geographic citizen science. In line with inclusion and the 'leaving no one behind' principles of the United Nation's 2030 Agenda for Sustainable Development, scientists are slowly realising the importance of understanding how people interact with their local environments, which is something that geographic citizen science currently supports with several of the examples discussed in this volume.

The fact that citizen science activities attract people of different ages, backgrounds and interests has its own design challenges. These include the need to design interfaces which help to attract and retain volunteers; to design user-friendly applications to enable volunteers with diverse skills and experiences to harness the potential of citizen science and collect accurate and high-quality data; and to generate user experiences that match those that users require and so on. Although still gaining ground, these and other design challenges have slowly started to attract the attention of the citizen science and human–computer interaction (HCI) research communities (Preece 2016).

The majority of citizen science and geographic citizen science activities have been focused on a specific demographic of Western, Educated, Industrialised and Developed (WEIRD) participants (Dourish 2015); and from a geographical point of view, most efforts have concentrated on urban centres located mainly in the Global North. This 'leaves behind' a significant proportion of the population, particularly less privileged citizens in non-urban centres of the Global South, who may benefit substantially from using these applications (e.g. in terms of taking ownership and addressing issues that are of significant local concern, but also in terms of promoting equality and improving scientific literacy to mention just a few). Yet, technological solutions developed in the Western world usually ignore the unique environmental, cultural, user and other contextual characteristics which influence successful technology adoption and utilisation in these areas.

Experiences and lessons learned from the context of geographic citizen science – especially those which focus on the design, development and evaluation of applications to support users with their tasks and how users interact with them – remain in their majority based on anecdotal evidence. The present book takes an anthropological and HCI approach to improve understanding of how geographic citizen science projects and their associated interfaces should be designed to maximise their anticipated impacts. The volume presents, discusses and reflects on case studies which engage diverse user audiences in both urban and non-urban contexts and for various purposes. Each chapter elaborates on the methodological principles, design decisions, interaction barriers and opportunities in specific contexts of use. By looking at the field through the lenses of specific case studies, this volume captures the current state of the art of research and development of geographic citizen science practices and provides important information to inform future technological innovation and research in this field.

## 3. Structure and content of this volume

Drawing on perspectives from geography, engineering, HCI and anthropology, the first part of this book outlines the theoretical, technological and methodological principles which underpin geographic citizen science and its design implications. Parts 2 and 3 of this volume provide a curated selection of geographic citizen science case studies being applied in various parts of the world, which are used to capture, present and effectively reflect on the differences, unique characteristics and design and interaction implications from using these applications. Part 2 presents case studies where the main users are located in urban areas, mainly in the Global North. Part 3 discusses case studies which are being implemented in non-urban areas, where geographic citizen science projects are used to engage with indigenous communities, mainly in the Global South. Case studies examine carefully the cultural context, user demographics and their characteristics (such as technology and literacy skills), issues of access and fitness for purpose, and provide further insight on interaction aspects and encountered barriers to conclude with a set of lessons learned which can be used to inform the design and development of future and existing geographic citizen science applications.

### 3.1 Part 1: Theoretical and methodological perspectives of geographic citizen science

Part 1 starts by setting the boundaries for geographic citizen science through a theoretical overview of the fields of citizen science and VGI, which is necessary to explain effectively how geographic citizen science emerges, its characteristics and what it entails. In Chapter 1, Haklay provides a historical overview and in-depth explanation of the congested terminology in both fields, and then goes on to discuss the necessary theoretical background, together with detailed practical examples from citizen science and geographic citizen science initiatives. Chapter 1 pays special attention to the societal and technical prerequisites of geographic citizen science which are essential in order for the reader to appreciate its underpinning meanings and complex relationships.

A common issue which is faced by most decision makers in geographic citizen science projects, from the early stages of setting up a new initiative, is that of choosing the most appropriate technological infrastructure to facilitate data collection, analysis and dissemination. This is usually accompanied by a lack of awareness on the technological front; that is, technological availability and most importantly the limitations and opportunities that specific technologies may bring to a project. These decisions are inseparable, and they are further destined to influence the way technology is eventually utilised by its end users and therefore they have a direct impact in the success (or failure) of the initiative. In Chapter 2, Antoniou and Potsiou introduce and discuss a range of technologies and the issues which are most likely to influence the success of a geographic citizen science project. The ultimate aim is to provide neither a checklist nor recommendations for a one-size-fits-all solution. Through a critical overview, the authors analyse the criteria and conditions which need to be taken into account when designing the technological infrastructure to fit specific contexts of use (e.g. ethical, legal, issues of data quality, geographic scale, number of participants, etc.)

Chapters 3 and 4 provide insight into methodological principles from HCI and anthropology, respectively, which can be used to support the design and development of successful geographic citizen science initiatives and applications. Specifically, in Chapter 3, Skarlatidou and Iglesias Otero provide an overview of popular HCI design approaches and methods which can be used to enable user involvement in the design, development and evaluation of citizen science applications suitable for projects that are implemented in both urban and non-urban contexts. The authors use actual examples from the geographic citizen science and wider citizen science contexts, where HCI methods have been used to extract user requirements and needs, obtain user feedback and evaluate user interfaces (UI). In doing so, Chapter 3 provides an overview of how HCI has so far been approached by citizen science and HCI practitioners and researchers, and it communicates the main lessons learned, as highlighted by the original studies.

In Chapter 4, Fryer-Moreira and Lewis discuss anthropological approaches to the development and implementation of geographic citizen science projects, which target non-urban contexts and indigenous populations, where sociocultural specificities must be considered in the design and development of their digital interfaces and the way the projects are designed and executed. The chapter provides an in-depth overview on methods such as participant observation, the development of the free, prior, informed consent process, the establishment of community protocols and approaches to co-designing user interface visualisations together with indigenous communities. The approaches and methods discussed in Chapter 4 have been used extensively by several of the authors in Part 3 of the volume. Therefore, this chapter provides readers with essential theoretical understanding before they go on to examine the case studies discussed in Part 3.

## 3.2 Part 2: Geographic citizen science case studies in the Global North

The second part of this volume presents and discusses five case studies which mainly target communities and individuals in urban areas, who are mostly literate, and where access to technological infrastructure is not a concern. All chapters in this part provide significant insight to understanding users' interaction barriers with the applications and the methods used for collecting this information and conclude with lessons learned, which citizen science practitioners may apply in similar contexts.

In Chapter 5, Feick and Robertson provide an overview of geographical expertise – that is, people's familiarity and knowledge of particular locales or with identifiable types of places – and they explain, using the example of three geographic citizen science applications, how consideration of geographic skills should inform the design of and improve interaction with citizen science tools and projects. In Chapter 6, Radicchi presents and discusses the Hush City app, a mobile app to collect data and map quiet areas in urban contexts. The author provides 15 people-centred recommendations to inform the design of citizen science mobile apps in soundscape research and public spaces studies.

In Chapter 7, Gibson presents the Global Forest Watch application, a data-collection mechanism and visualisation interface to provide the global community with information about the current state of our world's forests – an essential step towards their effective monitoring and sustainable management. The author discusses how a mixed-methods approach, using analytics but also an extensive consultation process with end users, can be applied to gain significant insights to improve interface design and the ways users interact with the application.

In Chapter 8, Pajarito Grajales et al. describe the Cyclist Geo-C mobile app, a geographic citizen science application which has been developed to involve citizens in the collection of open cycling data and feedback on their journeys. The authors emphasise the role of evaluation and user testing and pay particular attention to the use and applicability of collaboration-based gamification features, which are increasingly popular in citizen science in this context. In Chapter 9, Tsampoulatidis et al. present and discuss ImproveMyCity, a geographic citizen science application which is used by ordinary citizens to collect data about non-emergency issues in their town and report them to their local authorities. The authors explain how the application creates a direct communication channel between citizens and public authorities and therefore highlight the importance of providing a technological solution which enhances openness and transparency rather than hindering it. Design choices for moderation, motivation mechanisms and interface design recommendations are further provided by the authors.

## 3.3 Part 3: Geographic citizen science case studies with indigenous communities in non-urban areas

The third part of this volume consists of seven case studies which are being utilised by indigenous people and communities in mainly non-urban areas of the Global South, with the exception of Chapter 10 which involves First Nation communities in Northern Canada. All chapters in this part pay particular attention to the unique cultural and contextual characteristics of their case studies, such as environmental constraints, participants' literacy levels and their familiarity with and access to technology, which create a set of distinct challenges and opportunities in the ways these projects and their applications are implemented. This part provides a unique insight into geographical citizen science design and its methodological perspectives to enable and encourage the involvement of marginalised communities and those who are usually excluded from the environmental sustainability debate.

Large-scale resource development projects in Northern Canada face a legal requirement and a duty to consult the First Nation communities which are mostly being affected by them, which has left these communities with a massive and growing number of impact assessment proposals they have to manage and review. To assist communities in this process, in Chapter 10, Corbett and Derrickson discuss the co-design process for the development of Gather, a geographic citizen science tool which is used to view information relevant to the referral process and contribute data related to the use of community land and resources. This case study is the only one in this part which involves the development of an interface to act further as a communication channel between multiple stakeholders (i.e. indigenous communities, government and industry). The existence of multiple users with different skills, experiences, needs and requirements in a politically contentious environment creates a set of additional implications and interaction barriers, which the authors discuss extensively in the chapter.

Baka Pygmies of northern Republic of Congo have experienced for years an unprecedented exploitation of their local forest and natural resources in ways which are clearly juxtaposed with the indigenous ways of interacting with them. The Sapelli interface was developed in 2013 with the aim of supporting local communities and in collaboration with local non-governmental organisations, and it has been used since then with non-literate people to collect data about illegal logging and poaching in the area. In Chapter 11, Vitos describes a case study in the Congo Basin where Sapelli is used to enable local communities to participate in socio-environmental monitoring schemes and the collection of Traditional Ecological Knowledge. For the development and evaluation of the interface, the author thoroughly describes a user-centred design process and provides rich insight into methodological implications, interface design and interaction barriers.

Chapters 12–15 also explore the implementation of geographic citizen science initiatives and offer a significant anthropological insight into the unique cultural characteristics and environmental conditions present and explore how these influence the initiation of a geographic citizen science project, as well as the design and development of the relevant technologies to support the data collection. All four chapters describe the use of Sapelli with:

- Baka communities in Cameroon – to collect data about illegal wildlife crime and animal monitoring, which at the moment is the only viable solution to obtaining reliable information to inform effective forest management plans (Hoyte, Chapter 12);
- Local communities in the Prey Lang forest in Cambodia – to collect data about illegal logging and forest resource management (Theilade et al., Chapter 13);
- Local fishers in the Pantanal wetland, Brazil – to support local populations in the collection of data about natural resources use and fishing strategies as a way to gather evidence subsequently to inform the development of effective conservation models and the establishment of relevant environmental regulation which takes into account the local context and the population needs (Chiaravalloti, Chapter 14).

- Ashaninka communities in the Brazilian Amazon – to collect data about illegal activities and land invasions in their territories which can then be used to inform and communicate with government authorities and enforcement institutions (Comandulli, Chapter 15).

This volume would not be complete without a case study from the OpenStreetMap project. With more than a million users, OpenStreetMap is the one of the most popular geographic citizen science projects. Its aim is to create an open-source map of the world. Humanitarian OpenStreet-Map focuses on humanitarian action and response in crisis events and disaster management and supports community development initiatives towards achieving the Sustainable Development Goals. In Chapter 16, Ward and Firth discuss a Humanitarian OpenStreetMap case study from Peru, with young children exploring gender issues using mapping interfaces and geographic information. This case study provides insight into another dimension of geographic citizen science – that of its educational impact – and explores how technology can be used to support youth expanding their skills and abilities, as well as identifying important interaction barriers with geographic interfaces when they are used for this purpose.

Last but not least, the editors of this volume provide an overview and synthesis of the main issues discussed in the various case studies and highlight directions for future research in the last part of this volume (Synthesis and Epilogue).

Together, the theoretical chapters and the case studies demonstrate that geographic citizen science is possible at different scales – from the very local to the global; in a range of locations that differ markedly due to their infrastructure and economic development; and in widely differing societal and cultural contexts. With the imperative of addressing the Sustainable Development Goals in the coming decades, the present collection demonstrates that it is possible to provide geographic citizen science approaches that will reach across gender, socio-economic, literacy and cultural divides – thus achieving the aim of leaving no one behind.

## References

Dourish, Paul. 2015. 'Forward'. In *At the Intersection of Indigenous and Traditional Knowledge and Technology Design*, edited by Nicola Bidwell and Heike Winschiers-Theophilus. Santa Rosa, CA: Informing Science Press.

Oxford English Dictionary (OED) Online. 2014. Citizen science. Accessed 13 September 2020. https://www.oed.com/view/Entry/33513?redirectedFrom=citizen+science#eid316619123.

Preece, Jennifer. 2016. 'Citizen science: New research challenges for human–computer interaction', *International Journal of Human–Computer Interaction* 32: 585–612.

# Part I
# Theoretical and methodological principles

# Chapter 1
# Geographic citizen science: an overview

Muki Haklay

## Highlights

- Geographic citizen science is the area where volunteered or crowd-sourced geographic information and citizen science coincide – scientific work undertaken by members of the general public where the data generated has a deliberate and explicit geographic aspect, such as capturing an ecological observation by recording its Global Positioning System coordinates.
- By examining typologies of volunteered geographic information and citizen science, we can delineate major characteristics of geographic citizen science, including the agency of the participants as well as the intentions and the aims of the project.
- Different activities within citizen science lend themselves to geographic citizen science in different degrees – while volunteer computing is mostly non-geographic, ecological observations are completely within this field.
- Special concern should be paid in the effort of making geographic citizen science inclusive and acknowledging its multiple exclusionary potentials – from access to technology to access to knowledge – and addressing them.

## 1. Introduction: defining geographic citizen science

The early 2000s witnessed the flourishing of new terms and phrases, which frequently happens when societal and technological changes come together, and journalists and scholars are rushing to describe and explain them. The Internet, the web, mobile computing, crowdsourcing, volunteered geographic information (VGI) and citizen science are but a few of the terms that emerged from this flurry, and while some of the terms fell out of favour (such as cyberspace or neogeography), others proved enduring. The intersection between two of these terms – VGI and citizen science – is at the centre of this book and forms a distinct area of geographic citizen science. However, before turning to their definitions, it is important to consider the reasons for the changes that took place around the turn of the millennium.

Technologically, the 1990s had seen the emergence of the Internet as a global telecommunication infrastructure and the rapid growth of the World Wide Web (or the web for short). The ability to share geographic information over the web, in the form of interactive maps, started five years after the web was created (Putz 1994). Within the services that the rapidly growing web provided, interactive maps became useful and popular websites, although the management and visualisation of geographic information are more challenging than text, static images and video (Haklay, Singleton and Parker 2008).

Another innovation that came with the web is that of user-generated content (UGC), which started early on, when systems such as GeoCities (launched in 1994) allowed people with relatively limited technical skills to create their own websites (Brown 2001). With further technological and interaction design advances, it became possible to create content with even less technical knowledge through weblogs (blogs), images, audio (podcasts) and video-sharing websites. However, until the early 2000s, recording location-based information in a digital form was undertaken in the office, and the ability to use computers in the field was limited.

The first devices to support mass mobile computing started appearing in the late 1990s and included the handheld PalmPilot. The ability to capture the location digitally – through the Global Positioning System (GPS) – for everyday applications also became possible at that time. This led to pioneering CyberTracker which can be considered as one of the very first examples of geographic citizen science. CyberTracker allowed trackers to participate in recording information about rhinos, without knowledge of writing and reading (Liebenberg et al. 1999). Until 2000, when the GPS signal became open to civilian applications, the ability to capture locations digitally in an accurate way required expensive equipment and a lengthy process. However, within a few years, following changes in the availability of signals for non-military applications, the costs of a GPS chipset became affordable to the degree that it could be integrated into mobile phones.

Societal trends are no less important, and chief among them is the increased focus in educational policy on science, technology, engineering and mathematics (STEM) and higher education. In the 1990s, primary-level education received much attention across the world, following the Jomtien World Conference on Education for All (EFA), with further attention received after being included in the Millennium Development Goals (MDGs; Rose 2005). At the other end of the educational spectrum, the number of students in post-secondary education increased from an estimated 51 million in 1980 to 139 million in 2006 (Teichler and Bürger 2008). These trends mean that by the turn of the century, there was a large pool of people who were capable of understanding the technical issues that relate to systematic data collection and analysis, and the tools that they needed to collect and share data while out and about were available and affordable (see Haklay, Singleton and Parker 2008).

This chapter pays special attention to the societal and technical prerequisites that different forms of geographic citizen science assume. After all, the picture that was drawn in the opening paragraphs obfuscates the uneven way in which societal and technical advances spread across the world, for which the range of case studies in this book provides vivid evidence. The early hand-held devices cost about US$400, with additional GPS receivers that cost about US$500. Even with the technological advances and the economies of scale that mass production brings, smartphones in the late 2000s cost about US$600. When adding to these the connectivity costs, these devices were out of reach for most of the population until the early 2010s. This is one of the multiple facets of the 'digital divide': the division of the population between those who are connected and able to use digital technologies and those who are not. The facets of the divide include economic barriers such as the ability to purchase the equipment and pay for the contracts for access to the Internet and mobile network. Technical barriers include the availability of a fast broadband Internet connection (which is still not available in many rural parts of the world) or mobile network coverage. Societal barriers also play their part, such as functional illiteracy which is a level of literacy that prevents a person from being able to participate fully in society and therefore being able to carry out more complex tasks with their mobile phones. There are also design barriers that are more subtle, such as the provision of instructions on the interface of an application in a language that is difficult to comprehend without – often technical – background knowledge (see also Sui, Goodchild and Elwood 2013).

Taken together, these factors influence who can participate in projects, where they can do that and at what times, and which literacies are required. The outcome is uneven information production that represents the views and interests of those on the active side of the digital divide. Consideration of digital divides can therefore be included in the design of the software and the hardware that will be used to engage participants in a specific activity.

This chapter focuses on the intersection of VGI and citizen science – both the result of the changes noted above. The intersection can be termed 'geographic citizen science'. Here, VGI is defined as digital geographic information that is generated and shared by individuals. VGI is part of UGC, which was mentioned above. Within VGI, geographic information is an integral part of the digital media object, for example coordinates are an integral part of the exchangeable image file format (Exif) element of a picture taken with a digital camera (Goodchild 2007). Citizen science, on the other hand, is defined by the Oxford English Dictionary (2014) as 'scientific work undertaken by members of the general public, often in collaboration with or under the direction of professional scientists and scientific institutions'. Citizen science, when recorded using computers, is also a type of UGC, and here the content is scientific facts, observations or analysis. Geographic citizen science can therefore be defined as scientific work undertaken by members of the general public where the data generated have a deliberate and explicit geographic aspect (Figure 1.3). It is frequently place-based activity, although not always. As we will see in the rest of this chapter, not all VGI is geographic citizen science, and not all citizen science is geographic. We will come back to the boundaries of geographic citizen science at the end of the chapter.

While the rest of the chapter explores the relationships between VGI and citizen science, some aspects of geographic citizen science can already be observed. A citizen science project that is concerned with recording an environmental observation by taking a geotagged picture with a smartphone is clearly producing VGI – this is one of the common examples of geographic citizen science. In contrast, a project that engages volunteers to map the location of all water sources in an informal settlement in the open digital database of OpenStreetMap is carrying out a systematic collection of facts, and therefore it can be considered as citizen science (and therefore geographic citizen science). It is also possible to delineate where activities clearly fall outside the parameters of geographic citizen science, such as when VGI is not concerned with recording information in a systematic and objective way. Opinions regarding restaurant quality that are recorded in TripAdvisor cannot be considered citizen science. In addition,

VGI that is done without an intention of producing scientific outcomes or purpose falls outside geographic citizen science. Finally, when a citizen science project is not concerned with the geographic location of the observations, it will not be classed as VGI. For example, a citizen science about classifying galaxies is not geographic. Naturally, there will be complexities in the definition of specific cases. Throughout, this chapter explores the intentions of volunteers in their act of participation, as well as the issue of power between the contributor and the technical and social systems that facilitate the contribution, especially the role of the project originator and owner – the person or group that designs and runs the project.

## 2. The challenge of terminologies

When approaching the areas of VGI and citizen science, it is important to consider the way that terminology influences the way we perceive an area and, in particular, the role and agency of participants. Different terms create a different picture of what the role is of the person who is involved in the process, and they can help when considering cases such as setting an app to record automatically, without ongoing intervention, information about the environment – does this amount to geographic citizen science? The analysis of the terminologies will help us to conceptualise and think about this case. Both VGI and citizen science are related to another concept within the societal and organisational realm: crowdsourcing. The term 'crowdsourcing' was coined by Howe (2006) to describe a process in which a large group of people are asked to perform business functions that are either difficult to automate or expensive to implement. Fundamentally, crowdsourcing allows an organisation to ask a large group of unremunerated or marginally remunerated people to carry out piecework tasks for which the organisation is the prime beneficiary. Although Howe (2006) focused on the business context of crowdsourcing and the above focuses on the business transactional element of this activity, the term is also used to describe other activities such as requests for volunteering time to assist a humanitarian or scientific effort through an open call (which means that it is open to anyone who wishes to respond) and addresses potentially a large group of participants (known as the 'crowd').

Of particular importance is the labelling of the purposeful activity that people can partake in and the use of terms such as 'volunteer', 'citizen', 'user' and 'crowd' to describe the participants (see Eitzel et al. 2017). We can see that all these terms have their limitations. While 'volunteer' and 'citizen' are loaded with meaning, 'crowd' and 'user' might seem at first glance neutral or simply descriptive. Yet, crowdsourcing has been criticised as an exploitative practice that reduces humans to automatons or machine parts (Silverman 2014) and therefore that the term 'crowd' is used to treat contributors as an anonymous, faceless (and potentially expendable) group. The term 'user', which is common in digital technology and in computing language, has also been criticised by Brenda Laurel (2001), who observed that user

> implies an unbalanced power relationship – the experts make things; everybody else is just a user. People don't like to think of themselves as users. We like to see ourselves as creative, energetic beings who put out as much as we take in' and goes on to suggest an alternative term: Partner – this person has agreed to work on something together with you.
>
> The idea of being in partnership with the people purchasing your products or on your site is not only emotionally attractive; it is quite literally true. (Laurel 2001, 49–50)

As for the 'volunteer' in VGI, this has received special attention from Sieber and Haklay (2015), who argue that the assumption of free-will volunteering, without any wish for personal gain, is not reflected in practices such as crowdsourcing where there is no explicit volunteering for a higher cause, and, conversely, instead of seeing volunteering as a reason to increase the trust in the participant, it is a source of concern about their motivations. Finally, as might be expected, the 'citizen' in citizen science has also raised a lot of questions, as demonstrated by Mueller, Tippins and Bryan (2011) who argue that the use of the term 'citizen' requires the linking of public participation in science to a strong concept of democratisation and citizenship, especially when citizen science projects are related to education. Further discussions about this appear in Wilderman (2007), Calabrese Barton (2012), Cooper (2012), Eitzel et al. (2017) and Strasser et al. (2019).

These are merely a few examples of a much wider literature that critiques and questions the use of these loaded terms to describe large-scale activities that have emerged in the past decade. Arguably, they are the result of the underlying tensions that are at the heart of VGI and citizen science, which are portrayed as altruistic collaborative efforts towards a common goal and a greater good, on the one hand, and as extracting free labour, in an exploitative way, where the benefits accrue to the entrepreneurs who have set up the system or have the knowledge and skills to exploit the resulting information, on the other. The reality is somewhere in-between, depending on the nature of the project and its dynamics. This requires careful consideration of the meaning, terminologies and their fit to the specific case as well as to the people who together contribute to the resulting outputs.

## 3. Citizen science and volunteered geographic information

Before turning to locate the boundaries of geographic citizen science, it is important to establish what type of activities fall under the umbrella term 'citizen science' and then to examine the area of VGI. This section introduces several core characteristics of these two fields, with illustrative cases (including those from this book) to demonstrate a range of activities.

The framework used as a basis for analysis is from Craglia, Ostermann and Spinsanti (2012) in which they suggest differentiating between volunteer and geographic content. In each of these, we can differentiate between implicit and explicit contributions. Explicit volunteering is when people are knowingly volunteering effort to a project. Implicit volunteering is when information is shared openly but without people knowing how their contribution will be eventually used. For example, carrying out bird observations and reporting to a shared database is considered explicit volunteering, while the reuse of all the georeferenced images of parks that are shared on a photo-sharing website (such as Flickr) to assess the level of interest in the green space is implicit volunteering (e.g. Gliozzo, Pettorelli and Haklay 2016), since the images were shared without this purpose in mind. By and large, implicit volunteering examples are not covered in this book, since our focus is on the design and implementation of explicit data collection and sharing. Yet, it is worth mentioning this mode of contribution, since it can contribute to citizen science efforts, for example by searching through photo-sharing websites for observations that are relevant to the topic and then integrating them with a data set of explicit contributions. Another helpful distinction can be made between whether the participant needs to contribute information actively and knowingly (e.g. use an app such as WideNoise to measure the level of noise; see Becker et al. 2013) or to share information passively (e.g. use a phone to sense the signal from different telephone masts and share this information via OpenSignal – http://www.opensignal.com).

To understand the landscape of citizen science, this chapter uses a combination of two typologies: Haklay's (2013) analysis of levels of

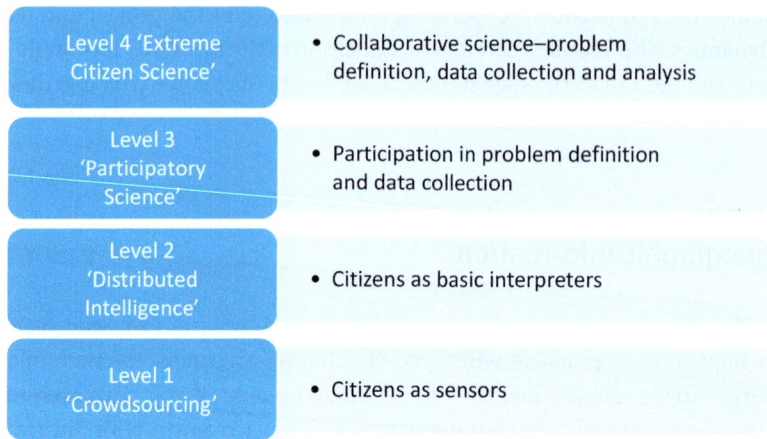

**Fig. 1.1** Different levels of engagement in citizen science (after Haklay 2013). Reprinted by permission from Springer: 'Citizen science and volunteered geographic information: Overview and typology of participation' by Muki Haklay, 2013.

engagement in citizen science (Figure 1.1) and Shirk et al.'s (2012) five models of public participation in scientific research (Figure 1.2). Both typologies address the balance between participants and scientists in terms of setting the aims of the project and executing it.

Haklay's (2013) typology offers four levels of engagement, according to the role of the participants and project owners. It particularly emphasises the agency of the participants in terms of carrying out tasks within the project and shaping the scientific process of setting the question, collecting and analysing the data and using the results. It categorically does not suggest that participation at a lower level is not a significant and meaningful experience from the point of view of the participants. It also does not suggest that limited agency in shaping the scientific process negates the agency of the participants in choosing to join a project and engage with it. At level 1, participation is focused on the provision of resources, and the cognitive engagement is minimal after setting in motion the sharing of resources. This is the level of crowdsourcing, where the participants contribute through a wide range of resources: computing, the locations that they visit on their daily routes or access to sensors that they have. Level 2 is distributed intelligence in which the participants are engaged in higher cognitive tasks such as data collection or interpretation. To do this, participants are asked to undertake training which can vary in length – from a few minutes to a preparatory workshop that might

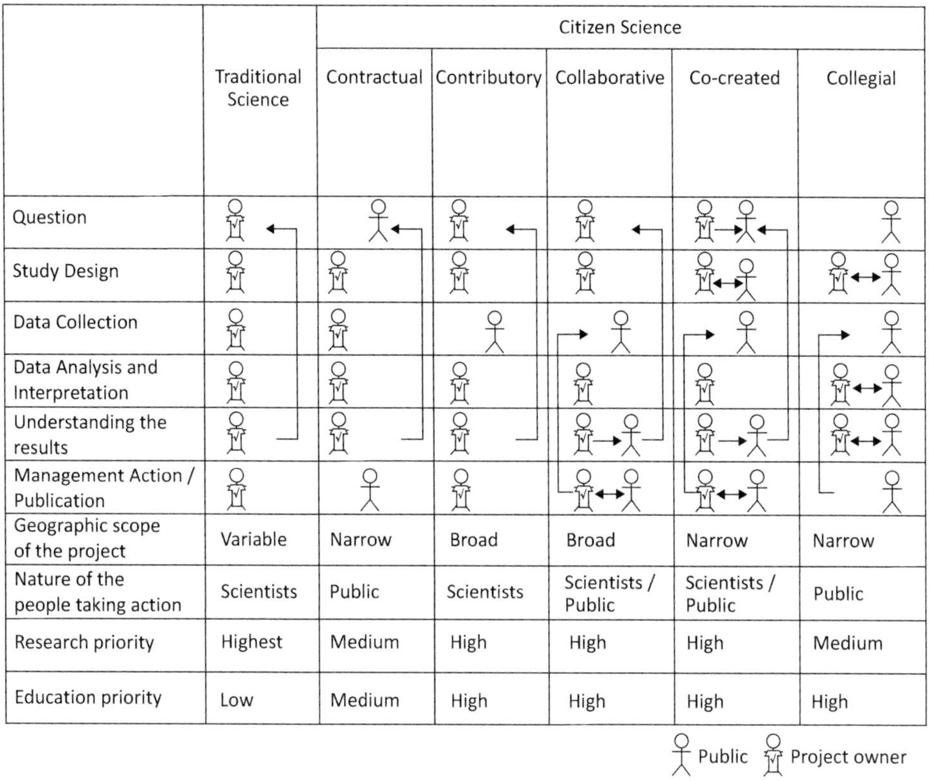

|  | Traditional Science | Citizen Science | | | | |
|---|---|---|---|---|---|---|
|  |  | Contractual | Contributory | Collaborative | Co-created | Collegial |
| Question |  |  |  |  |  |  |
| Study Design |  |  |  |  |  |  |
| Data Collection |  |  |  |  |  |  |
| Data Analysis and Interpretation |  |  |  |  |  |  |
| Understanding the results |  |  |  |  |  |  |
| Management Action / Publication |  |  |  |  |  |  |
| Geographic scope of the project | Variable | Narrow | Broad | Broad | Narrow | Narrow |
| Nature of the people taking action | Scientists | Public | Scientists | Scientists / Public | Scientists / Public | Public |
| Research priority | Highest | Medium | High | High | High | Medium |
| Education priority | Low | Medium | High | High | High | High |

Fig. 1.2 The Five Cs model of participation.
Source: After Cooper et al. 2007.

take a day or more. At these two levels of engagement, the project owner is setting out the goals of the project and the methodology in which data will be collected, and the participants are following instructions on how to participate. This does not make them passive participants, and the literature records highly inventive ways in which people become engaged (see Cooper 2016). However, the projects that utilise these levels frequently conceive participants' participation as restricted and that they have little role to play in the design of the scientific task. It is also important to note that for many participants, the structures in place match their motivations and interests. At level 3, participatory science, there is increased input from participants, including involvement in problem definition, and in consultation with scientists and experts, a data collection method is devised. The participants are then engaged in data collection, but they frequently require the assistance of the experts in analysing and interpreting the results. The participants are not involved in the analysis of the results of their effort – perhaps because of the level of knowledge that is required to infer scientific conclusions from the data. Finally, at level 4, extreme citizen science, the participants can choose their level of engagement and potentially can be involved in all the stages of the process – from identifying the scientific question to the analysis and publication or utilisation of results. This form requires scientists to act mostly as facilitators, in addition to their role as experts. It is also possible that an extreme citizen science project can take place without any involvement from professional scientists.

The Shirk et al. (2012) set of models is somewhat similar and can be termed 'the Five Cs' typology, as it uses five models of relationships between project owners (assumed to be scientists) and the public: contractual, contributory, collaborative, co-created and collegial (Figure 1.2). They describe them as:

> *Contractual* projects, where communities ask professional researchers to conduct a specific scientific investigation and report on the results;
> *Contributory* projects, which are generally designed by scientists and for which members of the public primarily contribute data;
> *Collaborative* projects, which are generally designed by scientists and for which members of the public contribute data but also help to refine project design, analyze data, and/or disseminate findings;
> *Co-created* projects, which are designed by scientists and members of the public working together and for which at least some of the public participants are actively involved in most or all aspects of the research process; and
> *Collegial* contributions, where non-credentialed individuals conduct research independently with varying degrees of expected recognition by institutionalized science and/or professionals. (Shirk et al. 2012, 4)

Figure 1.2 is useful to understand the five types of relationships. As it is partially based on the work of Cooper et al. (2007), it reflects the situation in environmental projects, which are the most common type of project in geographic citizen science. The figure depicts a typical scientific process on the left. The process starts with setting the research question, followed by study design, then data collection, analysis, interpretation and finally action, which can be an environmental management action or, in the case of scientific research, a publication. It also examines the geographic scope of the project and who is supposed to act. Finally, it is helpful to look at the balance between the research priority: producing scientific results that can be used for management or a scientific publication versus the education priority, which might include increasing awareness, increasing the education of participants, reaching out to new audiences and so on. If we look at the columns from left to right, we first see the traditional science process in which all the actions are carried out by scientists. The research priority is the highest, and there is little (if any) education and engagement priority. The scope is variable – it might be a small area, or a whole continent, depending on the interest, funding and focus of the scientists. In any case, the public is out of the process, and in case of scientific publications that are not open access, they cannot see the results of this work.

The second column explains the contractual model. This model of public engagement in research can happen when a community group engages scientists (either paid or pro bono) to research an issue that concerns them, for example local concern about watershed issues such as water quality which are then handled by researchers and students at a local university – in a model akin to the 'science shop' which also provides such services (Wilderman 2007). In this type of project, the geographic coverage is usually narrow, and the research and education elements are both at the medium level. However, the members of the public are expected to use the results in a way that suits their goals.

The contributory model is the most common in citizen science, and while Figure 1.2 shows that the public is involved mainly in data collection, there are projects where they are also involved in basic analysis tasks, such as examining and classifying an image. In this form of research, both the research and education aims are a priority, and the geographic coverage can be extensive. The results are mostly used by scientists.

The relationships in the next model, the collaborative model, are more complex. The scientists are setting the question and, most of the time, design many elements of the study. However, participants are involved not only in data collection but also in understanding the results, which might lead to refining the research questions, for example during a local air quality study (Evans-Agnew and Eberhardt 2019). The collaboration also extends to the use of the results, with participants taking part in the resulting action. As with the contributory model, the research and education priorities are high, and the geographic scope can be broad.

In co-created projects, the participants are involved in setting the questions and in the data analysis through an ongoing dialogue between the project owners and the participants. However, the scientists still maintain control over the goals of the project and lead on the design of the data collection and analysis part with high research and educational elements, while the geographic scale is commonly narrow. As for the results, they can be used by both scientists and participants, although the use can be different, for example addressing a local problem versus publishing a paper.

Finally, the collegial model can be carried out without scientists, and it requires that the entire research process is undertaken by participants. In this model, the scientists are providing expert advice (if needed), most commonly on study design and data interpretation. For example, the bottom-up project in which young parents carried out scientific research as part of the 'Parenting Science Gang' was supported in certain areas by asking scientists questions that they wanted to explore (Collins 2019). The education element of such a model is high, while the research output is less critical. As with the previous model, because of the need for close interaction between participants, the geographic scope is narrow.

Notice that the two typologies can be related to each other. The first two levels in Haklay (2013) describe mostly contributory projects, while level 3 describes a co-created project and level 4 relates to a specific type of collegial project.

Equipped with these typologies, we can now turn to the discussion of different forms of citizen science and see how geography interacts with the project aims, goals and data.

## 3.1 Citizen science

The following will briefly look at the types of activities that are included in citizen science (for a more in-depth examination, see Haklay 2013; Haklay, Mazumdar and Wardlaw 2018). Seven types of citizen science are discussed here: (1) passive sensing, (2) volunteer computing, (3) volunteer thinking, (4) environmental and ecological observations, (5) participatory sensing, (6) community/civic science and (7) do-it-yourself (DIY) science.

(1) Passive sensing relies on participants in the project providing a resource that they own for automatic sensing and information sharing. The information that is collected through the sensors is then used by scientists for analysis. The projects can involve stationary sensors – such a home-based weather station that is linked to the Weather Underground network (https://www.wunderground

.com/), with the data being used by meteorologists in forecasting. A mobile example is provided by the Contagion! experiment, in which mobile phone software was used to provide data for a pandemic propagation model (Klepac, Kissler and Gog 2018). Looking at our two classifications, this activity falls under crowdsourcing in Haklay (2013) and contributory in Shirk et al. (2012).

(2) Volunteer computing is a method in which participants share their unused computing resources, on their personal computer, tablet or smartphone, and allow scientists to run complex computer models during the times when the device is not in use. An example of this is the Climate Prediction network (https://www.climateprediction.net/) in which participants share computing resources to allow scientists to run multiple climate models. There are rare exceptions to the scientist-led project, such as the Rechenkraft effort in which a group of participants set systems and set projects by themselves (see Haklay 2015), which is an example of a more bottom-up-led project – although by necessity, the volunteers do not have much control over the project when they are carrying out the computing tasks. As with passive sensing, this activity falls under crowdsourcing in Haklay (2013) and contributory in Shirk et al. (2012).

(3) Volunteer thinking uses what Clay Shirky (2010) termed 'cognitive surplus', which is the cognitive ability of people not used in passive leisure activities such as watching TV. In this type of project, the participants contribute their ability to recognise patterns or analyse information that will then be used in a scientific project. For example, GeoTag-X (http://geotagx.org/) recruits volunteers to help with the classification of images as part of humanitarian efforts (Smith 2017). These projects are at the distributed intelligence level (level 2) and are usually contributory. A few projects in this area started in a more collegial manner, such as the OpenStreetMap (http://www.openstreetmap.org/) project in which volunteers are creating a digital map of the world.

(4) Environmental and ecological observation focuses on monitoring environmental pollution or observations of flora and fauna through activities such as bio-blitz in which a group of volunteers study a site thoroughly, using their phones to record and share observations. This is the area of activities where the contractual, contributory, collaborative, co-created and collegial models exist (for examples, see Cooper et al. 2007; Shirk et al. 2012), while in Haklay (2013) it covers the three top levels. However, surveys and analysis of projects in this area (e.g. Pocock et al. 2017) demonstrate that the

majority fall under the distributed intelligence and contributory categories.

(5) Participatory sensing is similar to the previous type of observation but gives the participant more roles and control over the process. While many environmental and ecological observations follow data collection protocols that were designed by scientists, in participatory sensing, the process is more distributed and emphasises the active involvement of the participants in setting what will be collected and analysed (e.g. the Hush City app, discussed in Chapter 6). In participatory sensing, because of the level of skills required to develop the software and hardware, the projects tend to be at the participatory science level (level 3) and are co-created projects, although some will be strongly contributory in their settings.

(6) Community/civic science, also known as bottom-up science, is initiated and driven by a group of participants who identify a problem that is a concern for them and address it using scientific methods and tools. Within this type of activity, the problem, data collection and analysis are often carried out by community members or in collaboration with scientists or established laboratories. We therefore find these projects at the participatory science level (level 3) and using the contractual or the collegial models.

(7) DIY science includes the creation of new devices and methodologies that are created by participants to explore their own questions. The type of questions that DIY science addresses can be environmental concerns, for example in the work of the Public Laboratory for Open Technology and Science (https://publiclab.org/) (Dosemagen, Warren and Wylie 2011) where people share experiences in developing tools and using them to find information about issues such as air or water quality. Another form of DIY science focuses on molecular biology and the manipulation of DNA, within DIY biology, or DIYBio (Strasser et al. 2019). Activities in DIY science fall under extreme citizen science (level 4) and the collegial model. They usually require a significant investment of time and some resources to participate in the activity.

These seven activities that are associated with citizen science show the breadth of the forms via which people can participate in scientific research today. The projects range from a complex effort to engineer bacteria to installing the app Vodafone DreamLab which supports cancer research while participating mobile phones are being charged at night (Pattnaik et al. 2018). This variety is causing frequent discussions about the boundaries of citizen science and the need to clarify its terminology (for a detailed discussion, see Eitzel et al. 2017). We can now turn to VGI.

## 3.2 Volunteered geographic information and geographic citizen science

Within the category VGI, some activities clearly fall outside the realm of citizen science, for example when people use their phone to 'check in' to a bar or provide a restaurant review on applications such as TripAdvisor (https://www.tripadvisor.co.uk/). This is an example of explicitly volunteering with an active and explicit geographic contribution. This is also a non-scientific example of a contributory project that focuses on distributed intelligence. VGI also includes contributions to Wikipedia (http://www.wikipedia.org/) that contain place names (e.g. an article about a historical figure mentioning a place that they travelled to) but are not explicitly geographic contributions, since the aim of the article is not to generate geographic information, although it is explicitly contributed.

Some VGI is more similar to citizen science in that it is concerned with recording geographic facts and observations. An app such as StreetBump (http://www.streetbump.org/) runs on a participant's phone and uses its sensors to detect bumps in the road while they are driving their car. Here, there is an explicit sensing of car movement associated with the geographic location from GPS. This is explicitly volunteered, passive and with an explicit geographic contribution. OpenStreetMap is another VGI example that can be identified as part of geographic citizen science, as it is concerned with recording facts about the world and measuring them accurately, and an outstanding example of a large-scale collegial project in which the participants are setting the goals of the project, creating and maintaining the software for it and using the resulting information.

When examining the overlap between citizen science activities and geography (Figure 1.3), we can see the range of citizen science activities that fall within VGI and are therefore geographic citizen science. The activities of environmental and ecological observations and civic/community science are inherently place based (Cooper et al. 2007; Haywood, Parrish and Dolliver 2016; Newman et al. 2017), and are therefore an integral part of geographic citizen science. Passive and participatory sensing are also frequently place based, and therefore part of geographic citizen science, although when passive sensing focuses on health issues, it might not use location information, and therefore it is only sometimes depicted as geographic citizen science.

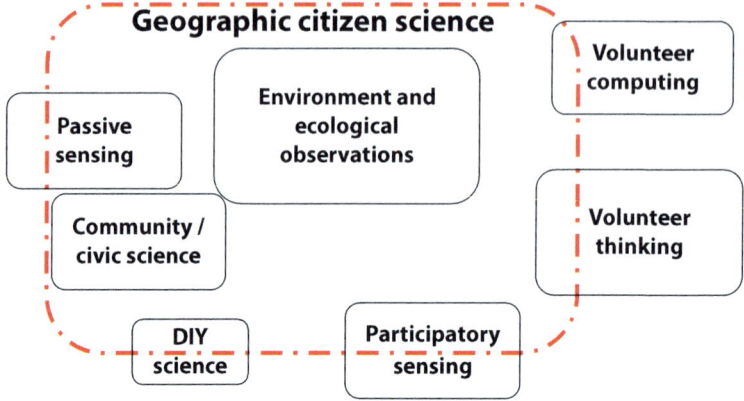

**Fig. 1.3** Conceptual overlap between volunteered geographic information and citizen science – the boundaries of geographic citizen science. DIY: do-it-yourself. Source: author.

The case is more complex with volunteer computing and volunteer thinking, where projects do not necessarily deal with geographic information and can be about analysing neurons or looking at images of galaxies. Here, only when the issue is explicitly geographic – as in classifying images from a camera trap in the Serengeti – is the result defined as geographic citizen science. It is likely that this is the minority of the cases for both these categories. Finally, within DIY science, it is possible to separate the environmental sensing activities, which are part of geographic citizen science, from activities such as DIYBio or other chemical and physical experiments, which will not be.

## 3.3 Digital technologies, interaction and geographic citizen science

Having mapped the landscape of citizen science and its geographic aspects, we can now turn to the need to understand its interactions with digital technologies. It is important to note that the different modes of participation and interaction can lead to widely different requirements regarding how the participant will interact and use digital technologies during a citizen science project. Let us look at two cases: a passive sensing application, such as the Contagion! experiment mentioned above, and an application for reporting non-emergency issues, such as ImproveMyCity (discussed further in Chapter 9).

In a passive sensing activity (crowdsourcing in Haklay 2013 and a contributory project in Shirk et al. 2012) with an implicit geographic contribution, we can see several implications for the design. There is a need to bring the existence of the mobile app and the time frame of the experiment to the awareness of the public, and to encourage them to download and activate the app. It is also important to encourage people in a specific geographic area to participate in the experiment, and therefore we need to have an idea about the phones that are commonly used in the area, as well as the app's compatibility with a range of operating systems. We also need to ensure that the app is not consuming too much power when recording its location, as this will inconvenience participants and might lead to the removal of the app. There is also a need to ensure that the Bluetooth communication is left active for the period of the experiment, since that is how the phone senses that another participant in the experiment is nearby. Finally, as this is happening in urban areas, there is an expectation of good mobile network coverage that will allow the information to be transmitted to the server, although as it will be used by people during commuting, we need to consider areas without coverage (e.g. while travelling on an underground train). We do not need to design many interactions with the participants, apart from alerting them when things are not working (e.g. if Bluetooth is switched off), and we do need to provide an indication when the experiment ends. Notice that because of the limited level of engagement, interaction with participants, such as answering questions and providing technical support, is limited.

In turning to the other case, the reporting of non-emergency issues is distributed intelligence and a contributory project, with an explicit and active geographic contribution. Moreover, we are expecting an ongoing contribution over time, with people reporting issues they see around them and that require handling by the local government. Similar to the other case, there is a need to make people aware of the existence of the app and to ensure that people within the specified locality are encouraged to download and use it. However, there is a more complex requirement, as we want participants to remember to open the app actively when they see an issue that requires intervention (e.g. a broken street light). Here, we also need to understand which mobile phones are in use by residents and if we can expect good mobile network coverage in all the places where the app will be used. We can assume that participants will expect to see a response to their reporting and indications of when the issue was resolved. The app interface and the experience of using it will also play an important role in the level of use.

We can see that each type of application will have some generic requirements, which can be inferred from the mode of operation and the typologies above, as well as application-specific requirements. Table 1.1 summarises the relationships between the type of activities, the frameworks and some design implications. This area is receiving attention within human–computer interaction studies (Preece 2016), and guidelines for designing mobile apps for citizen science are emerging (Sturm et al. 2018). The case studies in this book provide detailed descriptions of design and implementation across the range that the typologies cover.

The case studies also provide a vivid demonstration of the diverging requirements between the core and the periphery in terms of digital connectivity and technological and social development. The cases that describe geographic citizen science applications in the urban part of the Global North (such as those discussed in the second part of this book) are operating in an environment where there are many preconditions that make the use of citizen science a possibility. The network coverage across built and open areas is usually good, and the speed of delivering information to and from mobile devices is high. In addition, the population that lives in these areas is technologically advanced, and most people will own mobile phones that can run geographic citizen science apps. The level of education of the population will also be relatively high, and there is an assumption that the majority will be literate. It is also safe to assume that a significant number will have a higher education degree, so they can understand the goals and the processes of the project. The project originators will also benefit from the availability of mass media and the potential to share information about the project through it, as well as recruiting participants through social networks and events that bring people together. All these preconditions mean that the main effort of the project originators is in developing the software (application and back-end) and in the recruitment of and engagement with the participants.

In contrast, the cases from the periphery (such as those discussed in the third part of this book) demonstrate when many of these preconditions do not apply, and the development and deployment of the project need to confront these issues. In such conditions, it is not reasonable to expect ubiquitous mobile network coverage, and therefore applications need to be able to store their data and to share it when the device is near a communication node. The use of devices, while increasing, is not ubiquitous, and therefore project originators need to consider how they can provide devices to participants, as well as training in how to use them. Also, the assumption of literacy cannot be taken for granted, and solutions that are inclusive need to find a way to engage people with different

**Table 1.1** Types of projects, typologies and some implications for design

| Type of activity | VGI framework (Craglia et al. 2012) | Levels of engagement (Haklay 2013) | Model of PPSR (Shirk et al. 2012) | Example of design implications |
|---|---|---|---|---|
| Passive sensing | Volunteered – explicit Geographic – implicit | Crowdsourcing (L1) | Contributory | Recruitment and set-up crucial; battery and data use |
| Volunteer computing | Volunteered – explicit Geographic – implicit | Crowdsourcing (L1) | Contributory | Maintaining engagement when participants change device; notification of processing and the scientific outputs |
| Volunteer thinking | Volunteered – explicit Geographic – implicit | Distributed intelligence (L2) | Contributory | Design of micro-tasks, encouraging participants to contribute |
| Environmental/ ecological observations | Volunteered – explicit Geographic – explicit | Distributed intelligence (L2), also participatory science (L3) and extreme citizen science (L4) | Mostly contributory, also contractual, collaborative, co-created and collegial | Connectivity at data-collection location, ensuring data quality |
| Participatory sensing | Volunteered – explicit Geographic – explicit | Participatory science (L3) | Co-created, some contributory | Identifying participants aims and needs during data-collection tasks |
| Community/ civic science | Volunteered – explicit Geographic – explicit | Extreme citizen science (L4) | Contractual, collegial, some co-created | Early engagement with participants, designing end-user customisation, ensuring visualisation of results that serve the purposes of the community |
| DIY science | Volunteered – explicit Geographic – explicit | Extreme citizen science (L4) | Collegial | Designing peer-to-peer learning, instructions for DIY tools, ensuring guidelines for data quality |

VGI: volunteered geographic information; PPSR: public participation in scientific research; DIY: do-it-yourself.

levels of literacy. Even the availability of electricity to charge the devices cannot be assumed without examination of the local conditions. As a result, the project originators need to consider all the aspects of the project: software, hardware, connectivity, energy, training and engagement.

## 4. Development and future directions

Even though VGI and citizen science (and therefore geographic citizen science) have much longer histories, most of the attention from policy-makers, researchers and businesses has only occurred in the last decade. Frequently, questions arise about the quality of the resulting information as well as the motivation of the participants (see Sieber and Haklay 2015). As more evidence emerges to confirm the quality of data is adequate and that participants' motivations are a contribution to shared knowledge, attention can turn towards the compilation of longitudinal data collection. In some VGI activities, the collected information is 'hyper-local', making it only relevant to a small area in both space and time, for example information about a traffic jam and its implications for navigational decisions. Yet, even this localised information has relevance on a wider scale. In most VGI data sets, and especially in the area of geographic citizen science, there is a need to understand how the information changes over time. Thus, the activities in these fields have the duality of describing a snapshot of the world (capturing an observation at a specific time and place and recording it). Yet, because of the continuous sharing of the information, the data set is always dynamic and in a state of change (see Perkins 2014).

The process in which the information is produced, controlled and shared demonstrates differences in the power relationship. Concurrently, the ability to maintain the repository of information over time should receive more attention. For example, OpenStreetMap servers require regular operating system updates and effort as well as resources to deal with hardware failures. Sustainability requires an organisation, institution or company to take responsibility in the long term. As a result, the control of the system (understood here in the wider sense and not just the hardware/software part of it) foregrounds issues of power, control and resources into these seemingly distributed non-hierarchical activities. In addition, the process of data quality assurance requires oversight and moderation of more experienced and knowledgeable participants who check the information provided by novices. Over time, power relationships reveal themselves in the case of geographic citizen science.

Within the frameworks presented here, there is scope to develop further guidance, according to the type of project and in a way that will support those who want to initiate and develop new geographic citizen science projects. As awareness of the potential of geographic citizen science increases, there will be an increasing need to understand what design elements will work best in each context and what are the appropriate guidelines for each case. This is a multidimensional problem that requires analysis of context, culture, technology and science.

## 5. What's next for geographic citizen science

As VGI and citizen science activities progress, questions regarding data quality, the longevity of engagement, incentives and motivation of volunteers, as well as the roles that participants take, persist. Some research examining differential power differences has begun (Sieber and Haklay 2015). Nonetheless, there is plenty of scope to study geographic citizen science critically and to consider how the lessons learned can be integrated into the design of new applications. For example, there are different levels of inclusiveness in terms of who is involved in data collection and the areas that are being monitored, which are tightly linked to the analysis of the core and periphery of digital connectivity. There is also much to gain from further investigating organisational practices and cultural influence with regards to the recruitment and ability to retain participants over time. Aspects of gender inequality are being discussed (Cooper and Smith 2010; Stephens 2013), while ethnic, socio-economic and age disparities have received less attention. There is also scope to understand how wider politics and economic incentives lead to outcomes, for example which thematic areas receive attention and funding and, more specifically, why the production of base maps is perceived as a valuable commercial activity while the recording of biodiversity is not. Understanding geographic citizen science as a socio-technical system, and giving due attention to social aspects might provide better insights into the nature of the spatial data being produced through these activities and what they tell us about the state of the world.

## Acknowledgements

Many thanks to Katerina Zourou and Susanne Hecker for their comments on an earlier version of this paper. The research that underlines this project was supported by the European Union's H2020 research and innovation programme Doing It Together Science (under Grant Agreement No. 709443), the ERC Advanced Grant project European Citizen Science: Analysis and Visualisation (under Grant Agreement No. 694767) and the Natural Environment Research Council (Grant Nos NE/R012067/1 and NE/S017437/1).

## References

Becker, Martin, Saverio Caminiti, Donato Fiorella, Louise Francis, Pietro Gravino, Mordechai (Muki) Haklay, Andreas Hotho et al. 2013. 'Awareness and learning in participatory noise sensing', *PloS One* 8: e81638.

Brown, Janelle. 2001. 'Three case studies'. In *Online Communities: Commerce, Community Action, and the Virtual University*, edited by Chris Werry and Miranda Mowbray, 33–46. Upper Saddle River, NJ: Prentice Hall.

Calabrese Barton, Angela M. 2012. 'Citizen(s') science. A response to "The Future of Citizen Science"', *Democracy and Education* 20: 12.

Collins, Sophie. 2019. Executive summary: Evaluation report. Accessed June 2019. http://parentingsciencegang.org.uk/evaluation/executive-summary-evaluation-report.

Cooper, Caren. 2012. 'Links and distinctions among citizenship, science, and citizen science. A response to "The Future of Citizen Science"', *Democracy and Education* 20: 13.

Cooper, Caren. 2016. *Citizen Science: How ordinary people are changing the face of discovery*. New York: Overlook Press.

Cooper, Caren, Janis Dickinson, Tina Phillips and Rick Bonney. 2007. 'Citizen science as a tool for conservation in residential ecosystems', *Ecology and Society* 12: 11.

Cooper, Caren, and Jennifer Smith. 2010. 'Gender patterns in bird-related recreation in the USA and UK', *Ecology and Society* 15: 4.

Craglia, Max, F. Ostermann and Laura Spinsanti. 2012. 'Digital Earth from vision to practice: Making sense of citizen-generated content', *International Journal of Digital Earth* 5: 398–416.

Dosemagen, Shannon, Jeffrey Warren and Sara Wylie. 2011. 'Grassroots mapping: Creating a participatory map-making process centered on discourse', *Journal of Aesthetics and Protest* 8.

Eitzel, Melissa, Jessica Cappadonna, Chris Santos-Lang, Ruth Duerr, Sarah Elizabeth West, Arika Virapongse, Christopher Kyba et al. 2017. 'Citizen science terminology matters: Exploring key terms', *Citizen Science: Theory and Practice* 2: 1–20.

Evans-Agnew, Robin A., and Chris Eberhardt. 2019. 'Uniting action research and citizen science: Examining the opportunities for mutual benefit between two movements through a woodsmoke photovoice study', *Action Research* 17: 357–77.

Gliozzo, Gianfranco, Nathalie Pettorelli and Mordechai (Muki) Haklay. 2016. 'Using crowd-sourced imagery to detect cultural ecosystem services: A case study in South Wales, UK', *Ecology and Society* 21: 6.

Goodchild, Michael F. 2007. 'Citizens as sensors: The world of volunteered geography', *GeoJournal* 69: 211–21.

Haklay, Mordechai (Muki). 2013. 'Citizen science and volunteered geographic information: Overview and typology of participation'. In *Crowdsourcing Geographic Knowledge: Volunteered geographic information (VGI) in theory and practice*, edited by Daniel Z. Sui, Sarah Elwood and Michael F. Goodchild, 105–22. New York: Springer.

Haklay, Mordechai (Muki). 2015. *Citizen Science and Policy: A European perspective*. Washington, DC: Woodrow Wilson International Center for Scholars.

Haklay, Mordechai (Muki), Suvodeep Mazumdar and Jessica Wardlaw. 2018. 'Citizen science for observing and understanding the Earth', In *Earth Observation Open Science and Innovation*, edited by Pierre-Philippe Mathieu and Christoph Aubrecht, 69–88. Cham, Switzerland: Springer.

Haklay, Mordechai (Muki), Alex Singleton and Chris Parker. 2008. 'Web Mapping 2.0: The neogeography of the GeoWeb', *Geography Compass* 2: 2011–39.

Haywood, Benjamin K., Julia K. Parrish and Jane Dolliver. 2016. 'Place-based and data-rich citizen science as a precursor for conservation action', *Conservation Biology* 30: 476–86.

Howe, Jeff. 2006. 'The rise of crowdsourcing', *Wired Magazine* 14: 1–4.

Klepac, Petra, Stephen Kissler and Julia Gog. 2018. 'Contagion! The BBC Four pandemic – the model behind the documentary', *Epidemics* 24: 49–59.

Laurel, Brenda. 2001. *Utopian Entrepreneur*. Cambridge, MA: MIT Press.

Liebenberg, Louis, Lindsay Steventon, Karel Benadie and James Minye. 1999. 'Rhino tracking with the CyberTracker field computer', *Pachyderm* 27: 59–61.

Mueller, Michael P., Deborah Tippins and Lynn A. Bryan. 2011. 'The future of citizen science', *Democracy and Education* 20: Article 2.

Newman, Greg, Mark Chandler, Malin Clyde, Bridie McGreavy, Mordechai (Muki) Haklay, Heidi Ballard, Steven Gray et al. 2017. 'Leveraging the power of place in citizen science for effective conservation decision making', *Biological Conservation* 208: 55–64.

Oxford English Dictionary (OED) Online. 2014. Citizen science. Accessed August 2014. https://www.oed.com/view/Entry/33513?redirectedFrom=citizen+science#eid316619123.

Pattnaik, Swetansu, Catherine Vacher, Hong Ching Lee, Warren Kaplan, David M. Thomas, Jianmin Wu and Mark Pinese. 2018. 'Network-aware mutation clustering of cancer', *BioRxiv* 432872.

Perkins, Chris. 2014. 'Plotting practices and politics: (Im)mutable narratives in OpenStreetMap', *Transactions of the Institute of British Geographers* 39: 304–17.

Pocock, Michael J. O., John C. Tweddle, Joanna Savage, Lucy D. Robinson and Helen E. Roy. 2017. 'The diversity and evolution of ecological and environmental citizen science', *PLoS One* 12: e0172579.

Preece, Jennifer. 2016. 'Citizen science: New research challenges for human–computer interaction', *International Journal of Human–Computer Interaction* 32: 585–612.

Putz, Steve. 1994. 'Interactive information services using World-Wide Web hypertext', *Computer Networks and ISDN Systems* 27: 273–80.

Rose, Pauline. 2005. 'Is there a "fast-track" to achieving Education For All?', *International Journal of Educational Development* 25: 381–94.

Shirk, Jennifer L., Heidi L. Ballard, Candie C. Wilderman, Tina Phillips, Andrea Wiggins, Rebecca Jordan, Ellen McCallie et al. 2012. 'Public participation in scientific research: A framework for deliberate design', *Ecology and Society* 17: 29.

Shirky, Clay. 2010. *Cognitive Surplus: Creativity and generosity in a connected age*. London: Penguin.

Sieber, Renée E., and Mordechai (Muki) Haklay. 2015. 'The epistemology(s) of volunteered geographic information: A critique', *Geo: Geography and Environment* 2: 122–36.

Silverman, Jacob. 2014. 'The crowdsourcing scam: Why do you deceive yourself?', *The Baffler* 26.

Smith, Cobi. 2017. 'A case study of crowdsourcing imagery coding in natural disasters'. In *Data Analytics in Digital Humanities*, edited by Shalin Hai-Jew, 217–30. Cham, Switzerland: Springer.

Stephens, Monica. 2013. 'Gender and the GeoWeb: Divisions in the production of user-generated cartographic information', *GeoJournal* 78: 981–96.

Strasser, Bruno J., Jérôme Baudry, Dana Mahr, Gabriela Sanchez and Elise Tancoigne. 2019. '"Citizen science"? Rethinking science and public participation', *Science and Technology Studies* 32: 52–76.

Sturm, Ulrike, Sven Schade, Luigi Ceccaroni, Margaret Gold, Christopher Kyba, Bernat Claramunt, Mordechai (Muki) Haklay et al. 2018. 'Defining principles for mobile apps and platforms development in citizen science', *Research Ideas and Outcomes* 4: e23394.

Sui, Daniel, Michael Goodchild and Sarah Elwood. 2013. 'Volunteered geographic information, the exaflood, and the growing digital divide'. In *Crowdsourcing Geographic Knowledge*, edited by Daniel Sui, Sarah Elwood and Michael Goodchild, 1–12. Dordrecht, The Netherlands: Springer.

Teichler, Ulrich, and Sandra Bürger. 2008. 'Student enrolments and graduation trends in the OECD area: What can we learn from international statistics', *Higher Education to 2030: Demography* 1: 151–72.

Wilderman, Candie C. 2007. 'Models of community science: Design lessons from the field'. Paper presented at the Citizen Science Toolkit Conference, Cornell Laboratory of Ornithology, Ithaca, NY, June 2007.

# Chapter 2
# Design and development of geographic citizen science: technological perspectives and considerations

Vyron Antoniou and Chryssy Potsiou

## Highlights

- Developments in information technology have a direct impact on the aims, goals and missions of different geographic citizen science projects, and create both challenges and opportunities.
- Geographic citizen science decision makers should always invest time and effort to increase their own technological awareness.
- It is imperative for technologies adopted to be interoperable with other technologies used for the project or collaborative projects and not work in their own silos.
- Technology should be an enabling factor for all and not create biases among participants.

## 1. Introduction

Over the past few years, a major shift in the technological arena has had a direct impact on geographic citizen science: the combination of significant developments in open-source software, do-it-yourself (DIY) hardware proliferation and the equipping of mobile devices with multiple sensors. This combination has enabled millions of citizens to be involved in geographic citizen science projects with minimum cost and, indeed, fuelled its expansion to every domain. However, this combination is neither always straightforward nor hassle free. Moreover, it is often a challenge to intertwine any of the possible technological options with the aims, goals and missions of different geographic citizen science projects. It is difficult to find and implement the best possible technological solution (given a project's characteristics and constraints, e.g. funds, time, resources, etc.), and this can be true irrespective of how technologically savvy project managers and participants are. There are several factors that contribute to this challenge, but perhaps the most important one is the pace of change and innovation on the technological front, from hardware and software that might appear or become obsolete, to prerequisites and requirements that might change, to the introduction of disruptive technologies that completely change everything.

Nevertheless, technological advances can be thought of both as a challenge and as an opportunity to combat existing problems, progress the aims and broaden the reach of geographic citizen science projects. For example, hardware is getting smaller and more efficient; new hardware is becoming available (e.g. drones); algorithms and software are evolving; free and open-source software now exist; multiple sophisticated sensors equip everyday devices (e.g. mobile devices, wristbands); and access is provided to what was until recently extremely expensive data that previously only governments or big corporations could afford and had access to (e.g. high-resolution satellite imagery). For example, as Pimm et al. (2015) note, technologies for species monitoring have evolved from expensive radio collars to satellite imagery or unmanned aerial vehicles (UAVs) and drones. Through this constant evolution, technology offers a new and diverse set of opportunities in geographic citizen science by opening new channels for public involvement, providing means to engage new audiences (Mazumdar et al. 2018) and for collaborative work (Mooney, Corcoran and Ciepluch 2013).

Today, technology is an indispensable part of both citizen science and geographic citizen science projects in almost every step of the process cycle. Some kind of technological solution is invariably applied to both small and large phases of a project in order to facilitate it. For example, mobile phones, DIY kits, sensors and drones are used to collect data; web and mobile applications are used to transfer data to central depots; databases are used to store, manage and disseminate data; and specific tools, hardware and software are used for the analysis of data and the dissemination of results. Technological solutions are used for controlling, managing and steering projects and much more.

However, a technological approach alone cannot solve all problems. Any technology adopted needs to be put to good use and be supported by effort and time, not least because technology itself can be a source of challenges and biases. Before adopting a specific approach, it is important to consider, among others, issues such as participant demographics, affordability, access and fitness for purpose (Mazumdar et al. 2018). A careful choice might boost the development of the project, while the adoption of a suboptimal technological approach can negatively affect the project in terms of both the financial support needed and functionality provided (Wiggins 2013). Thus, any technological consideration and research for a solution should keep a broad view and examine the entire ecosystem in which a technology is developed (Antoniou and Skopeliti 2017). A case in point is OpenStreetMap (OSM) where the entire project needs to be understood and examined (i.e. wiki specifications, editing processes, updates, versioning, medium used, number of developers, usability, etc.) before a decision can be made on which OSM editor to select in order to collect homogeneous data. In other cases, the granularity of the observations or outcomes that is required in each step might dictate the need for specific technology or vice versa; that is, if only a specific technology can be used, for example due to financial constraints, then the outcome will depend on the capabilities of this technology.

It is clear that the challenges are many, of varied nature and can appear in any step or process of a project. Thus, this chapter's target audience ranges from newcomers to the geographic citizen science domain with little or no experience in technological issues to experienced geographic citizen science administrators who battle on a daily basis with technology-related challenges. The aim of this chapter is to introduce and discuss a range of technological issues and their impact on the life cycle of a geographic citizen science project. In a nutshell, the aim is to make the life of geographic citizen science decision makers easier by providing a heads-up analysis of the technological front. The bottom line is that there is no one clear how-to method – a universal solution that will apply to all cases. The variables of each geographic citizen science project are often so numerous, so complicated and intricately intertwined that is next to impossible to suggest a one-size-fits-all solution. Nevertheless, in the following sections, the reader will find several technological aspects that need to be considered when designing, running or maintaining a geographic citizen science project. While this chapter will try to broaden the sphere of the discussion in order to touch upon many aspects of the use of technology in the geographic citizen science context, it cannot be considered exhaustive, not least because innovations and interruptions are generic issues with any technological discourse. What applies today can very soon be obsolete.

## 2. Main technological considerations

Before starting a discussion about technology and its relationship with geographic citizen science, the context needs explaining. This chapter will not cover the role of ubiquitous or underpinning technologies, but rather aims to provide thoughts and recommendations about technological advances that are challenging for geographic citizen science shareholders to grasp, use and maximise to their full potential. Thus, the chapter will not discuss directly issues such as the World Wide Web, the Internet or mobile networks (although there are geographic citizen science projects that need to battle the absence of such ubiquitous technologies, such as the case studies in Part 3 of this book). Furthermore, the chapter will not discuss specific technologies (although several examples will be provided), but rather present a generic discussion about what needs to be considered when geographic citizen science participants are faced with technological challenges.

Geographic citizen science decision makers should always invest time and effort to increase their own technological awareness. They need to monitor technological developments and how other projects overcome similar problems and challenges. This will give them the necessary orientation and confidence to search for the right solutions, which in many cases might already be in use. Developing a technological approach that is innovative is an option, but it may lead to unnecessary expenditure of effort, resources and time or to outcomes that do not fulfil the purpose of the project or are far from users' expectations.

Irrespective of technological choices, certain issues will always need to be considered, as they are generic in any type of technology used. Examples include issues such as human–computer interaction and user experience design (Preece 2016), which are extensively discussed in Chapter 3 and also throughout the book. Simply put, these issues deal with how humans interact with computers and how computers interface their capabilities to their users. Geographic citizen science decision makers should think carefully about factors that have an effect on their choices: from how technological savvy their participants will be, to how complicated the tasks are, to what the learning curve of each technological solution will mean for their geographic citizen science project. Of course, there might be technological solutions that do not require computer involvement (e.g. the use of a DIY kit), but still the overall concept of user-friendly interfaces that maximise usability should be considered.

Another group of issues, irrespective of the technological approach adopted, includes security, privacy, legal and ethical considerations. All these need to be planned carefully and, when necessary, communicated to geographic citizen science participants in a comprehensive and transparent way. For example, geographic citizen science decision makers should not take for granted that all participants will be noble and well behaved during their contributions. Tight security policies, especially in digital technologies, will contribute towards project viability and data integrity by protecting geographic citizen science projects from malevolent users. Similarly, respecting and protecting user privacy is not only crucial for a project's reputation, it can also be a legal obligation. Finally, ethical issues should also be carefully considered. Examples include intellectual property rights and access rights that users can have over the data they create and the results that come out of a geographic citizen science initiative. What belongs to whom, who has access to what and how this access will be provided (API, bulk downloads, etc.) need to be transparent from the outset of a geographic citizen science initiative. Of course, each decision on the above can have a different effect on overall processes adopted and the infrastructure needed.

In any case, the adoption of a proper technological approach should follow a generic but long-term plan. This does not mean that a geographic citizen science project should lose its flexibility to follow evolving technologies, but rather that the adoption of technological solutions should not be short-sighted and only solve ephemeral problems (which sometimes can be very important) such as more battery life, greater signal coverage or mobile app development. Available technologies should be seen not as a pool of tools to choose from but as an ecosystem that will enable geographic citizen science participants to address current and (most importantly) future challenges. Geographic citizen science decision makers should be aware from the outset of a project that most, if not all, technological solutions come with a number of drawbacks and restrictions. Drawbacks can include the learning curve of a technological solution or the cost of acquiring and maintaining the necessary hardware and software. Restrictions can exist in the use of specific technology. For example, the use of UAVs can be restricted by no-fly zones around airports or other protected areas. Technological continuity, future support, compatibility and interoperability of the adopted technology with other solutions used or to be used in the course of the project should also be carefully examined. Especially regarding software development, the update and release cycles are very important, as applications can become obscure if they fail to keep up with technological trends.

Thus, it is paramount to avoid viewing technology as a quick fix for transitory problems and start embracing the technological element of geographic citizen science projects as the key enabler and the primary medium through which citizens will engage with the cause of the initiative and interact with the natural or digital environment. It allows information and data to flow both horizontally and vertically, data integrity to be preserved, scientific tasks to be concluded, decisions to be made and the future steps of the project to be designed.

In this context, experimenting with existing and emerging technologies is crucial (Newman et al. 2012) for both the current and future needs of a geographic citizen science project. Instead of exclusively tying down the future and the potential of a project to the fate of a specific technology, decision makers should actively seek technological improvements whenever possible. Similarly, changes in the design and processes of a geographic citizen science project should not be avoided, but rather sought after when there is evidence of emerging and disruptive technologies that can improve and strengthen the project. Examples can be found in the adoption of social networking and gamification that can provide a totally different context of user engagement and participation or in the use of cloud computing to meet data management and computational challenges for global projects.

In practice, technological apparatuses can be used in different ways and combinations, for different purposes and in various contexts inside the geographic citizen science domain. Thus, appropriate considerations should also be made for each individual case. For example, technological needs are different for projects run by individuals or small groups compared to local, national or global ones. Even for projects of the same geographical scope, the needs differ based on how decentralised each project is. For example, a centralised project (e.g. OSM), where all participants contribute to the same database, needs a different technological infrastructure compared to a project that focuses on several specific places in the world (e.g. forest monitoring in several countries). Another important factor is the sought level of human–technology interaction of each project. Will technology have a dominant presence in the entire process cycle of the project, or will it be used in the background as a supporting element to allow for more human initiative? Moreover, different technological approaches can be used for certain individual tasks such as data gathering, metadata recording, data integrity, quality control and data management. Finally, technological solutions should provide options of modularity and extension. For example, it is not uncommon for a technological approach not to cover the entire spectrum of needs for a specific task, and thus some in-house development could be also needed; as the needs evolve, so should the solution adopted.

In all these cases, it is imperative for the individual technological solutions to be interoperable with other technologies used for the project or collaborative projects and not work in their own silos. This can be achieved by using technologies that comply with international standards and do not function as black boxes relying solely on proprietary formats, hardware, software or other prerequisites. Moreover, the adoption of a 'selfish' technology means that the project's evolution will depend on how this specific technology progresses. Meanwhile, other options can become better, faster, cheaper or easier to use. Adopting technologies that apply and stick to standards makes it easy to switch or extend to other technologies, if needed, with minimum cost while securing continuity, connectivity and interoperability with other technologies. Thus, standards and interoperability principles allow following technological developments at will and building modular project with the best of what technology can provide.

Standards, interoperability and openness of projects is also pursued from the top down. For example, Mazumdar et al. (2018) refer to guidance issued on crowdsourcing and citizen science by the Director of the US Office of Science and Technology Policy which suggests that 'federal agencies should design projects that generate datasets, code, applications and technologies that are transparent, open and available to the public, consistent with applicable intellectual property, security, and privacy protections' (Holdren 2015, 2). Similarly, in 2015, the European Commission's Horizon 2020 Framework Programme issued a call for the Coordination of Citizens' Observatories and Initiatives from 2016–17 in order to promote standards and ensure interoperability (EC 2015).

## 3. Other considerations

One of the most compelling characteristics of geographic citizen science projects is the fact that they are carried out in varying contexts and socio-economic conditions. Naturally, these variations pose different challenges to each initiative. Challenges and needs that are considered fundamental for certain initiatives in the developing world (e.g. literacy and technologically savvy participants) might seem trivial and pointless for citizen science projects in the developed world and vice versa (e.g. explain the magnitude of deforestation and its impact on the natural environment and local population). The second and third parts of this book attempt to capture and reflect on some of the challenges that geographic citizen science initiatives are facing based on their background context, the locations where they are being implemented, their user audiences and so on.

In any kind of geographic citizen science project, two of the main goals are user engagement and participation. Geographic citizen science derives its strength and efficiency from participants who voluntarily contribute their time, effort, resources and money to the causes of the initiative. Thus, it is imperative initially to engage with the citizens to participate and then to sustain, enhance and expand this participation. In Chapter 1 of this volume, Haklay discusses a typology of the different levels of participation and engagement in citizen science projects. As he explains, they range from crowdsourcing, where participation requires limited cognitive engagement or resource contribution, to extreme citizen science, which refers to collaborative science as a completely integrated activity. Another common need for geographic citizen science projects is how to enable and guide participants into actually performing the required tasks correctly. This includes how to train volunteers to be prepared to complete the tasks successfully, how to interface the entire process cycle with individuals and what guide/controlling mechanisms will be present in order to prevent user errors. Furthermore, coordination and supervision of citizen scientists is fundamentally different from what researchers normally deal with inside a research lab (Antoniou 2018), and thus proper actions should be sought.

Also, for geographic citizen science projects, it is desirable that there are no biases in the participants' pool and that all parts of society are represented. Biases can spring from various sources. In terms of participation, projects should be ideally open to participants irrespective of race, gender and socio-economic status. Furthermore, a lack of education and technical skills or limited availability of resources and time, which usually correlate with not so advanced societies and economies, might drive unprivileged contributors away (Antoniou 2018). Thus, measures and actions through the appropriate use of technological solutions should be sought in order to democratise participation.

In addition, for geographic citizen science, it is particularly desirable that participants are spatially distributed in the area of interest, especially if there is a need for in situ observations. Even more important than the spatial distribution of the participants is the issue of observations. The sheer number of observations in geographic citizen science is not indicative of the efficiency or productivity of a project. Spatial and temporal biases can infiltrate the data and thus severely affect any outcome or result (Antoniou 2018; Antoniou and Schlieder 2018). Then, there is the need to collect, store and manage spatial data. The famous axiom 'spatial is special' (Anselin 1989) is still valid and needs to be addressed accordingly by geographic citizen science projects. The data itself need special tools to be collected (e.g. Global Positioning System–enabled devices), special storage mechanisms (e.g. spatially enabled databases), special algorithms and tools for analysis (e.g. geographic information systems (GIS)), special processes for quality assurance (e.g. topological rules) and special tools for the visualisation and communication of results (e.g. cartography-enabled software). Naturally, all these need the appropriate level of spatial understanding and knowledge, first from the project managers and then – according to their level of involvement – from the participants.

Equally important to the management of data collected is the management of the project itself through a series of decisions that need to be made. For geographic citizen science in particular, these decisions need to be based: (1) on the understanding of the underlying geography, the spatial relationships between the phenomenon observed and the geography, as well as the constraints and opportunities of the geography; and (2) on the current state of the project as well as the spatial and temporal distribution of the observations at hand. Thus, the plans and decisions for everyday management as well as for the future development of a project should be based upon extensive introspection of the project through the lenses of geography, data and spatial analysis.

## 4. Intertwining geographic citizen science and technology

Several types of technological solutions can be considered to address various challenges in a geographic citizen science project. In what follows, several groups of technological options will be highlighted (although the list is not exhaustive), and examples will be drawn mainly from the geographic citizen science domain.

An important phase of each project is data collection, especially in geographic citizen science initiatives; data and observations should bear their spatial footprint. This combination requires a position recording mechanism to be in place that will provide coordinates with the required accuracy (i.e. positional accuracy). In turn, depending on the medium(s) that will be used to collect data, for example from dedicated Global Positioning System devices to smartphones, or to head up digitising over satellite imagery in a computer display, positional accuracy can vary significantly from a few centimetres to a few metres. Fusing and/or overlapping data with different positional accuracies can create multiple problems, leading to confusing results. Likewise, data collection with higher or lower positional accuracy than required can also lead to problems. In the former case, data storage and management issues can arise, while in the latter, poor results can be generated in terms of accuracy. Thus, decision makers should be aware of what is actually required regarding the aims of the project in terms of positional accuracy and then decide on the type and capabilities of the technology that should be used.

Another closely related issue is the prominent medium(s) of human–technology interaction for the specific project. The use of mobile, PC, tablet or DIY devices – or a combination of those – will also influence the requirements for technological interfaces. Which programming language or how the interaction design will be implemented needs to be carefully considered and designed for each medium separately; but it should also be kept in mind that the look and feel, the processes and the functionality provided should be the same for each medium used in a specific project so that it provides a seamless user experience. For example, if the task requires the recording of a point location, adding metadata and uploading a photo of an observation, all these functionalities should be feasible to perform in both mobile and desktop applications, and the steps that lead to the final outcome should be similar (if not identical) for all mediums. Real-life examples of this challenge can be found in the OSM project and the various editors created (Antoniou and Skopeliti 2017). Only a few of the many mobile apps for data collection managed to survive; they became the default data-capturing mediums, while the rest either have disappeared or are only used occasionally. This is a waste of both time and energy for the OSM community, while at the same time it is questionable what the quality and completeness is (in terms of the features' attributes) of data provided by the obscure editors which still co-exist in the OSM database alongside data from more prominent editors.

As discussed in the previous section, a constant need of all geographic citizen science initiatives is first to make them visible to potential participants by promoting their aims and goals, then to engage users with the project so they can actively participate and finally to maintain user engagement and participation. In all these fundamental steps, the correct use of technology can make a huge difference. For example, an obvious solution is to exploit the power of social networks and set up publication campaigns that will help in the promotion of initiatives. However, social networking technologies can have many more applications in the project's process cycle. Keeping participants interconnected and allowing them to interact with several aspects of the project by pursuing their individual preferences and capabilities might prove to be very beneficial to the project. Embedding social networking functionality inside a project enables participants to create subgroups for specific tasks, facilitates communication and coordination of activities and allows micromanagement such as organising meetings, answering questions or clarifying grey areas. At the same time, it enables project managers to have a clear view of how participants think and behave. In short, they can touch the pulse of the project. In turn, this gives decision makers the opportunity to act proactively and guide the overall project accordingly. Another example can be found in the use of gamification. Developing a gamification strategy for several project processes can increase participant engagement and continuous involvement, but it also provides solutions to challenges such as quality control and quality assurance. Examples include Foldit (https://fold.it/portal/), which is a multiplayer online game that helps scientists to use crowdsourced input in order to predict protein structures (Cooper et al. 2010), or Geo-Wiki (https://www.geo-wiki.org/ ) (Fritz et al. 2012), which is an online game where, through geo-tagging images, global land cover is improved. Furthermore, Antoniou and Schlieder (2018) examined how gamification can tackle spatial and temporal participation biases in OSM and thus increase the overall quality of the data created.

Similarly, when it comes to data collection, technological solutions can move things more quickly and efficiently in almost every sense. In addition to what has been discussed previously regarding mediums and interfaces, proper technologies can help in coherent and standardised data collection when the most appropriate solution is used and may help overcome some of the most important obstacles such as cost and time. For example, relatively inexpensive Raspberry Pi and Arduino devices can be the heart of DIY sensors or devices capable of collecting data for water, air, noise or other observation-based tasks. In particular, DIY devices can be affordable versions of commercial off-the-shelf hardware for multiple purposes without compromising accuracy and reliability. Similarly, the use of drones can be a fast and relatively inexpensive solution for mass aerial coverage of large areas that would have been cumbersome and costly to achieve without this type of technology. Alternatives would have been the purchase of satellite imagery or aerial photography.

One of the most common challenges in the digital age is data storage, management and security. Issues such as loss or corruption of data, proper data management and the risk of hacking need to be constantly addressed by project managers. For geographic citizen science projects, data storage is confined by the presence of coordinates that dictate the use either of spatially aware formats (e.g. shapefile) or spatial databases (SDBs) that can host coordinates and apply spatial functions to data. Geospatial information, when the area of interest is large or when the data are captured in high granularity, can quickly become huge in volume. Today, there are both proprietary (e.g. ESRI GDB, Oracle Spatial, etc.) and open-source (e.g. Postgis, SQLite, etc.) SDBs that can hold and efficiently manage large volumes of data (as their limitations correlate with the limitations of the hardware or operating system). Specialised algorithms have been created and are embedded into SDBs for indexing (vector), tilling (raster) and querying spatial data. Organisations such as the Open Geospatial Consortium and ISO/TC211 have worked to develop processes and standards that all major providers are following, thus harmonising the development environment of geo projects.

The data, as valuable as they are, are just the beginning of a scientific journey. Data-analysis tools and methods are needed to make the next step – transforming data into useful information that can lead to meaningful and effective results. The analysis of the data, especially for big projects, needs thorough planning. Two common approaches involve either in-house analysis of the entire volume of data or the splitting of data into small parts and crowdsourcing the analysis by exploiting the idle computational power of the crowd. This volunteered computing was introduced as early as 1999 by the SETI@home project which used volunteers' computing power to analyse data from a radio telescope searching for extraterrestrial intelligence (Anderson et al. 2002). Each of these approaches poses its own technological challenges. In the former case, the bigger and more successful the project is, the greater the need for tools that have the capability to analyse huge volumes of data. The magnitude of each of the five Vs in the definition of big data (i.e. velocity, volume, value, variety and veracity) will have an impact on the infrastructure needed to achieve the analysis, visualisation and communication of data and results. In the latter case, the challenges are located in how to split and merge data and results seamlessly and how to share the algorithms and tools used for the analysis (especially if these are proprietary ones).

The advances in geographic information science and the proliferation of open-source GIS software, for both standalone desktop applications and GIS web services, enables complicated spatial analyses on massive volumes of data with very little cost. Exploratory data analysis (Anselin 1999) on a cartographic backdrop conducted by GIS tools is a powerful combination that can reveal hidden relationships and spatial correlation that would be extremely difficult to detect otherwise. Moreover, data and results visualisation and communication are equally important to the data analysis. GIS can now provide several ways to visualise data, from small overview maps inside a report, to paper maps of different scales, to interactive online maps and services, up to three-dimensional globes with fly-through options. All or a selection of these can be included in the arsenal of a geographic citizen science project to visualise any stage or aspect of the project, either internally or for end users.

Finally, since a geographic citizen science project should be anything but introverted, decision makers should make sure that participants have the necessary feedback on what is happening with their contributions and effort. After all, it is their own free time, money and effort that supports the cause. Informing them about the progress of the project is a minimum requirement, and an appropriate mechanism to do so needs to be in place as soon as the project begins. The cartographic capabilities of GIS software can once again help in visualising the results and in presenting the outcomes, successes, failures and needs of a project at each stage in order to communicate them to the target audiences. This in turn can inform, motivate and guide leaders and participants towards project goals yet to be achieved.

Another important part of a geographic citizen science initiative is the management of the project itself. Here, help can be found in other domains too. There are multiple tools that allow a project manager or a team leader to monitor and help the work of their colleagues. Many of these options are now online, and they are provided as either free or freemium services. A simple Internet search can reveal the available options as well as comparisons between the options that can help to determine the best solution for a project. Interestingly, these online tools are easy to learn and thus can be used almost immediately by an entire team. Also, as is the case with any project, the management of citizen science projects often requires planning and action for future steps. Especially for geographic citizen science projects, GIS is almost an imperative technology to support this task. Apart from their ability to store and manage geographic information, GIS tools can depict what-if scenarios, run simulations, project future data, present data in timelines and, in general, provide a broad range of products and services that can help stakeholders to make the correct decisions based on concrete evidence versus intuition.

Apart from managing the day-to-day business of a geographic citizen science project, it is also important for decision makers to keep on their radar developments on the technological front in both geographic citizen science and other domains. They need to be aware of new practices and strategies that might improve existing activities and tasks or introduce completely new ones to achieve the project aims better and to meet user needs. Examples can be found in the developments of virtual and augmented reality software and hardware that can be helpful in educating and training participants (Albers, de Lange and Xu 2017) in order to perform the tasks required over the course of a project. Also, there are the developments regarding the Internet of things (Ashton 2009) which, despite the heterogeneity in the approaches and the varying standards so far, can blend smoothly with input from human participation (Antoniou et al. 2017) and organise, improve and extend the reach of geographic citizen science projects in new ways. Similar is the impact of the advances in wearable technology which transform everyday portable and wearable things into smart sensors capable of collecting a variety of data. Although many breakthroughs in wearables and citizen science have taken place in the health-care domain (Vesnic-Alujevic, Breitegger and Pereira 2018), there are also examples where wearable technology has been used to map data to the physical environment and analyse it through a geographic lens (Mazumdar et al. 2018).

## 5. Future outlook: artificial intelligence and machine learning

It is hard to predict the future. The list of those who have tried and failed is long. However, perhaps the safest way to approach the future – in our case, the future of technology in geographic citizen science – is to examine where we stand now, how other disciplines have evolved and where they are headed in terms of technological development. Then, we need to uncover the driving forces behind these evolutions and whether and how these factors will affect the practices and developments of geographic citizen science. While the appearance of disruptive technologies can undermine such predictions, it is useful to contemplate future developments.

If there is one technological breakthrough that needs attention when looking at the future of technology and geographic citizen science, it is the combination of artificial intelligence (AI) and machine learning (ML). In almost every known domain, AI and ML are breaking ground and providing solutions to all sorts of challenges that were either extremely difficult or unimaginable to cope with before. It is just a matter of time before AI and ML are being used extensively in geographic citizen science projects. For example, it would become easy to classify photos of species uploaded by users automatically to the correct class or to spot outliers in existing data sets with extreme accuracy. Land use and land cover maps which currently require huge amounts of work and in situ input from citizens would be straightforward to produce through satellite or drone imagery classification with AI.

However, these new technologies have their own requirements. AI and ML often need huge volumes of cleansed and curated training sets to develop their models. For example, object recognition models need accurate delineation of the borders of the object to train the model properly. Although such databases exist for experimentation and testing algorithms and models (https://en.wikipedia.org/wiki/List_of_datasets_for_machine-learning_research/), each domain needs to create its own training data sets to accrue the benefits of AI and ML.

## 6. Conclusion

As noted in the introduction, there is no one-size-fits-all solution to technological challenges in geographic citizen science projects. The variables of each project are so many, complicated and intertwined that it is difficult to suggest a simple how-to guide. However, a few mandatory points have been highlighted.

The first point is a thorough and unbiased inspection of the geographical citizen science initiative itself. Decision makers should have a clear understanding of the aims of the initiative, the stakeholders, the participants and volunteers; how the project will evolve; the resources and constraints of the initiative; and any other topic that will affect decision processes. Then, a high-level broad understanding and monitoring of current technological advances is fundamental. This should not be restricted to the world of geographic citizen science. Other domains can also provide paradigms to follow. Next is the intertwining of these two worlds: geographic citizen science initiatives and technology. For any challenge faced by a geographic citizen science project, a quest for the best technological solution should be undertaken. However, it is imperative that two main principles are respected: (1) the technologies adopted should follow international standards and thus be interoperable; and (2) the technological ecosystem decided upon should create an environment where initiatives and ideas can grow without being constrained by technological dependencies.

## References

Albers, Bastain, N. de Lange and Shaojuan Xu. 2017. 'Augmented citizen science for environmental monitoring and education', *International Archives of the Photogrammetry, Remote Sensing and Spatial Information Sciences* 42: 1–4.

Anderson, David P., Jeff Cobb, Eric Korpela, Matt Lebofsky and Dan Werthimer. 2002. 'SETI@home: An experiment in public-resource computing', *Communications of the ACM* 45: 56–61.

Anselin, Luc. 1989. What is special about spatial data? Alternative perspectives on spatial data analysis (89-4). UC Santa Barbara: National Center for Geographic Information and Analysis. Accessed 15 September 2020. https://escholarship.org/uc/item/3ph5k0d4.

Anselin, Luc. 1999. 'Interactive techniques and exploratory spatial data analysis', *Geographical Information Systems: Principles, Techniques, Management and Applications* 1: 251–64.

Antoniou, Vyron. 2018. 'A chimera of VGI, citizen science and mobile devices'. In *Mobile Information Systems Leveraging Volunteered Geographic Information for Earth Observation*, edited by Gloria Bordogna and Paola Carrara, 133–49. Cham, Switzerland: Springer.

Antoniou, Vyron, and Christoph Schlieder. 2018. 'Addressing uneven participation patterns in VGI through gamification mechanisms'. In *Geogames and Geoplay*, edited by Ola Ahlqvist and Christoph Schlieder, 91–110. Cham, Switzerland: Springer.

Antoniou, Vyron, Linda See, Giles Foody, Cidália Costa Fonte, Peter Mooney, Lucy Bastin, Steffen Fritz, Hai-Ying Liu, Ana-Maria Olteanu-Raimond and Rumiana Vatseva. 2017. 'The future of VGI'. In *Mapping and the Citizen Sensor*, edited by Giles Foody, Linda See, Steffen Fritz, Peter Mooney, Ana-Maria Olteanu-Raimond, CidÁlia costa Fonte and Vyron Antoniou, 377–90. London: Ubiquity Press.

Antoniou, Vyron, and Andriani Skopeliti. 2017. 'The impact of the contribution microenvironment on data quality: The case of OSM'. In *Mapping and the Citizen Sensor*, edited by Giles Foody, Linda See, Steffen Fritz, Peter Mooney, Ana-Maria Olteanu-Raimond, CidÁlia costa Fonte and Vyron Antoniou, 165–96. London: Ubiquity Press.

Ashton, Kevin. 2009. 'That "Internet of things" thing', *RFID Journal* 22: 97–114.

Cooper, Seth, Firas Khatib, Adrien Treuille, Janos Barbero, Jeehyung Lee, Michael Beenen, Andrew Leaver-Fay, David Baker and Zoran Popović. 2010. 'Predicting protein structures with a multiplayer online game', *Nature* 466: 756–60.

European Commission (EC). 2015. TOPIC – coordination of citizens' observatories initiatives. Accessed 19 October 2019. http://ec.europa.eu/research/participants/portal/desktop/en/opportunities/h2020/topics/sc5-19-2017.html.

Fritz, Steffen, Ian McCallum, Christian Schill, Christoph Perger, Linda See, Dmitry Schepaschenko, Marijn Van der Velde, Florian Kraxner and Michael Obersteiner. 2012. 'Geo-Wiki: An online platform for improving global land cover', *Environmental Modelling and Software* 31: 110–23.

Holdren, John P. 2015. Memorandum to the heads of executive departments and agencies. Executive Office of the President Office of Science and Technology Policy. Accessed 15 September 2020. https://obamawhitehouse.archives.gov/sites/default/files/microsites/ostp/holdren_citizen_science_memo_092915_0.pdf.

Mazumdar, Suvodeep, Luigi Ceccaroni, Jaume Piera, Franz Hölker, Arne Berre, Robert Arlinghaus and Anne Bowser, eds. 2018. *Citizen Science Technologies and New Opportunities for Participation*. London: UCL Press.

Mooney, Peter, Padraig Corcoran and Blazej Ciepluch. 2013. 'The potential for using volunteered geographic information in pervasive health computing applications', *Journal of Ambient Intelligence and Humanized Computing* 4: 731–45.

Newman, Greg, Andrea Wiggins, Alycia Crall, Eric Graham, Sarah Newman and Kevin Crowston. 2012. 'The future of citizen science: Emerging technologies and shifting paradigms', *Frontiers in Ecology and the Environment* 10: 298–304.

Pimm, Stuart L., Sky Alibhai, Richard Bergl, Alex Dehgan, Chandra Giri, Zoë Jewell, Lucas Joppa, Roland Kays and Scott Loarie. 2015. 'Emerging technologies to conserve biodiversity', *Trends in Ecology and Evolution* 30: 685–96.

Preece, Jennifer. 2016. 'Citizen science: New research challenges for human–computer interaction', *International Journal of Human–Computer Interaction* 32: 585–612.
Vesnic-Alujevic, Lucia, Melina Breitegger and Ângela Guimarães Pereira. 2018. '"Do-it-yourself" healthcare? Quality of health and healthcare through wearable sensors', *Science and Engineering Ethics* 24: 887–904.
Wiggins, Andrea. 2013. 'Free as in puppies: Compensating for ICT constraints in citizen science'. In *Proceedings of the 2013 Conference on Computer Supported Cooperative Work*, 1469–80. New York: ACM.

# Chapter 3
# Design approaches and human–computer interaction methods to support user involvement in citizen science

Artemis Skarlatidou and Carol Iglesias Otero

## Highlights

- Citizen science activities attract, and are relevant to, people of all ages, backgrounds and interests. Participants' education and literacy skills, access to technological infrastructure, their familiarity with technology and local environmental conditions present various obstacles in the successful adoption and utilisation of geographic citizen science applications.
- Human–computer interaction (HCI), the discipline which investigates how humans interact with computerised systems and other technological artefacts, provides methodologies and techniques to inform the design and development of geographic citizen science applications.
- Different design approaches (e.g. user-centred design and participatory design) and methodologies reviewed in this chapter can be used to explore local cultural contexts and develop user-friendly and useful geographic citizen science applications that meet user needs in culturally appropriate ways, and which generate a positive user experience.

## 1. Introduction

Human–computer interaction (HCI) is the multidisciplinary discipline which investigates how humans interact with computerised systems and other technological artefacts. HCI made its appearance in the 1970s, although the term was coined later in the 1980s and matured quickly, mainly driven by continuous technological developments and the new modes of interactions that these introduced. HCI had a massive impact on our lives, with the development of not only technologies such as the personal computer and the World Wide Web, but also communication and entertainment technologies. Apple's iPhone and Google's search box are popular products of HCI research and design (Preece 2016). Nevertheless, the impact of HCI on the design, development and evaluation of citizen science applications is still at an infant stage. According to Preece (2016, 588), 'this is largely because biologists, other scientists, and HCI specialists do not often work together'. Despite the lack of extensive research and evidence from this context, there are still a few cases which demonstrate the successful implementation of HCI methodologies and techniques to inform the design and development of citizen science applications. To make it practical and easy to grasp, some of these cases are used as examples throughout this chapter.

Citizen science activities attract, and are relevant to, people of all ages, backgrounds and interests. Although this itself is a massive design challenge for HCI, so far, citizen science and geographic citizen science have attracted a limited demographic profile that is Western, educated, industrialised, rich and democratic (frequently described using the acronym WEIRD; Dourish 2015). In other words, the majority of such activities take place in the developed world and urban environments, such as those described in the second part of this book, where education skills as well as basic access to and familiarisation with technologies are usually taken for granted. On the other hand, citizen science initiatives, such as those described in the third part of this book, mainly target communities in developing countries, where education and literacy, access to technological infrastructure and familiarity with technology, as well as the local environmental conditions, present various obstacles in their successful adoption and utilisation. It is therefore common, due to contextual differences, to use different design approaches to work with users in designing and developing functional, usable and useful applications.

The purpose of this chapter is to improve the understanding of how HCI methods and tools can be applied in real geographic citizen science contexts. We provide an overview of design approaches and methods which can be used to enable and support user involvement at various stages of designing and developing geographic citizen science applications. The aim of using these methods is to understand local cultural contexts, meet user needs in culturally appropriate ways and develop

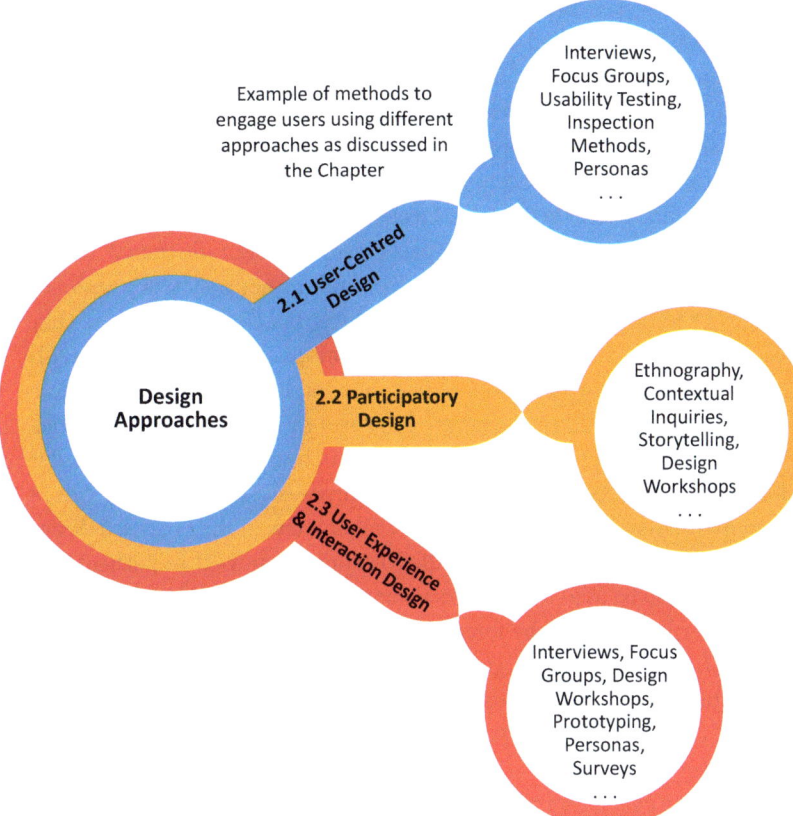

**Fig. 3.1** Overview of design approaches and methods used to engage users in the design, development and evaluation of citizen science applications. This conceptual model is rather generalised, and it captures methods as they are currently being implemented in similar initiatives. Methods can be modified and designed to address different aims and purposes, and therefore they can be used in different contexts supported by all or few of the proposed design approaches.
Source: authors.

products which are user-friendly and useful and which potentially generate a positive user experience. Figure 3.1 provides a summary of the design approaches and the methods that are discussed in this chapter.

The conceptual model (Figure 3.1) captures design approaches and methods as they are currently being implemented in citizen science initiatives and therefore is not by any means exhaustive; more design rationales which enable the involvement of the end user can be found in the HCI literature (e.g. see Bekker and Long 2000). For example, user-centred design (UCD), discussed in Section 2.1, requires user engagement and involvement at all stages of the design and development process iteratively. Therefore, this design approach assumes that users are in close proximity and available to interact with the development teams at all times, which is harder to apply when citizen science initiatives take place in developing countries and remote geographic locations, such as those discussed in the third part of this book. Instead, participatory design, discussed in Section 2.2, is more suited to studies which aim to not only work with and involve users in developing functional products, but also to explore the characteristics of cultural contexts and local conditions of living in ways that substantially inform their design and development. Respectively, the methods which we discuss in this chapter are informed by these design approaches. Yet, it should be kept in mind that methods can be modified and designed in ways which address different aims and purposes. Therefore, they can be used in different contexts supported by all or a few of the proposed design approaches. For example, storytelling is commonly used in participatory design but can be used in other design approaches discussed herein.

A term which will be frequently encountered in this chapter is usability. The proposed ISO ergonomics definition explains that 'usability is a measure of the effectiveness, efficiency and satisfaction with which specified users can achieve specified goals in a particular environment' (Brooke et al. 1990, 357). Usability therefore is not a qualitative attribute (i.e. how easy a user thinks it is to complete a set of tasks), but usability can in fact be measured empirically in terms of learnability, efficiency, effectiveness, error rate and user satisfaction (Nielsen 1993). There is a set of methods (e.g. usability testing, prototyping, focus groups) that can be used to evaluate usability or identify potential usability barriers and also a set of 'discount' methods (e.g. heuristic evaluation) which can be used for the same purpose when full-scale evaluation is either too complicated or too expensive (Nielsen 1993).

## 2. Design approaches for the development of geographic citizen science

### 2.1 User-centred design

UCD is one of the most popular design approaches in HCI. The term was coined and made its appearance in the 1980s with the publication of the book *User-Centered System Design: New Perspectives on Human–Computer*

*Interaction* (Norman and Draper 1986). UCD is a design philosophy and a set of methods which place the user right at the centre of the design process. The focus of UCD is on what users need and how an interface should be designed to accommodate these needs so that it meets users' expectations and is usable.

The role of the researcher or designer in UCD is to facilitate the user interaction in such a way that meaningful information can be extracted about the users' 'needs, behavior, likes, and dislikes' (Preece, Rogers and Sharp 2015, 454) and other human and contextual factors which influence the usability of the end product. This information is then used to inform interface design, development and implementation. UCD requires the involvement of the actual (or a representative sample of) end users. Users are involved at various stages in an iterative design and development process; that is, from product conceptualisation – where user requirement elicitation can be particularly useful – to interface evaluation and subsequent redesign to address user issues and improve overall interaction. According to Sanders (2002), in UCD, the roles of the researcher, the designer and the user are distinct yet interdependent – something which further differentiates this design rationale from other design approaches, which are discussed in the next sections.

In citizen science, there are a few examples from the way UCD approaches have been so far applied in practice. For example, Woods and Scanlon (2012, 3) describe a 'light-touch user-centred design approach' for the development of a geographic citizen science application which aims at improving volunteers' understanding of natural history by identifying and submitting geolocated observations of nature using the iSpot (iSpotnature.org) platform. The authors describe a UCD process which includes two iterations. The first focuses on conceptual design; it uses the methods of storytelling (Section 3.5) and interviews (Section 3.1) with existing iSpot users to elicit their requirements and subsequently develop functional requirements to inform the design of the application. This process is followed by some preliminary testing using some user-based scenarios and a review of the application's competitors. The authors explain that testing at this stage helped them to realise that 'we had been creating a service that largely mimicked website navigation and we therefore had to completely redesign the navigation and layout for the mobile app' (Wood and Scanlon 2012, 4). The second iteration includes usability testing (Section 3.6) of the redesigned interface with 10 users, as well as the incorporation of popular accessibility standards in the interface. The authors, when describing their future plans, explain that usability testing with mobile eye tracking is being considered prior to the application's final release, and they suggest that a UCD approach has proven beneficial for improving the quality of the final product by ensuring that user input is incorporated after each iteration.

In another example, Newman et al. (2010) use a UCD approach for the design and development of a geographic citizen science application for the collection of invasive species data. Although less information is provided about the exact procedure that it is followed, the authors explain that 'Our software development lifecycle included iterative investigation, design, requirements specification, development, implementation, testing, and maintenance' (Newman et al. 2010, 1854). They describe a usability testing (Section 3.6) experiment in the lab with 16 users, followed by a survey to evaluate web usability concepts further, assess the overall user experience and also determine users' skills and familiarity with the use of similar applications. Continuous feedback obtained from user interviews (Section 3.1) in subsequent iterations reveals that they 'overlooked many tasks citizen scientists and volunteer coordinators need to accomplish online' (Newman et al. 2010, 1860). This highlights the importance of continuing to collect user input, even after the application is launched.

## 2.2 Participatory design

Early HCI research embraced approaches that place special attention on the user as an information-processing mechanism influenced by a set of factors which have to be taken into account in the design process. In these approaches, 'the notion of the user – a very limiting term – as an active actor in the process was missing' (Bannon 2011, 52). The growing maturity of the field led to the realisation that there was a need to 'create new settings and experiences that can assist computer professionals to work in partnership with diverse users in improving both computer technologies and the understandings that make computer technologies successful in real use' (Muller and Druin 2012, 1051). Influences from the Scandinavian participatory design movement and the field of Computer Supported Collaborative Work resulted in an HCI shift from a psychological to a more sociological perspective, which expanded its methodological toolkits to incorporate field observations, not only lab-based studies (Bannon 2011).

Participatory design in HCI involves a 'set of theories, practices and studies related to end-users as full participants leading to software and hardware computer products and computer-based activities' (Muller and Druin 2012, 1051). Participatory design draws from other research and design approaches, such as UCD discussed in the previous section, but also from disciplines such as psychology, anthropology, sociology and political science. Its practices are motivated by democratic ideals inspired by a need to support the incorporation of 'multiple voices in knowledge production, but [also] in the production of technologies as knowledge objectified in a particular way' (Suchman 2002, 93), which is particularly important in terms of empowering marginalised and disempowered groups. This creates a 'third space' in HCI where users and developers work together in a process of mutual learning, understanding and respect which, according to Fowles (2000), enables the transformation from 'symmetries of ignorance', among end users and developers, to 'symmetries of knowledge' through participation. Participatory design methods do not just enable an understanding of what the users need; the emphasis is less on the outcomes and more on the design process which aims to challenge both end users and designers to enter each other's world and learn from each other. Simultaneously, this process can create and shape new products in ways which transcend traditional thinking and knowledge structures.

Participatory design is frequently encountered in citizen science literature. Several studies, as discussed below, describe its use in the development of citizen science applications. Considering the roots, values and meanings that surround participatory design, there are several studies which unfortunately use the term incorrectly as an umbrella term to describe rationales of end-user involvement. Those studies that follow a participatory design approach highlight its value in empowering communities and users to solve local issues or issues of a higher cause and in creating mutual reciprocity.

This book provides several examples of participatory design in the third part, and it is more extensively discussed from a methodological point of view in Chapter 4. For example, in Chapter 14, Rafael Chiaravalloti uses Sapelli, a geographic citizen science application, to collect data which subsequently supports natural resource use and management in the western border of the Pantanal in Brazil. The ultimate aim of his study is to understand the implications that local activities have for local conservation and environmental sustainability. The study engages deeply with local communities and fishermen at all stages of the project: from understanding their needs and goals in natural resource use and management, as well as how they can benefit from the implementation and use of Sapelli, to working with them in the development of appropriate icons to use in Sapelli's interface design while respecting all of their concerns in the way the study is implemented at all stages. For the purposes of this research, Chiaravalloti explains that he used ethnographic methods which required living with local communities for long periods of time to build a bond and helping them in their daily activities, where he had a chance to observe and understand how the technology was used and the barriers to take into consideration in subsequent design improvements.

## 2.3 Interaction and user experience design

Interaction design and user experience (UX) design are two popular terms that are now commonly encountered in HCI literature, which represent the growth of the field over the years. The distinction between the two is not usually clearly defined, either in terms of their definitions or in terms of the practices they embrace (Interaction Design Foundation 2019).

Interaction design, as a design discipline, aims to explore the design of interactive digital products and services, and focuses on their behaviour to ensure that they satisfy user needs and desires while taking into account the contextual settings in which these are being used (Cooper, Reimann and Cronin 2007). UX design focuses on the design of experiences that digital products and services generate. Such experiences 'emerge from the integration of perception, action, motivation and cognition' (Hassenzahl 2011, 3). As Norman, in a commentary on Hassenzahl's text, puts it:

> Design has moved from its origins of making things look attractive (styling), to making things that fulfil true needs in an effective understandable way (design studies and interactive design) to the enabling of experiences (experience design). Each step is more difficult than the one before each requires and builds upon what was learned before. (Norman 2011, 8)

UX design therefore focuses on non-utilitarian aspects of user interactions, which include dynamic concepts such as 'emotional, affective, experiential, hedonic, and aesthetic variables' (Law et al. 2008, 2396), whereas interaction design focuses more on how the digital products behave. To get an in-depth understanding of the broader spectrum of UX research, the reader may refer to an early work by Hassenzahl and Tractinsky (2006) which captures the questions the field was trying to address in its early days and its challenges, which are still relevant today.

Both interaction and UX design draw from the growing popularity of ethnographic methods in the design and development of interactive systems (Forlizzi and Battarbee 2004), and their methodological approaches emphasise: understanding the user (both approaches), understanding the product (both approaches), understanding the interaction between the user and the product (interaction design) and the experience this generates (UX design). There are different approaches, methods and tools to guide and support interaction and UX design. For example, Cooper, Reimann and Cronin (2007) describe methods which cover the whole spectrum of the interaction design process, and which include: understanding the users and their requirements (e.g. interviews, contextual inquiries); modelling users and their goals (e.g. personas and scenarios); design principles and theories for defining the design and its interaction (e.g. interface paradigms, visual design, design values) and evaluation of the final product (e.g. usability testing). Since both approaches draw from various disciplines, there is a plethora of theories and methods, with some used more frequently than others, to support them (for more information on this topic, the reader may refer to Rogers 2004).

The previous sections (2.1 and 2.2) provided examples to demonstrate how design approaches – mainly UCD and participatory design – have been slowly incorporated in citizen science to inform how end users interact with and use these applications in various contexts, mainly in terms of incorporating user needs and improving their usability. Yet, the way citizen science technologies (such as wearables, do-it-yourself and gamified citizen science applications) are currently utilised, as well as their potential future capabilities, require extending our understanding and research to start incorporating interaction and UX design principles. There is already evidence (Skarlatidou, Ponti et al. 2019) that these principles may address critical concerns in citizen science which extend beyond the design of simple and easy-to-use technologies. For example, there is a need to design applications which are motivating and fun; applications which encourage behaviour change and a better understanding of the scientific process for those who participate; applications which have the ability to connect end users with the physical environment that surrounds them; as well as applications which enable the collection of data trusted by end users themselves and by the wider community.

# 3. User-based methods to assist the development of geographic citizen science applications

The design approaches reviewed in Section 2 come with a set of theories and methods that can be used to guide the design process. In this section, we will review some of the most popular methods that are used for these purposes while providing examples from the context of citizen science.

## 3.1 Interviews

Interviews are used as a qualitative method in a wide range of disciplines such as sociology, anthropology and HCI research. The method consists of asking participants a series of questions in order to obtain in-depth information that often cannot be acquired through other means (Adams, Lunt and Cairns 2008). In preparation for an interview, a template is often constructed, which can range from a fixed script to a schedule of targeted topics. Frequently interviews use a combination of open and closed questions (Wilson 2013). Open questions allow the interviewees to phrase the answers with their own words, while closed questions push the interviewee to select from a number of potential choices (Wilson 2013).

In addition to deciding on questions or topics to be addressed, preparation for an interview template also requires determining the right order of questions (Wilson 2013; Weller 1998). Moving from the general to the specific is often a good strategy, since it allows the interviewee to understand the research direction (Weller 1998), and it explores the interviewee's opinions without imposing any concepts or ideas which would not have been initially triggered by the question. Also, it has been shown that using questions that are too specific at the beginning of an interview can lead to the user getting carried away with anecdotal information and lead to biases (Weller 1998). In that sense, interviews can be less or more structured, leading to different degrees of replicability and richness (Berg 2009).

Given the prevalence of interviewing as a method, it is not surprising that there are numerous examples of utilising this method in various citizen science projects and for various purposes. Woods and Scanlon (2012) recount using interviews for the development of the iSpot website, which can be used to support volunteers collect and visualise their observations of plants and animals. The researchers used interviews at an intermediate stage; that is, after the collection of some preliminary requirements but before developing and testing the interface which they developed during the first iteration of their UCD approach. According to the authors, the main reason for using interviews was to obtain feedback from using the application, which was also the goal in another study by Newman et al. (2010) – to inform the development of the Citsci.org (https://www.citsci.org/) application. Here, the implementation of the method aims to acquire preliminary user feedback and perceptions to form the requirement specification, while additional interviews – which are conducted in later phases of the application's development following usability testing (Section 3.6) experiments – aim to obtain feedback about specific usability barriers experienced by participants.

Similar examples demonstrate that in most citizen science cases, interviews are mostly used in combination with other methods, and they are particularly helpful when applied reiteratively to collect user information at various stages and subsequently guide other methodologies. For example, Michener et al. (2012) describe using direct interviews in preparation to develop detailed personas (Section 3.8), the traits of which were based on a compilation of the traits of those who were interviewed. Moreover, the act of interviewing itself can be enhanced through the use of other methodologies. For instance, Fledderus (2016) applies storytelling (Section 3.5) and scenario strategies to inform the preparation of semi-structured interviews and then uses them to steer the conversation topic towards the participants' prior experience with birdwatching. Subsequently, Fledderus (2016) used affinity diagrams in order to compile the results and data from the interviews and to create personas as representations of the interviewed target audience.

## 3.2 Ethnography and contextual inquiries

Ethnography is a qualitative methodology that aims to understand cultural or human characteristics linked to a site or setting, which may include tacit modes of knowledge, workflows, attitudes, habits and experiences (Randall, Harper and Rouncefield 2007). Typical ethnographic methodology mixes situated, long-term observation and thick description, which is characterised by rich and textured accounts of the observations (Mannik and McGarry 2017). While ethnographic methods originated in cultural anthropology and sociology, they are now being used in numerous other fields, including psychology, education, business and design. The practice of observing how different cultural groups or audiences interact with various technologies and devices is now quite widespread, and many large technology corporations, such as Intel and Apple, hire anthropologists and HCI experts to investigate consumer needs and demands.

In the field of HCI, especially within the context of participatory design approaches (Section 2.2), ethnography is now frequently using methods such as contextual inquiries, which are used to observe 'users' actual living situations and behaviors' (Kurosu 2013, 73). Contextual inquiry is a method that combines immersive observation and interview techniques (Cooper, Reimann and Cronin 2007). This method was pioneered by Hugh Beyer and Karen Holtzblatt, who explain that it should be based on a master–apprentice model where the interviewer asks questions of the user 'as if she is the master craftsman' (quoted in Cooper, Reimann and Cronin, 58). Contextual inquiries take place situationally, allowing the participant to interact with the technology as the conversation moves along (Holtzblatt and Beyer 2015). This is beneficial because it allows the researcher to observe interaction with the technology and workflow patterns. Contextual inquiries are mostly carried out in a space where the activity under analysis would normally take place. This allows the researcher to understand the users' needs and motives better through immersion in the very situations where the technology will be used. In a sense, contextual inquiry can be understood as an HCI version of ethnographic participant observation, where the researcher partakes in and shares the use of technology with the users. Contextual inquiries can often be improved by preceding them with ethnographic interviews, which allow the researcher to identify user goals and priorities (Cooper, Reimann and Cronin 2007).

Since contextual inquiry allows the researcher to observe and interview stakeholders situationally, it is a good strategy to address different groups of users and stakeholders in the process. For instance, Kim et al. (2011) recount meeting with a variety of organisations and participants, including scientists, volunteers, environmental activists and water control boards, in order to carry out contextual inquiries for the development of Creek Watch (i.e. a mobile app and website that enable volunteers to collect and share water flow and trash data from rivers and creeks). Since the application needs to meet both scientists' and volunteers' needs, contextual inquiries addressed the situations and tacit needs of both groups and further aimed to understand data flows and data standards. Here, the method demonstrates that every group has a different reason for wanting to engage with this citizen science project, and the information acquired through the method was used to build an early version of the application.

## 3.3 Questionnaires and surveys

Questionnaires and surveys consist of a prepared compilation of questions which aim to obtain objective and subjective information about the users themselves (e.g. demographics), a product or service they use and their experiences. Questionnaires and surveys are perhaps one of the most popular methods used in a variety of disciplines to engage with a much larger population sample of participants in order to obtain qualitative and quantitative information to meet the purposes of the study (Ozok et al. 2008). They can be implemented digitally or by paper and pencil using both quantitative and qualitative scales (Ozok et al. 2008). Because they can be administered to multiple users at the same time, questionnaires and surveys are a much less time-intensive method to obtain user data from a large pool of people.

Their results can be compared using simple or more advanced statistical methods. It is a faster approach if one seeks to detect particular trends as well as the requirements or opinions about the usability of a product (Adams, Lunt and Cairns 2008). Questions need to be fixed in advance, which makes the design and preparation process extremely important for the effective implementation of the method (Dix et al. 2004) – something which less experienced researchers usually overlook. Moreover, arriving at the right set of questions for a survey requires a higher level of familiarity with the targeted audience, their potential needs or expectations as well as a good understanding of the design and use of the method (e.g. when it is necessary to avoid exhausting participants by asking too many questions and, as a consequence, collecting a small number of responses). Some types of questions (e.g. demographic surveys) can be particularly helpful at earlier stages of citizen science design to get an idea of the characteristics that define a user group (Adams, Lunt and Cairns 2008).

The malleability of questionnaires means there are numerous examples that use them in citizen science, with questionnaires being extremely popular in gathering data about citizen scientists' motivation and attitudes. Raddick et al. (2010) use a quantitative questionnaire in order to survey the motivations of more than 10,000 users from Galaxy Zoo, a citizen science project that asks volunteers around the world to classify galaxy images. For the survey, Raddick et al. (2013) conceptualise motivation around 12 constructs, and for each construct, the survey asks participants to rate the application using a seven-point Likert scale that ranges from 'not motivating' to 'very motivating'. The survey is implemented as an online questionnaire – a fitting decision for a citizen science project that takes place entirely online. In addition to the constructs addressing motivation, the survey includes a demographics section with questions aimed at gathering data about respondents' age, sex, country of residence and education level. While the results of this survey demonstrate that a desire to contribute to science was the greatest motivating factor driving participation in this citizen science project, the survey does not include any questions which aim to understand any qualitative relations to the interface design of the application.

Another survey, carried out by Nov, Arazy and Anderson (2011) with participants of Stardust@home, also focuses on user motivation. The volunteers, who in this case use the application to contribute to tracing interstellar dust, also state the importance of their contribution to science and personal enjoyment as primary motivations (Curtis 2015). In this case, the survey was only distributed to volunteers who had been active 'in the 30 days prior to the survey date' (Nov, Arazy and Anderson 2011). Perhaps because of this, the response rate, at a 27.1 per cent, was much higher than the one achieved in the Galaxy Zoo survey which was sent to all email addresses of registered participants (Nov, Arazy and Anderson 2011; Raddick et al. 2013). The respondents to the survey were also asked to rate the relevance of different motives on a seven-point Likert scale.

It is noteworthy that the findings of both surveys result in realising key implications for the design of the citizen science projects that they examine. For example, Nov, Arazy and Anderson recommend that 'designers and leaders . . . focus their recruiting and retention efforts on motivational factors that are more salient', and if contribution to, and involvement in, scientific pursuits is a key motivation, designing for 'communicating the project's mission, achievements, and its scientific contribution' should be considered an important goal or design requirement (Nov, Arazy and Anderson 2011, 72). Although the authors do not provide any examples of how this knowledge may support the design of specific features and functions of their applications, this is valuable knowledge to support a further exploration of the 'features and designs [that] work well in attracting and retaining participants' (Curtis 2015, 726).

## 3.4 Focus groups

Focus groups, a popular market research method, consists of open-ended, semi-structured group interviews which involve one (or more) moderator(s) and a selected group of usually 6–12 participants. Focus group participants should be selected to match the target audience (Cooper, Reimann and Cronin 2007, 69); that is, a group representative of the intended or predicted user population for a product. The focus group is most often used to gather opinions and feedback on technologies as well as initial reactions to a product.

Focus groups require a moderator to lead the activities and discussions in order to encourage conversation between all participants and to ensure that research goals are fulfilled. This often gives the method a benefit over individual interviews. As Adams, Lunt and Cairns (2008) explain, discussions held among participants often provide useful data on a specific topic for a broader set of participants in a shorter period of time than individual interviews. Moreover, while individual interviews may pose challenges in trying to generalise responses, focus groups provide the researcher with an idea of shared requirements and expectations beyond personal preferences. On the other hand, focus groups, unlike contextual inquiries (Section 3.2) and usability tests (Section 3.6), do not necessarily involve direct interaction with a technology. Nevertheless, a challenge to consider is the fact that focus groups tend to end in consensus, where the loudest opinion may 'win'. With this in mind, focus groups are considered particularly useful for reflection on or during the execution of collaborative activities such as prototyping (Adams, Lunt and Cairns 2008).

Applications of focus groups in citizen science design varies. Sometimes, a focus group is used once, at the beginning or the end of the activity, to assess group experience and expectations. There are also examples where repeated sessions are scheduled in order to obtain 'extensive feedback . . . on results, experiences and usability' (D'Hondt, Stevens and Jacobs 2013, 682). For example, a participatory noise mapping project in Antwerp, which entailed collaboration with a group of 13 volunteers from a citizen-led activist group, lasted for several months and included several focus group activities (D'Hondt, Stevens and Jacobs 2013). In this example, focus groups were used for various purposes, such as introducing the experiment, taking group decisions on the experimental conditions (e.g. the geographic focus and measurement dates), training the participants and sharing good practices to improve data quality, as well as obtaining feedback (D'Hondt, Stevens and Jacobs 2013).

## 3.5 Storytelling

Storytelling involves the creation of narrative forms in order to convey ideas, emotions or events through a linguistic act, written or oral, which can at times be accompanied by other pictorial or sound-based media. As a form of social activity that has existed for centuries, storytelling can pertain to a variety of realms, including art, anthropology, psychology, history and many others. In the design context, it involves the use of narrative and narrative fiction to explore potential users' responses and interactions with a design solution (Spaulding and Faste 2013). The method is especially popular with participatory design practices. The central characteristic of storytelling is the capacity to create 'a reality behind the design that reminds the designer of the real people that will touch and interact with their design' (Hunsucker and Siegel 2015, 2). Stories achieve this complexity by creating context about underlying conditions (such as attitudes, tacit knowledge and others) that determine decisions such as the usability of a design. In order to populate stories and storytelling with situational context and key actors, often they are used in conjunction with scenarios and personas (Spaulding and Faste 2013; Section 3.8).

Just like a work of fiction, storytelling in design processes also requires the key elements of narrative: plot, characters, arcs and resolution (Spaulding and Faste 2013). Through this combination, stories offer a description of events or actions, and not only (as scenarios do) of the situation where they could take place (Quesenbery and Brooks 2010). However, unlike in fiction writing, storytelling in HCI is 'based on data from listening and observing in formal and informal settings' (Quesenbery and Brooks 2010, 50). Furthermore, storytelling can be used through all phases of the development of a design: 'research, ideation, prototyping, and presentation' (Hunsucker and Siegel 2015, 2). For instance, in early stages of the design, stories can be gathered from potential users and compiled into a single narrative that helps communicate design directions (Hunsucker and Siegel 2015). During ideation and prototyping, storytelling may be used to communicate and visualise usability problems and user needs in a way that allows designers to imagine the design solution in context. For example, the main character in the story can encounter the design solution and attempt to use it, potentially running into pain points or difficulties in the process (Hunsucker and Siegel 2015). Alternatively, instead of presenting users with a prepared story, they can be asked to elaborate on real or imagined stories in order to describe their interactions with a product (Cooper, Reimann and Cronin 2007). Finally, since narrative allows for both creativity and collective sharing, storytelling is very effective for communicating a design solution or idea to stakeholders (Cooper, Reimann and Cronin 2007).

In citizen science design, storytelling is often used in combination with other methods. For example, Woods and Scanlon (2012) utilise a storytelling process followed by a series of interviews. The stories here are used to develop 'authentic' scenarios that are presented to a group of 10 experienced mobile phone users to guide the evaluation of the first prototype in a realistic context (Woods and Scanlon 2012). Storytelling in this way is used to summarise and communicate user experiences and knowledge into scenarios or briefs that are then used to guide a user during testing or discussion. Another example comes from Phillips and Baurley's (2014) usage of a storyboarding process to design and construct citizen science sensor devices, in which storytelling allowed the participants to move beyond data collection to designing and fabricating the monitoring devices. Phillips and Baurley streamlined findings from user research studies into a toolkit made up of 150 printed cards that gave parameters such as constraints, goals, images and user traits. During the discussion, participants picked up different cards and 'interpreted [them] into narratives to engage parties unfamiliar to the process of design' (Phillips and Baurley 2014, 4). The authors concluded that the storyboarding process was useful because it allows participants to capture different viewpoints and create a rounded narrative.

Beyond the process of design in itself, storytelling can also be used to discover, convey and record information from design processes and experiences. For anthropologists and HCI professionals involved with a variety of users, storytelling can be a key tool to move from case studies to the recognition of patterns, relations and recurrent challenges. Finally, when deciding to use storytelling as a method, it is crucial to be aware that storytelling is a practice of great technical and cultural importance in many indigenous cultures around the world (Fernández-Llamazares and Cabeza 2018). Most importantly, in this context, stories – often passed down as oral histories generation after generation – have a collective character and can reveal 'conceptualizations of nature–culture interrelations that differ from Western epistemologies' and ontologies (Fernández-Llamazares and Cabeza 2018, 3). For citizen science projects carried out in collaboration with indigenous communities, storytelling can help researchers understand indigenous world views relevant to the collaborative design of the project in a way that allows for a more profound 'approach to knowledge coproduction' (Fernández-Llamazares and Cabeza 2018, 3). Readers interested in this method may refer to Truna (2015) for a thorough discussion and critical approach to storytelling development.

## 3.6 Usability testing

Usability testing can facilitate a better understanding of the users' difficulties and, in turn, facilitate the development of design solutions that can overcome usability deficiencies and often create novel solutions. The focus here is on usability. Although there are several HCI methods which support the engagement and involvement of end users in usability assessments, the most popular is usability testing, as the best way to understand how users interact with a system is by observing how they use it. Therefore, the method aims to identify usability problems through observation of users performing real tasks using the system under investigation and subsequently inform the development of functional products which are easy to use (Rubin and Chisnell 2008).

In usability testing, users are provided with a set of tasks, which they are then asked to accomplish. Therefore, a good understanding of the system, any problematic features which may pose difficulty to users or particularly strong features that the users may enjoy are essential in choosing representative tasks and designing effective usability testing experiments. The observer and note taker may be physically present in the same room with the user while the testing session takes place, or they may be sitting behind a one-way mirror in the more traditional usability lab setting. In usability testing, the observer is not allowed to intervene and answer any task-related questions. A usability testing experiment is not a training session to demonstrate how an interface can be effectively utilised, but rather a method to understand where it fails so that others can use it without requiring the physical presence and support of an instructor. The purpose of usability testing is not only to observe whether the user will succeed in completing the given tasks, but also to observe the strategies the users develop to overcome any problems they encounter and the reasoning behind the difficulties they experience. To gain this insight, users are asked to 'think aloud', which involves 'verbalizing their thoughts as they move through the user interface' (Nielsen 2012). Although the process may initially lead to an unnatural situation, constantly reminding and encouraging the users to think aloud quickly helps overcome any uneasiness. According to Nielsen (1993), thinking aloud is 'the single most valuable usability engineering method'.

Apart from taking notes from the observation and users' thinking aloud data, other common metrics which are gathered during a usability testing experiment include success rates and task completion times. A success rate is a bottom-line usability metric which can be used to communicate the usability performance of a system. It reflects a percentage of the tasks which were completed successfully (Nielsen 2001). Completion times are also very effective in providing an indication of how easy it is for a user to complete a task; for example, if a user needs five minutes to record a single observation, that might not only be unrealistic and frustrate users but might also mean that the interface design can be improved to make the task easier to complete and increase user speed. For this purpose, it is recommended to pilot the tasks and completion times before the testing session. The users may be interrupted and asked to move to the next task if it takes them more time than anticipated to complete them.

Another critical decision in usability testing concerns the optimal number of users who should be invited to participate so that an in-depth insight is gained into its potential usability problems. In Nielsen's (1989) paper 'Usability Engineering at a Discount', the author argues that testing with five users can give you enough usability insight while getting the maximum cost–benefit ratio from using the method. Nielsen and Landauer (1993) have further demonstrated that after the fifth user, usability findings are repetitive, while you only need a maximum of 15 users to identify all usability problems. Another thing to consider is that the only usability problems that will be identified in a usability testing experiment will relate to the tasks that are provided during the experiment. If additional features need to be evaluated, then repeating the experiments with a different set of tasks and perhaps a different set of users should be considered, depending on the significance of minimising any potential learning bias.

The citizen science literature has several usability testing examples to demonstrate their potential. For example, Idris et al. (2016) evaluated the performance of a system which aims to engage with local indigenous communities in collecting and mapping the ecotourism community assets in the Royal Belum reserve forest in Malaysia. In their study, the authors recruited 40 users to carry out a field-based usability testing experiment while video recording each experiment to gather observational data. During the experiment, they provided the users with five tasks while they observed their performance. Note that a second experiment took place which focused on testing the condition of whether the provision of training and support before and during the session had a positive impact on participants' overall performance, but this was not part of the usability testing experiment, since it violated the condition of not intervening and interrupting participants. The experimental protocol they followed required participants to sign a consent form, and demographic data were also collected via a pre-experiment interview session. Apart from observational data, the authors gathered further performance metrics such as task completion times and error rates.

Although Idris et al. (2016) provide a well-structured field-based usability testing experiment with indigenous communities, it should be noted that the method has been shown to be somewhat problematic in similar settings. Pejovic and Skarlatidou (2020) interviewed nine researchers who use citizen science with indigenous communities to investigate interaction design challenges in the implementation of mobile extreme citizen science initiatives. With respect to usability testing, the authors report that their interviewees 'commented that [their participants] struggled in understanding these hypothetical situations, which are introduced by usability testing tasks, and providing the feedback the research were seeking' and that

> several of our interviewees mention the lack of honest and constructive feedback from their users as this is not in line with the way their societies and belief systems work. Users lacking formal education

> seldom criticize projects . . . criticism is avoided as it goes against the principles of the egalitarian society that these users belong to.
> (Pejovic and Skarlatidou 2020, 265)

To overcome these methodological problems, in some of the projects the interviewees describe, they had to be creative with the provided tasks and incorporate their experiment into a treasure hunt game, as discussed further in Chapter 11. In other cases, the experience of the researchers in conducting ethnographic studies with the same communities proved very useful in better understanding and interpreting the results of usability testing experiments.

## 3.7 Design workshops and prototyping

User-led design workshops, commonly used in participatory design (where different terms can be found to describe the same or similar methods) are activities which support a group of participants in proposing ideas and working together with designers or researchers in order to co-design solutions, products and interfaces (Ahram and Falcão 2019). The broader aim of these workshops in participatory design is to enable participants to commit to shared goals, strategies and outcomes (e.g. designs; Muller and Druin 2012). Workshops usually include up to 10 participants (but also can be smaller groups of people or in some cases whole communities – as in the case studies discussed in the third part of this book), one or more facilitators and the researcher(s). The workshops are organised around specific questions and goals, which can be extracted using other methods such as interviews or surveys during initial exploration or discovery processes (Spinuzzi 2005).

Design workshops, rather than simply seeking feedback or personal accounts from the participants (e.g. as in a focus group), ask participants to identify and imagine their own product solutions and improvements, which may be described by drawings on whiteboards, sticky notes, paper or other means, depending on the context and the skills of the participants. A common output is therefore the development of low- or high-fidelity prototypes. Prototyping, which may be also used in other contexts (i.e. in UCD or UX activities), is an important tool, and therefore it is discussed next.

Prototypes are defined as artefacts that simulate some or all features of the technological solution that is being designed. They serve to develop creative solutions, but also test and evaluate intermediate design stages before arriving at the final product, and can be used as throwaways (i.e. knowledge is acquired through the test but the prototype is discarded), incrementally (i.e. each prototype stands for a component of the final design) or evolutionarily (i.e. each prototype is used as the basis for a following iteration; Dix et al. 2004). Prototypes can be done on paper or digitally. They can be 'anything from a paper-based storyboard through [to] a complex piece of software, and from a cardboard mock-up to a moulded or pressed piece of metal' (Preece, Rogers and Sharp 2015, 386).

Paper prototypes require the researchers and participants to sketch drawings, dialog boxes, screens or other interactive features of a user interface, which represent or simulate the electronic device's processes, as users interact with the paper interface (Carter and Hundhausen 2010). These prototypes are also known as 'low-fidelity' prototypes. These are valuable sources of information, not only because they may contain creative and novel understandings and visualisations to represent participants' knowledge structures and needs, but also as they require no coding, they can be used to 'demonstrate the behavior of an interface very early in the development' at a low cost (Rettig 1994, 22). Digital prototyping, also known as 'high fidelity', includes transferring this information in digital forms to build demos and digital interfaces, which usually offer some functionality and interactivity options via a digital device. These are usually used to demonstrate and test technical solutions, and therefore they may be used in usability testing experiments (Section 3.6) for evaluation purposes or in focus groups (Section 3.4) to obtain further feedback.

There are a few examples in citizen science which describe the use of design workshops and prototyping for various purposes. Phillips et al. (2014) use design workshops in the Bee Lab project implementation – a citizen science project which develops toolkits to support beekeepers in constructing their own monitoring devices and to collect data from hives. They use design workshops at various stages of the project, initially as a method to collect participants' concepts for monitoring, which they then use to develop tangible monitoring kits. In the second stage, they use design workshops to evaluate preliminary designs of the Bee Lab kits, where participants try to assemble the kits and discuss their functionality with respect to their contextual needs and usability.

In another citizen science–inspired example, Bowser et al. (2014) explore the development of a gamified citizen science application to report plant phenology data, called Floracaching. The authors provide an excellent description of how prototyping can be used as part of an iterative co-design process with end users while considering and building on contextual characteristics. Since Floracaching is a 'location-based app and game', the authors propose a novel approach: Prototyping for Location,

Activities and Collective Experiences (PLACE), which is based on a set of principles for comprehensively applying the method in a way that fits a particular context. The authors suggest 'that the prototyping process should organize sessions so that they accurately reflect the way actual users will experience time in the app or game' (Bowser et al. 2013, 1520), and this involves taking into account and mimicking environmental and social conditions, which are representative of place, time and other contextual characteristics. In their study, the authors further describe how the application uses different types of prototyping incrementally in a co-design process with end users, combined with other techniques such as designed-centred focus groups and contextual inquiry.

### 3.8 Personas

Personas are a popular tool in HCI which have been used in the context of citizen science, albeit not extensively. The method allows design teams to create fictitious characters, the so-called personas, that represent the needs of the intended user audiences, and scenarios which are used to describe user needs and how these can be met. Although the method is not designed to obtain new user input directly, it is used to communicate user findings from other methods (e.g. interviews, storytelling activities, focus groups, etc.), but it may also be used in combination with other HCI methods (e.g. see inspection methods in Section 3.9) which support the design team in mimicking user actions in order to understand how well these are met and whether there are usability deficiencies. Blomkvist (2002) explains that a persona may not be a description of a particular existing user, but rather a collection of patterns of user 'behaviour, goals and motives, compiled in a fictional description of a single individual' (Blomkvist 2002, 1). The lifelike quality of personas can help product developers make user needs more 'tangible and alive' (Blomkvist 2002, 1). Personas are situated in similarly archetypical 'scenarios', which are also built relying on information gathered during initial research. Personas and scenarios are used together in order to imagine and communicate how a design will operate in the hands of potential users.

An example of a persona constructed for the purposes of evaluating the Gender and Tech Magazines citizen science application is provided in Figure 3.2. The application collects data from users who scroll through digitised tech magazine pages and answer a set of questions about how and how frequently women are represented there. The persona in this example is 'Claire Thomson', a scientist and an ambassador of science, technology, engineering and mathematics (STEM) subjects, who offers

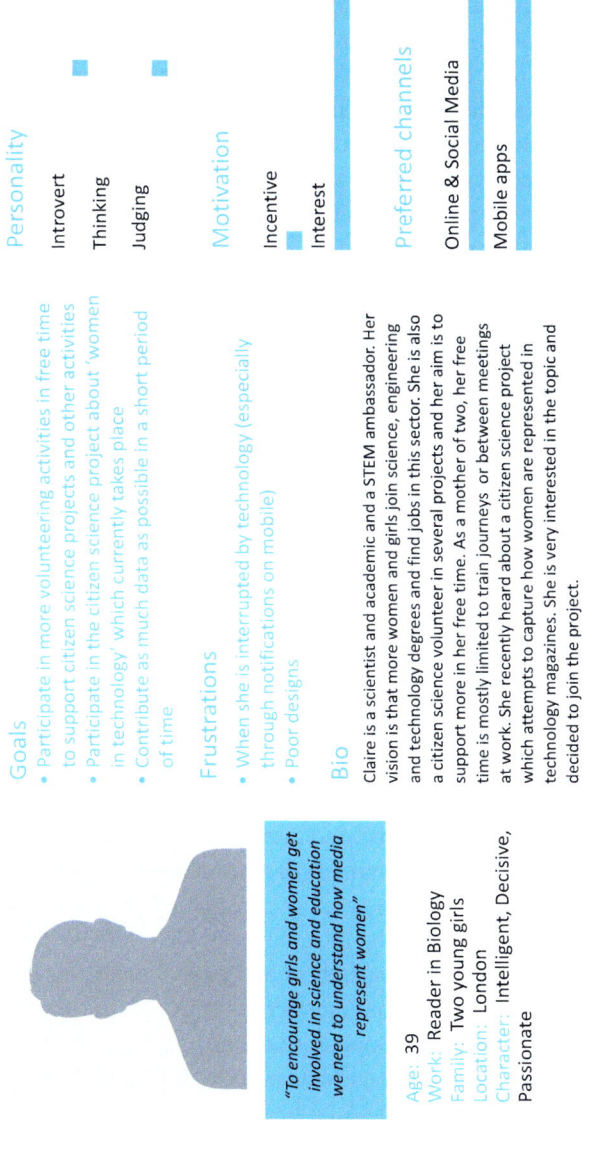

**Fig. 3.2** Persona example used for an evaluation of the Gender and Tech Magazines citizen science application (created with Xtensio.com). Source: authors.

her time and enthusiasm to demonstrate the importance of these subjects, especially in women's lives and careers. The persona includes details about Claire's age, job role, location and personality traits. It further provides a short biographical description to bring the persona to life. It provides a brief description of her goals, frustrations and motivations, as well as her preferred channels and technologies for participating, which are particularly important when the method is used as a tool to evaluate or design an application. For example, Claire's time is limited, and therefore it would only be reasonable to assume that most appealing to this user would be applications which support the quick completion of tasks.

Although there are different types of personas and ways to represent them, a common mistake that occurs in persona design is that the representations of potential users are often dry and vague. These do not inspire those who use them, and it does not help them empathise with the persona(s), which is one of the goals of using this method in the first place. Realistic designs, such as the one provided above, may help achieve these aims.

Although few examples of the method can be found in citizen science literature, those which exist sufficiently capture and describe how the method can be applied, mostly in combination with other methods. Getto and Moore (2017) describe the use of personas in a usability assessment of the North Carolina Coastal Atlas – an online tool used to show changes to the coastline, used by researchers, coastal managers and educators as well as ordinary citizens. Their study provides significant insight into UX design (Section 2.3), with a combination of lab-based and field experiments (i.e. contextual inquiries), and it shows how the method can be used not only for data representation, but also for data collection, analysis and visualisation to support complex geographic tasks. In this study, through a series of interviews and usability testing experiments with actual users of the application, the authors explain that they found that technical expertise and familiarity with geographic information systems (GIS) strongly influences user interactions with the application, and therefore the four initially constructed personas are used to represent potential users who have different job roles (i.e. GIS experts, coastal planners, coastal researchers and coastal educators). They also explore how the four personas interact not only with the application but also with each other (via the application) and how this influences its overall functionality. What it is particularly interesting is that continuous engagement of users with the Atlas team during their evaluation process had a pedagogic character, as it helped them 'to make several important links between UX, rhetoric, and networked communication . . . they were able to grasp UX in a concrete sense by trying out several methods on an actual application by interacting with real, live users' (Getto and Moore 2017, 27). This in turn resulted in users constructing themselves an additional persona – that of the 'citizen scientist' – which was also incorporated in the evaluation to ensure that this type of user finds the application useful and usable.

### 3.9 Inspection methods

Inspection methods are used to expose interaction problems, mainly usability related, associated with the user interface design. Although they only involve expert evaluators rather than the actual end users in the evaluation of an existing application, not only are these methods popular in wider HCI research, but they also appear in citizen science studies. Inspection methods are easy to apply, effective in terms of the number of problems which may be detected and time efficient, especially when compared to usability testing (Hollingsed and Novick 2007). The focus of this section will be on heuristic evaluation – one of the most popular inspections methods used in HCI (Molich and Nielsen 1990; Nielsen 1993).

Heuristic evaluation involves a set of evaluators, usually three to five – to minimise subjectivity and ensure that most problems are detected (Nielsen 1992) – inspecting the user interface of an application based on a list of heuristics (Nielsen 1994). These heuristics are usually a set of well-established principles or guidelines, which are established after extensive user testing in specific contexts of use (e.g. e-commerce applications, online citizen science applications, etc.). Each violated heuristic therefore presents a usability issue, which – depending on the problem – should be redesigned to improve interaction. A severity rating can be assigned to each violated heuristic based on criteria such as the frequency of the problem's occurrence, its impact and persistence (Nielsen 2003), which can help prioritise the redevelopment work.

The HCI field provides several lists of heuristics, the most popular being those introduced by Molich and Nielsen's (1990) 'Ten Usability Heuristics for Interface Design' and Shneiderman and Plaisant's (2004) 'Eight Golden Rules of Interface Design'. Although these heuristics cover a broad set of interaction principles to improve the usability of a wider range of websites, they do not particularly target geographic and citizen science applications which have their own specific characteristics. Some guidelines that would be perhaps more relevant to the specific context of citizen science are: Jennett and Cox's (2014) 'Eight Guidelines for Designing Virtual Citizen Science Projects' and Skarlatidou, Hamilton et al.'s (2019) best practice guidelines which were developed after a systematic literature review of user-based studies from the broader citizen science context. There are not yet guidelines which focus explicitly on the geographic component of citizen science, although GIS guidelines from the context of public web mapping (e.g. see Nivala, Brewster and Sarjakoski 2008) can be beneficial, since some of these applications may share similar characteristics.

Wald, Longo and Dobell (2015) use the method of heuristic evaluation to assess 20 virtual citizen science projects with six evaluators who received training in using the method. Their aim is to identify design principles which support this type of application in order to 'compete more effectively for volunteers, increase productivity of project participants, and retain contributors over time' (Wald, Longo and Dobell 2015, 562). Their list of heuristics comprises of principles from UX design (Section 2.3), as well as lessons learned from their own experience in setting and managing citizen science projects, guided by themes such as engagement, retention and usability. The severity ratings given by each of the six evaluators are averaged, and the authors attempt to identify correlations among specific heuristics, although this is not the common approach in other studies, given the low number of evaluators who usually perform a heuristic evaluation. In another study, Reed et al. (2013) also use an inspection method carried out by two expert evaluators, which resembles that of heuristic evaluation, in order to assess nine Zooniverse projects against specific usability principles which refer to: content, ease of use, made for the medium, emotion and aesthetics. Their list is inspired by Microsoft's Usability Guidelines but has been modified to fit the context of virtual citizen science. The authors emphasise less the violated usability principles and redesign suggestions which are usually aims of a heuristic evaluation and suggest that their generated usability index can be very helpful for creating systematic comparisons across a selection of projects.

It is clear from the previous examples that heuristic evaluation provides enough flexibility for adaptation, so that better supports the reasons behind conducting the evaluation in the first place. These decisions need to be carefully thought and well-justified, as a poor experimental design will lead to poor results and subsequently to ineffective solutions that may not help improve interaction but can have exactly the opposite results.

## 4. Discussion

Citizen science applications target a wide spectrum of end users, with different skills, expertise, experiences and interests to allow them to participate and volunteer their time and efforts in collecting data for scientific purposes. In a large survey conducted by West, Pateman and Dyke (2016) to identify motivations for participating in citizen science projects in the UK, they found that that the majority of respondents were mainly white middle-aged men with high incomes. In a different survey, by Geoghegan et al. (2016), the authors found that citizen science volunteers are more likely to be males in either the 25–34 or 55–64 age groups, with the majority of them having more than three years of experience in participating in similar projects. Other studies further highlight the advanced academic qualifications of volunteers (e.g. Crall et al. 2013; Wright et al. 2015). This is already a diverse user audience to design for, but as citizen science digital applications expand to target specifically children and teenagers, students, older people and underrepresented groups in science, such as indigenous communities, the importance of understanding user audiences and sufficiently incorporating their needs and requirements in the design of the final product becomes a critical concern. This will determine not only the successful implementation of the project but also the future of the field.

Rather than assuming that 'you know what your users want', which experience shows leads to problematic designs, we believe that geographic and citizen science needs more user-based studies to improve our understanding of how to design for these diverse user audiences. To improve understanding of how this can be achieved in practice, this chapter offers a review of a set of design approaches and methods, mainly from the HCI context, which can be used to support the engagement of the end user at various stages of the design, development and evaluation of citizen science applications, with the subsequent goal of improving how users interact with them. The methods that are reviewed in the previous sections can be used at different stages (e.g. from conceptualisation to evaluation) – iteratively, in combination, or coupled with other methods found in HCI and social sciences literature.

Whether you decide to follow a UCD approach or a participatory design approach to work with your end users, it is certain that you will gain a significantly better insight about your project as a whole and the specific features of the technology which will be implemented. Nevertheless, it should be always kept in mind that 'we can't just add users and stir'. A careful selection and thoughtful experimental design plan is absolutely essential to ensure that the project's aims are met without introducing further assumptions and biases. Methods need to be selected and applied at the appropriate stage and in ways that follow methodological protocols, and the users involved need to be representative of the target audiences. Feedback should be always sought, either from HCI specialists directly or from other professionals in geographic and citizen science and other disciplines who have experience with the implementation of similar methods. Last but not least, findings should no longer remain anecdotal evidence in the hands of development teams. It is essential that we learn from each other – that we share our success stories and failures to improve methodological understandings and design issues that can support the development of usable and exciting applications which will attract, rather than put off, much larger audiences of volunteers in citizen science.

## Acknowledgements

Many thanks to Marcos Moreu, Fabien Moustard and Judy Barrett for their comments on an earlier version of this chapter. This work is supported by the European Union's H2020 research and innovation programme Doing It Together Science (under Grant Agreement No. 709443) and the ERC Advanced Grant project European Citizen Science: Analysis and Visualisation (under Grant Agreement No. 694767).

## References

Adams, Anne, Peter Lunt and Paul Cairns. 2008. 'A qualitative approach to HCI research'. In *Research Methods for Human–Computer Interaction*, edited by Paul Cairns and Anna Cox, 138–57. Cambridge: Cambridge University Press.

Ahram, Tareq Z., and Christianne Falcão. 2019. *Advances in Usability and User Experience: Proceedings of the AHFE 2019 International Conferences on Usability and User Experience, and Human Factors and Assistive Technology, July 24–28, 2019, Washington, DC, USA*. Cham, Switzerland: Springer.

Bannon, Liam. 2011. 'Reimagining HCI: Toward a more human-centered perspective', *Interactions* 18: 50–7.

Bekker, Mathilde, and John Long. 2000. 'User involvement in the design of human–computer interactions: Some similarities and differences between design approaches'. In *People and Computers XIV – Usability or else!*, edited by Sharon McDonald, Yvonne Waern and Gilbert Cockton, 135–47. London: Springer.

Berg, Bruce L. 2009. *Qualitative Research Methods for the Social Sciences*, 7th ed. Long Beach, CA: Pearson.

Blomkvist, Stefan. 2002. 'Persona – An overview (extract from the paper 'The user as a personality. Using personas as a tool for design')'. Position paper for the course workshop

Theoretical Perspectives in Human–Computer Interaction at IPLab, 3 September 2002, 1–8. Stockholm: KTH-Royal Institute of Technology.

Bowser, Anne, Derek Hansen, Jennifer Preece, Yurong He, Carol Boston and Jennifer Hammock. 2014. 'Gamifying citizen science: A study of two user groups'. In *Proceedings of the 17th ACM Conference on Computer Supported Cooperative Work*, 137–40. CSCW Companion '14. New York: ACM.

Bowser, Anne, Derek Hansen, Matthew Reid, Jocelyn Raphael, Ryan Camett, Yurong He, Carol Boston, Dana Rotman and Jennifer Preece. 2013. 'Prototyping in PLACE: A scalable approach to developing location-based apps and games'. In *CHI '13: Proceedings of the 31st Annual SIGCHI Conference*, 1519–28. New York: ACM, 2013.

Brooke, John, Nigel Bevan, Fred Brigham, Susan Harker and David Youmans. 1990. 'Usability statements and standardisation: Work in progress in ISO'. In *Proceedings of the IFIP TC13 Third International Conference on Human–Computer Interaction*, 357–61. Amsterdam: North-Holland Publishing Co.

Carter, Adam S., and Christopher D. Hundhausen. 2010. 'How is user interface prototyping really done in practice? A survey of user interface designers'. In *2010 IEEE Symposium on Visual Languages and Human-Centric Computing*, 207–11. Leganes, Spain: IEEE.

Cooper, Alan, Robert Reimann and David Cronin. 2007. *About Face 3: The essentials of interaction design*. Indianapolis, IN: John Wiley.

Crall, Alycia W., Rebecca Jordan, Kirstin Holfelder, Gregory J. Newman, Jim Graham and Donald M. Waller. 2013. 'The impacts of an invasive species citizen science training program on participant attitudes, behavior, and science literacy', *Public Understanding of Science* 22: 745–64.

Curtis, Vickie. 2015. 'Motivation to participate in an online citizen science game: A study of Foldit', *Science Communication* 37: 723–46.

D'Hondt, Ellie, Matthias Stevens and An Jacobs. 2013. 'Participatory noise mapping works! An evaluation of participatory sensing as an alternative to standard techniques for environmental monitoring', *Pervasive and Mobile Computing* 9: 681–94.

Dix, Alan, Janet E. Finaly, Gregory D. Abowd and Russell Beale. 2004. *Human–Computer Interaction*. Harlow, UK: Pearson/Prentice-Hall.

Dourish, Paul. 2015. 'Forward'. In *At the Intersection of Indigenous and Traditional Knowledge and Technology Design*, edited by Nicola Bidwell and Heike Winschiers-Theophilus. Santa Rosa, CA: Informing Science Press.

Fernández-Llamazares, Álvaro, and Mar Cabeza. 2018. 'Rediscovering the potential of indigenous storytelling for conservation practice', *Conservation Letters: Journal of the Society for Conservation Biology* 11: e12398.

Fledderus, Tom. 2016. 'Creating a usable web GIS for non-expert users: Identifying usability guidelines and implementing these in design'. Master's diss., Uppsala University.

Forlizzi, Jodi, and Katja Battarbee. 2004. 'Understanding experience in interactive systems'. In *DIS '04: Proceedings of the 5th Conference on Designing Interactive Systems: Processes, practices, methods, and techniques*, 261–8. New York: ACM.

Fowles, R. A. 2000. 'Symmetry in design participation in the built environment: Experiences and insights from education and practice'. In *Collaborative Design*, edited by S. A. R. Scrivener, L. J. Ball and A. Woodcock, 59–70. London: Springer.

Geoghegan Hillary, Alison Dyke, Rachel Pateman and Sarah West. 2016. *Understanding Motivations for Citizen Science*. Final report on behalf of UKEOF, University of Reading, Stockholm Environment Institute (University of York) and University of the West of England. Swindon, UK: UKEOF.

Getto, Guiseppe, and Christina J. Moore. 2017. 'Mapping personas: Designing UX relationships for an online coastal atlas', *Computers and Composition* 43: 15–34.

Hassenzahl, Marc. 2011. 'User experience and experience design'. In *The Encyclopedia of Human–Computer Interaction*, edited by Mars Soegaard and Rikke F. Dam. Aarhus, Denmark: Interaction Design Foundation.

Hassenzahl, Marc, and Noam Tractinsky. 2006. 'User experience – A research agenda', *Behaviour and Information Technology* 25: 91–7.

Hollingsed, Tasha, and David G. Novick. 2007. 'Usability inspection methods after 15 years of research and practice'. In *SIGDOC '07: Proceedings of the 25th Annual ACM International Conference on Design of Communication*, 249–55. New York: ACM.

Holtzblatt, Karen, and Hugh R. Beyer. 2015. *Contextual Design: Evolved*. San Rafael, CA: Morgan & Claypool.

Hunsucker, Andrew, and Martin Siegel. 2015. 'Once upon a time: Storytelling in the design process'. Paper presented at LearnXDesign, Chicago, IL, 28–30 June 2015.

Idris, Nurul Hawani, Mohamad J. Osman, Kasturi D. Kanniah, Nurul Hazrina Idris and Mohamad H. I. Ishak. 2016. 'Engaging indigenous people as geo-crowdsourcing sensors for ecotourism mapping via mobile data collection: A case study of the Royal Belum State Park', *Cartography and Geographic Information Science* 44: 113–27.

Interaction Design Foundation. 2019. What is the difference between interaction design and UX design. Accessed 13 September 2020. https://www.interaction-design.org/literature/article/what-is-the-difference-between-interaction-design-and-ux-design.

Jennett, Charlene, and Anna L. Cox. 2014. 'Eight guidelines for designing virtual citizen science projects'. Paper presented at Second AAAI Conference on Human Computation and Crowdsourcing, Pittsburgh, PA, 2–4 November 2014.

Kim, Sunyoung, Jeffrey S. Pierce, Christine Robson and Thomas Zimmerman. 2011. 'Creek watch: Pairing usefulness and usability for successful citizen science'. In *CHI '11: Proceedings of the SIGCHI Conference on Human Factors in Computing Systems*, Vancouver, CA, 2125–34. New York: ACM.

Kurosu, Masaaki. 2013. *Human–Computer Interaction. Human-Centred Design Approaches, Methods, Tools, and Environments: Proceedings of the 15th International Conference, HCI International 2013, Las Vegas, NV, USA*. Berlin: Springer.

Law, Effie, Virpi Roto, Arnold P. O. S. Vermeeren, Joke Kort and Marc Hassenzahl. 2008. 'Towards a shared definition of user experience'. In *CHI '08 Extended Abstracts on Human Factors in Computing Systems*, 2395–8. New York: ACM.

Mannik, Lynda, and Karen McGarry, eds. 2017. *Practicing Ethnography: A student guide to method and methodology*. Toronto, Canada: University of Toronto Press.

Michener, William K., Suzie Allard, Amber Budden, Robert B. Cook, Kimberley Douglass, Mike Frame, Steve Kelling, Rebecca Koskela, Carol Tenopir and David A. Vieglais. 2012. 'Participatory design of DataONE – Enabling cyberinfrastructure for the biological and environmental sciences', *Ecological Informatics* 11: 5–15.

Molich, Rudolf, and Jakob Nielsen. 1990. 'Heuristic evaluation of user interfaces'. In *Proceedings of the SIGCHI Conference on Human Factors in Computing Systems*, edited by Jane C. Chew and John Whiteside, 249–56. New York: ACM.

Muller, Michael J., and Allison Druin. 2012. 'Participatory design: The third space in HCI'. In *Human–Computer Interaction Handbook*, edited by Julie A. Jacko, 1051–68. Boca Raton, FL: CRC Press.

Newman, Greg, Don Zimmerman, Alycia Crall, Melinda Laituri, Jim Graham and Linda Stapel. 2010. 'User-friendly web mapping: Lessons from a citizen science website', *International Journal of Geographical Information Science* 24: 1851–69.

Nielsen, Jakob. 1989. 'Usability engineering at a discount'. In *Proceedings of the Third International Conference on Human–Computer Interaction on Designing and Using Human–Computer Interfaces and Knowledge Based Systems*, edited by Gavriel Salvendy and Michael J. Smith, 394–401. New York: Elsevier Science.

Nielsen, Jakob. 1992. 'Reliability of severity estimates for usability problems found by heuristic evaluation'. In *Posters and Short Talks of the 1992 SIGCHI Conference on Human Factors in Computing Systems*, 129–30. New York: ACM.

Nielsen, Jakob. 1993. *Usability Engineering*. San Francisco, CA: Morgan Kaufmann.

Nielsen, Jakob. 1994. 'Heuristic evaluation'. In *Usability Inspection Methods*, edited by Jakob Nielsen and Robert L. Mack, 25–62. New York: John Wiley.

Nielsen, Jakob. 2001. Success rate: The simplest usability metric. Accessed 10 February 2009. https://www.nngroup.com/articles/success-rate-the-simplest-usability-metric/.

Nielsen, Jakob. 2003. Usability 101: Introduction to usability. Accessed 14 March 2019. https://www.nngroup.com/articles/usability-101-introduction-to-usability/.

Nielsen, Jakob. 2012. Thinking aloud: The #1 usability tool. Accessed 7 April 2014. https://www.nngroup.com/articles/thinking-aloud-the-1-usability-tool/.

Nielsen, Jakob, and Thomas K. Landauer. 1993. 'A mathematical model of the finding of usability problems'. In *Proceedings of ACM INTERCHI '93 Conference in Amsterdam, The Netherlands, 24–29 April 1993*, 206–13. New York: ACM.

Nivala, Annu-Maaria, Stephen Brewster and Tiina L. Sarjakoski. 2008. 'Usability evaluation of web mapping sites', *The Cartographic Journal* 45: 129–38.

Norman, Donald A. 2011. 'Commentary on Marc Hassenzahl's user experience and experience design'. In *The Encyclopedia of Human–Computer Interaction*, edited by Mars Soegaard and Rikke F. Dam. Aarhus, Denmark: Interaction Design Foundation. Accessed 13 July 2020. https://www.interaction-design.org/literature/book/the-encyclopedia-of-human-computer-interaction-2nd-ed/user-experience-and-experience-design.

Norman, Donald A., and Stephen W. Draper. 1986. *User Centered System Design: New perspectives on human–computer interaction*. Hillsdale, NJ: Lawrence Erlbaum Associates.

Nov, Oded, Ofer Arazy and David Anderson. 2011. 'Dusting for science: Motivation and participation of digital citizen science volunteers'. In *Proceedings of the 2011 iConference*, 68–74. New York: ACM.

Ozok, A. Ant, Dana Benson, Joyram Chakraborty and Anthony F. Norcio. 2008. 'A comparative study between tablet and laptop PCs: User satisfaction and preferences', *International Journal of Human–Computer Interaction* 24: 329–52.

Pejovic, Veljko, and Artemis Skarlatidou. 2020. 'Understanding interaction design challenges in mobile extreme citizen science', *International Journal of Human–Computer Interaction* 36: 251–270.

Phillips, Robert, and Sharon Baurley. 2014. 'Exploring open design for the application of citizen science; A toolkit methodology'. Paper presented at Design Research Society, Umea, Sweden, 16–19 June 2014.

Phillips, Robert, Jesse Blum, Michael A. Brown and Sharon Baurley. 2014. 'Testing a grassroots citizen science venture using open design, the Bee Lab Project'. In *CHI '14 Extended Abstracts on Human Factors in Computing Systems, CHI EA-14*, 1951–6. New York: ACM.

Preece, Jennifer. 2016. 'Citizen science: New research challenges for human–computer interaction', *International Journal of Human–Computer Interaction* 32: 585–612.

Preece, Jennifer, Yvonne Rogers and Helen Sharp. 2015. *Interaction Design: Beyond human–computer interaction*, 4th ed. Chichester, UK: John Wiley.

Quesenbery, Whitney, and Kevin Brooks. 2010. *Storytelling for User Experience: Crafting stories for better design*. New York: Rosenfeld Media.

Raddick, M. Jordan, Georgia Bracey, Pamela L. Gay, Chris J. Lintott, Carie Cardamone, Phil Murray, Kevin Schawinski, Alexander S. Szalay and Jan Vandenberg. 2013. 'Galaxy Zoo: Motivations of citizen scientists', *Astronomy Education Review* 12.

Raddick, M. Jordan, Georgia Bracey, Pamela L. Gay, Chris J. Lintott, Phil Murray, Kevin Schawinski, Alexander S. Szalay and Jan Vandenberg. 2010. 'Galaxy Zoo: Exploring the motivations of citizen science volunteers', *Astronomy Education Review* 9.

Randall, David, Richard Harper and Mark Rouncefield. 2007. *Fieldwork for Design: Theory and practice*. London: Springer.

Reed, Jason, M. Jordan Raddick, Andrea Lardner and Karen Carney. 2013. 'An exploratory factor analysis of motivations for participating in Zooniverse, a collection of virtual citizen science projects'. In *Proceedings of the 2013 46th Hawaii International Conference on System Sciences (HICSS '13)*, 610–19. Washington, DC: IEEE Computer Society.

Rettig, Marc. 1994. 'Prototyping for tiny fingers', *Communications of the ACM* 37: 21–7.

Rogers, Yvonne. 2004. 'New theoretical approaches for human–computer interaction', *Annual Review of Information Science and Technology* 38: 87–143.

Rubin, Jeff, and Dana Chisnell. 2008. *Handbook of Usability Testing: How to plan, design, and conduct effective tests*, 2nd ed. Indianapolis, IN: John Wiley.

Sanders, Elizabeth B. N. 2002. 'From user-centered to participatory design approaches'. In *Design and the Social Sciences*, edited by Jorge Frascara, 18–25. London: CRC Press, Taylor and Francis.

Shneiderman, Ben, and Catherine Plaisant. 2004. 'Eight golden rules of interface design'. In *Designing the User Interface: Strategies for effective human–computer interaction*, 4th ed. Boston, MA: Addison Wesley.

Skarlatidou, Artemis, Alexandra Hamilton, Michalis Vitos and Muki Haklay. 2019. 'What do volunteers want from citizen science technologies? A systematic literature review and best practice guidelines'. *Journal of Science Communication* 18: 1–8.

Skarlatidou, Artemis, Marisa Ponti, James Sprinks, Christian Nold, Muki Haklay and Eiman Kanjo. 2019. 'User experience of digital technologies in citizen science', *Journal of Science Communication* 18: 1–8.

Spaulding, Eric, and Haakon Faste. 2013. 'Design-driven narrative: Using stories to prototype and build immersive design worlds'. In *CHI '13: Proceedings of the SIGCHI Conference on Human Factors in Computing Systems*, 2843–52. New York: ACM.

Spinuzzi, Clay. 2005. 'The methodology of participatory design', *Technical Communication (Washington)* 52: 163–74.

Suchman, Lucy. 2002. 'Located accountabilities in technology production', *Scandinavian Journal of Information Systems* 14: 91–105.

Truna. 2015. 'African gamer: Whose story is it anyway?'. In *At the Intersection of Indigenous and Traditional Knowledge and Technology Design*, edited by Nicola J. Bidwell and Heike Winschiers-Theophillus, 35–66. Santa Rosa, CA: Informing Science Press.

Wald, Dara M., Justin Longo and A. R. Dobell. 2015. 'Design principles for engaging and retaining virtual citizen scientists', *Conservation Biology* 30: 562–70.

Weller, Susan C. 1998. 'Structured interviewing and questionnaire construction'. In *Handbook of Methods in Cultural Anthropology*, edited by H. Russell Bernard and Clarence C. Gravlee, 343–90. London: Rowman & Littlefield.

West, Sarah, Rachel Pateman and Alyson Dyke. 2016. *Data Submission in Citizen Science Project*. Report for Depfra PH0475. York, UK: University of York.

Wilson, Chauncey. 2013. *Interview Techniques for UX Practitioners: A user-centered design method*. Waltham, MA: Morgan Kaufmann.

Woods, Will, and Eileen Scanlon. 2012. 'iSpot Mobile – A natural history participatory science application'. In *Proceedings of Mlearn 2012*, Helsinki, Finland, 15–16 October 2012.

Wright, Dale R. Les G. Underhill, Matt Keene and Andrew T. Knight. 2015. 'Understanding the motivations and satisfactions of volunteers to improve the effectiveness of citizen science programs', *Society and Natural Resources* 28: 1013–29.

# Chapter 4
# Methods in anthropology to support the design and implementation of geographic citizen science

Raffaella Fryer-Moreira and Jerome Lewis

## Highlights

- The successful implementation of geographic citizen science projects in diverse cultural, environmental and infrastructural contexts requires a renewed attention to local specificities, and careful consideration of how these specificities can inform the design of geographic citizen science tools and strategies for implementation.
- Anthropological methods (e.g. ethnographic observation; negotiating a free, prior and informed consent process; and the development of community protocols) are key to the participative design of tools capable of translating indigenous ecological knowledge into data sets that can be placed in dialogue with current scientific conservation and policy models.

## 1. Introduction

This chapter outlines anthropological approaches to the development and implementation of the geographic citizen science projects that are discussed mainly in the third part of this book (Chapters 11, 12, 14 and 15). It highlights the importance of ethnographic methods, particularly in non-urban contexts and among indigenous populations, where sociocultural specificities must be considered in the design and development of digital interfaces and implementation approaches. Two key pillars of engagement are described here: free, prior and informed consent (FPIC) and community protocols (CPs). Theoretical approaches in anthropology will also be discussed, which place indigenous knowledge practices on an epistemic par with scientific knowledge practices, highlighting the importance of participatory forms of knowledge production which challenge the exclusivity of 'expert' status. By engaging with the epistemologies of others, our own scientific frameworks may be broadened and enriched, drawing on ethnographic diversity to adapt or develop new methods and concepts better able to address contemporary challenges in science and society.

The value of geographic citizen science approaches to environmental monitoring and conservation efforts is increasingly being recognised by the international scientific community (Haklay 2013), leading to a proliferation of such projects around the world. Not only do community members and scientists produce similar results in data quality and quantity when documenting the status of and trends in local species and natural resources (Danielsen et al. 2014), but in indigenous contexts, they also introduce forms of ecological knowledge that have historically been excluded from scientific and conservation discourses. The contributions in this volume are testimony to the diversity of social, cultural and ecological contexts in which such projects are being implemented, and they highlight the insights – as well as the challenges – that these diverse contexts offer. The interdisciplinary perspective this volume proposes is essential to addressing and overcoming the challenges faced in such work, and the methodological and theoretical approaches offered by anthropology and extreme citizen science are crucial to the design, implementation and evaluation of sustainable geographic citizen science projects. Anthropological perspectives are particularly pertinent in 'extreme' geographic citizen science contexts, where indigenous and local communities living in remote regions present unique cultural, environmental and infrastructural challenges to urban citizen science strategies.

Researchers must negotiate between the need to design context-specific technological assemblages that are appropriate to cultural and environmental particularities, and the need to produce data that are sufficiently standardised and robust to be interpreted and given due consideration by international actors, institutions and other stakeholders. Anthropological methods are key to the participative design of tools capable of translating Traditional Ecological Knowledge (TEK) into data sets that can be placed in dialogue with current scientific conservation and policy models (see Chapter 12). As contemporary conservation models fail to prevent environmental destruction and degradation on a global scale, indigenous ecological knowledge – and the approaches to conservation it informs – is playing an increasingly important role in global environmental policy, offering new ways of conceiving of human–environment relations which may help inform new conservation models (Lewis 2019) and new frameworks for scientific enquiry. In this context, anthropological methods and theoretical frameworks play a key role in making geographic citizen science projects in extreme conditions feasible, and contribute towards the diversification of conservation discourses and practices.

In this chapter, we will outline the key ways in which anthropological methods inform geographic citizen science projects, highlighting the role they play in community engagement, project design and implementation. We will go on to explore the theoretical contribution anthropology offers to geographic citizen science projects, and how geographic citizen science itself is theorised. Finally, we will discuss the importance of anthropological methods and theory for the development of successful geographic citizen science projects.

## 2. Ethnographic methods: anthropological approaches in geographic citizen science

Anthropological knowledge begins with the premise that we do not understand the communities with whom we conduct research: this is the purpose of ethnography. It is only through extended ethnographic engagement with a community – involving observation of and participation in daily social life, interviews with individual community members and groups, extensive co-habitation, careful observation of local technical practices – that we are able to gain glimpses into the cultural particularities presented by a given community. The successful implementation of geographic citizen science projects in diverse cultural, environmental and infrastructural contexts requires a renewed attention to local specificities and careful consideration of how these specificities can inform the design of geographic citizen science tools and strategies for implementation. All the case studies in this volume identify the importance of local involvement in the design, planning and implementation of successful projects. While some authors in this volume (e.g. see Chapters 12, 14 and 15) draw on their anthropological background to highlight the ways in which ethnographic methods shaped the process at every stage, authors from human–computer interaction and other disciplinary backgrounds (e.g. see Chapter 11) highlight the ways in which ethnographic approaches facilitated the evaluation of usability challenges and the development of solutions.

The success of any (geographic) citizen science project depends on the volunteers' interest and willingness to participate, and in many contexts, this will also depend on support from the wider local community. This makes building trust with communities an essential starting point for any project and will usually involve spending time with communities in a way that is not directly related to the project itself, including eating together, learning vocabulary and participating in social events and festivities while living in the community. All human societies have their own unique forms of relating with newcomers and establishing relationships of alliance and trust. Therefore, local norms must be understood and adhered to if community participation and consent is to be established.

## 2.1 Participant observation and observant participation: meeting local needs and incorporating local knowledge

If projects are to be sustainable and successful, they must meet the existing needs of local communities and draw on local knowledge frameworks. This entails starting with the premise that the researchers involved do not fully understand the existing needs of local communities or the knowledge frameworks through which these needs are locally articulated. Therefore, preliminary community engagement and research must be conducted in order to establish the project's aims and scope collaboratively.

Participant observation is a core component of ethnographic research, and involves an extended time observing and participating in community activities and the practices of local life (Descola 2005). It is only through immersion in the daily activities of a community – usually involving extended periods of co-habitation – that the researcher can gain sufficient understanding of the social and cultural specificities to inform the design, implementation and successful adoption of geographic citizen science projects. Community meetings and the debates which populate them can often provide valuable insights into the concerns and difficulties faced by local people and the divergent ways in which both problems and potential solutions are understood. Researchers must pay attention to the ways in which problems are posed, the impacts they have on people's lives at the level of both individuals and the group, and how the communities themselves may have attempted to address these challenges in the past.

If the geographic citizen science project in question aims to provide new approaches to environmental monitoring, for example, then it is important that project design is informed by both local and established forms of environmental monitoring, paying heed to the kinds of data that are usually gathered and considered important by local participants and decision makers. New activities will be understood by local peoples and decision makers in reference to existing activities. The more researchers understand about the range of local environmental monitoring practices, the more likely it is that the new practices proposed by the project will be understood in continuity with previous practices and, as a result, understood to be valuable by all. At the same time, researchers who understand the context in which environmental monitoring already takes place are less likely to propose activities which have been tried before and shown to be ineffective. Observation of and participation in existing environmental monitoring practices may be an important place to begin any geographic citizen science project which proposes to introduce new methods, tools and practices for this purpose.

The extent to which a researcher observes or participates in community activities will depend on the specific context, and while it may be appropriate to play the role of an external observer in some situations, active participation may be called for in others. Ethnographers of social movements and protest groups (Krøijer 2015) have often commented on the impossibility of accessing certain spaces without engaging as an active participant, as passive forms of participation are regarded with either suspicion or disdain. Similarly, other anthropologists (e.g. Wacquant 2004) have pointed to the kinds of bodily knowledge that are only acquired through practice, where external observation is insufficient to gain the level of understanding required to make sense of the activities involved. Approaches which place greater weight on researcher participation in activities have sometimes been described as observant participation (Wacquant 2004; Krøijer 2015) rather than participant observation. The method most suitable for a research project must be decided upon in reference to its context.

In-depth interviews are also a core component of ethnographic research, enabling researchers to gain insights into the way people understand their own communities, societies, culture and environmental relations. Interviews can be: (1) structured, where the interviewer has a precise set of questions which remain largely unchanged throughout the interview; (2) semi-structured, where the interviewer has a set of questions which guide the interview, but which leave room for unplanned lines of enquiry which may emerge during the interview itself; or (3) unstructured, where the researcher conducts the interview more like an informal conversation, allowing the participant and the ensuing conversation to shape the line of enquiry. While the semi-structured interview is the format most commonly used by anthropological researchers, the method that is most suitable depends on the interests and priorities of the research, as well as the specificities presented by participants in particular ethnographic contexts. Interviews enable researchers to learn more about the people with whom they work through listening to participants themselves, revealing ideas and perspectives which are often not visible through participant observation (Hockey and Forsey 2012, 71).

## 2.2 Free, prior and informed consent (FPIC)

While acquiring FPIC is a prerequisite for any scientific project with human subjects, when working with communities that are culturally distinct (from the researchers in question), additional considerations must be considered to ensure the process is sensitive to local cultural frameworks and local understandings of what constitutes consent (Lewis and Nkuintchua 2012, 155). Community members must thoroughly understand the objectives of the project and any potential benefits and risks, as well as be confident in their ability to withdraw from the project (and withdraw consent for their data) at any point. The process of soliciting consent must follow local protocols and, while being inclusive, also respect local hierarchies. Researchers must be patient with the time frames involved in local decision-making processes too. In short, if community involvement is to be established and maintained, projects must adhere to both the frameworks of scientific research ethics and local ethical frameworks.

Negotiating FPIC is a process consisting of informing the affected persons about planned activities and their potential positive and negative impacts. It involves verifying that the information provided has been understood before explicit consent can be requested. If people refuse, their decisions must be respected. Whatever outcome results, the process must be formalised in both scientific and locally meaningful ways.

Gaining FPIC is an ongoing iterative process based on the following key elements:

- Prior. All elements of the FPIC process must be completed before any data are collected by communities.
- Consent. This 'is required from people in situations where any externally initiated activity, by state agencies, private enterprises or NGOs, may impact on the lives and livelihoods of individuals and communities' (Lewis 2012, 175). Consent is culturally specific, and the ways that local participants understand consent must be docu-

mented and incorporated into this process. In the case of extreme citizen science, the FPIC discussion is the basis for planning the project with the community, and as much time is given as is needed. By the end of these discussions, participants should understand the purpose and main objectives of the project, understand the intended benefits and how these will be achieved and understand the potential risks and what strategies are in place to avoid them. On this basis, participants are able to make informed decisions on the terms under which they will participate, and with whom the different data sets they collect can be shared. They should understand their right to withdraw fully or partially from the project at any time, and that they have the right to request that their data contribution is deleted (Lewis and Nkuintchua 2012, 157).

- Free and informed. Free means that consent is offered without bribery, duress or coercion of those concerned. Informed is more complex because it requires culturally appropriate and effective strategies that account for literacy levels, as well as sociocultural and linguistic differences that are developed to inform people fully of potential positive and negative consequences. Participants' understanding needs to be independently tested wherever possible before FPIC is requested and formalised.

Each of these elements is crucial if a project or intervention is to be just and sustainable. There are numerous guides to FPIC (some examples include FAO 2014, FAO 2016 and Oxfam 2010), but here are some key steps to achieve an equitable FPIC in the context of citizen science:

(1) Identify which communities will be affected.
(2) Strengthen institutional capacities of both parties as required (languages, gender, tools/equipment, etc.).
(3) Develop culturally appropriate communication and information strategies.
(4) Identify appropriate decision-making processes with those concerned.
(5) Co-develop the project's objectives with those concerned, including strategies to enhance positive outcomes and strategies to reduce negative outcomes.
(6) Co-design the pictogram-based decision tree and the methodologies needed to achieve these objectives with those concerned (see below).
(7) Agree on how the project will be run and any benefit sharing that may accrue (CPs – see below).

(8) Record and formalise the process of obtaining consent both formally and in culturally appropriate ways.
(9) Maintain the relationship on which the consent is based.

## 2.3 Ethnographic approaches for designing suitable interfaces for interaction with devices

The case studies, especially those in the third part of this volume, highlight the importance of ethnographic methods to inform the design of the digital technologies and interfaces themselves, as well as informing evaluations in the field to improve usability and ensure end users can fully utilise the intended geographic citizen science applications.

Anthropological approaches to technical practice may also be useful here, as they can help us understand the existing technical contexts through which new technologies will be locally understood.

Ethnographic observation can help to structure the design process so that it is iterative, paying attention to challenges presented by local infrastructure and environmental conditions, as well as the specific forms of social organisation which will shape the ways in which new devices are received and engaged with. Anthropological approaches to technology have highlighted the ways in which all technical practice is embedded within social relations, how such practices are organised in relation to age and gender, and the broader social contexts in which they are situated. Ethnographic forms of investigating these practices can include producing detailed operational sequences (Lemonnier 1993) of technical activities or mapping out the networks of associations which particular technical processes both require and reproduce (Latour 2005).

Every context is unique, and while some case studies in this volume offer examples of challenges that can arise at different stages of the design and implementation process (e.g. Chapters 11, 12 and 14), the solutions offered in each case must emerge from their specific context and may or may not be transferrable. Participant involvement in the development of technologies and interfaces is essential if such technologies are to prove useful to the communities involved, and assumptions surrounding design and usability must be regularly questioned in extreme citizen science contexts.

### *Pictogram development and decision trees*

The first stage in the design process is to develop a prototype decision tree. Decision trees set out a flow diagram of the choices or decisions to

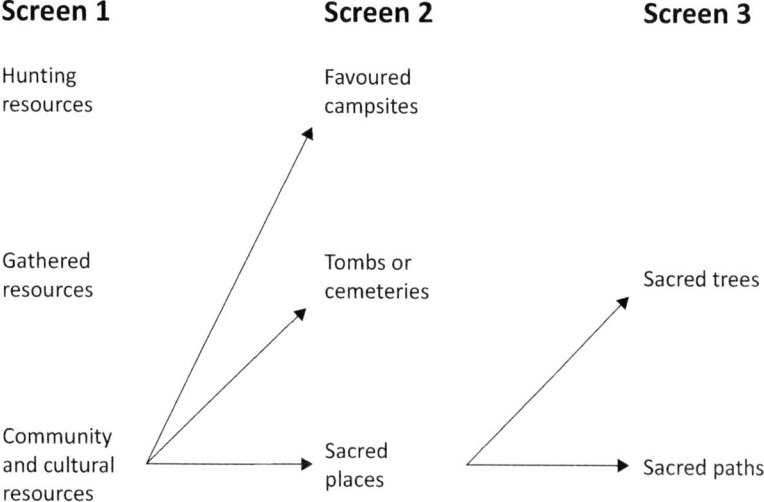

**Fig. 4.1** Example of decision tree flow diagram.
Source: Lewis, J. 2012. Technological leap-frogging in the Congo basin, pygmies and global positioning systems in central Africa: What has happened and where is it going?, *African Study Monographs*, Suppl. 43: 15–44.

be made to collect specific data points and the options required to get there. Figure 4.1 is an example provided by Lewis (2012, 33).

The decision tree forms the basis of the structure or logic for coding an XML project. The new Sapelli Designer, currently under development, removes the need to code, but the decision tree will still need to be designed. On the Sapelli website (http://www.sapelli.org/building-a-simple-decision-tree-with-sapelli-xml), a tutorial for creating a decision tree for programming in XML is provided.

Depending on the educational background of the community, it may be that facilitators offer suggested pictograms to represent the different nodes on the decision tree, or participants develop these using paper, by drawing on the ground or taking photographs. It is often a case of both parties combining their talents to do this collaboratively.

In the case of non-literate people, co-design processes around the development of pictograms to structure data collection may require more imaginative approaches. We have often found that preparing A4 prints/copies of individual pictograms is the best way of exploring design possibilities. We start by showing people the pictograms without saying what they are intended to mean. Instead, we ask the participants to tell us what they represent. If the answers are quick and accurate, the pictogram is a good one. However, if responses take time or the pictogram is understood in different ways, it requires further work. Participants should then take the lead in advising and suggesting how the pictogram can be adapted to make its meaning unambiguous.

Once all the pictograms that people want are on sheets of A4 paper, these can be arranged on the ground in the order that they would appear in the decision tree. The more participatory this process is, the more participants will understand what they are doing in the future, and this will improve accuracy and data quality when they are collecting data points independently.

When the preliminary decision tree and pictograms have been designed, they are made into Sapelli projects and uploaded to smartphones. Different segments of the community (men, women, youth, etc.) involved in the project test the decision tree by taking the prototype out on a phone into the local area to start mapping key resources or other items on the decision tree. As limitations, confusions or additional data points to be collected are identified, they are progressively addressed and incorporated into the decision tree, and a corrected Sapelli project is uploaded and tested again in a similar manner. In our experience, stability is generally achieved after three iterations of this process. Further issues tend to require longer to identify, and it is important to review the decision tree at regular intervals throughout a project's life.

## 2.4 Community protocols (CPs)

To sustain the project over its lifetime, we recommend the development of a CP that formalises the solutions collectively agreed by the community participating in the work.

The CP process is a method for setting the expectations of participants in the conduct of the project. By discussing in turn each element of project activity and how it will be done, the CP provides the opportunity to address problems before they emerge and to put into place solutions agreed upon by all.

The community members are central to planning and deciding how they will engage with project activities and how they will care for and use any equipment left with them, and deciding and naming who is responsible for each of the tasks the project requires. If any remuneration is needed, this should be decided publicly. A schedule for work needs to be agreed by key actors. A process for validating data collected and for sharing it is required. Who will check and who will receive the data collected? With whom can the data be shared and under what circumstances?

Every project will have different implications for those participating. It is during the CP discussions that potential impacts are explored and discussed, and strategies for ensuring that results are as intended are co-developed. This often results in strategic decisions being made that should be documented as part of the CP. If communities require support or capacity building in order to benefit from the project, these should be planned and agreed in the CP.

As with all participatory engagement and design activities, the quality of relations between facilitators and participants matters. Participants should feel empowered and confident that their views will be listened to without criticism or judgement. Facilitators should avoid 'leading the discussion' and should ask open questions with regular encouragement to all present to speak. In situations where a segment of the population is unlikely to talk in public (e.g. women, minorities, lower castes or classes), alternative meeting spaces where they can express themselves freely are required. By holding such additional meetings simultaneously with those in which dominant groups are participating, other voices are more likely to be heard.

Aspects from the FPIC discussion will inform the CP process, and working through a series of questions, such as those listed below, will start the process of who, what, when and where. The first part of the protocol consists of questions and answers about the functioning of the project such as:

- Who collects the data?
- When will they go to collect the data?
- How will they collect the data?
- How will they check the data?
- With whom will they share their data?
- Who is responsible for the equipment?
- What risks are there when collecting the data?

This is then followed by a session on the technical and methodological support, another about the logistics support and one on the data-sharing protocols.

- Technical and methodological support. This includes what equipment is needed, what charging facilities there are (e.g. generator, solar, nearby 'station') and what the level of connectivity is for data transmission. Further refinement/updating of the methodology will be required as part of the ever-changing nature of fieldwork.

- Logistics support. This incorporates the detail of what resources, where and when.
- Data-sharing protocols. For each data item, risks and benefits associated with sharing are discussed and the who, when and where of data sharing is decided upon. How data are shared (e.g. memory cards, secure wireless/relay transmission), in what contexts and with whom are documented.

In cases where the community needs to present their data to government officials or others in positions of power and responsibility in order to lobby for change, they may require additional support. In close consultation with the community and local actors with experience of such processes, a capacity-building programme can be developed to ensure participants can maximise the opportunities for using the data they have collected to make a positive difference to their circumstances. In general, this is best achieved through peer-to-peer learning. Such opportunities may need to be organised by facilitators. Identifying comparable communities that have experience of such lobbying is the first step. Facilitating visits by key players from the participating community to share and learn from their peers can be organised (Lewis and Nkuintchua 2012, 158–62, provide more detail).

## *Good practice*

The CP itself is a changing and iterative process defined by the local team. Further changes, such as inviting new members and defining their roles, are decided by the local team and need to be documented in the CP.

The project team must keep copies of the CP for each village/community and ensure they are each up to date as the work progresses and the project develops. Where possible, pictograms should be incorporated to assist the non-literate or those with limited literacy to understand the document.

## 2.5 Participative evaluation

The success of geographic citizen science projects must be evaluated by community actors themselves, who are best positioned to establish whether the project meets existing local needs. At the same time, researchers must be attentive to local cultural behavioural norms which may inhibit open criticism of a project introduced by outsiders for fear of causing offence or embarrassment or for negatively impacting the acquisition of future resources. Cultural sensitivity must be exercised when designing evaluative activities which pay attention to local codes of conduct, local forms of evaluation and training, and local conventions of critique.

Anthropological methods entail that every step of the process must be subjected to ethnographic investigation. This approach is key if we are to move away from projects which seek to impose Western solutions to problems defined in Western terms, with little attention paid to the communities they claim to support. It also allows for the possibility that we do not understand the social and ecological relationships at stake, the problems that are experienced in these relationships or the solutions that local communities propose. By starting from this premise of ignorance, we are not at a disadvantage – for not only do we have the opportunity to learn from the communities we work with about their own environmental challenges and solutions, but we also have the opportunity to develop an understanding of global environmental challenges that are ethnographically informed and sensitive to cultural specificity while developing cross-cultural frameworks for solutions to these urgent issues. Rather than trying to solve the problems of others, without fully understanding the way in which these problems are experienced, the approach adopted by extreme citizen science proposes to develop tools which enable the conditions in which local design solutions can emerge (e.g. see Chapters 14 and 15).

## 3. Theoretical frameworks

Grounded on the anthropological premise that any aspect of social life may be subject to cultural variation (Wagner 1981), the ethnographic methodologies used by the researchers in this volume have enabled them to implement geographic citizen science projects that both anticipate and design for cultural difference. Every stage of the design and implementation process is enriched by an ethnographic attention to detail which is drawn from a specific theoretical model in anthropological thought. In what follows, the theoretical models which inform each stage will be briefly summarised, with indications towards literature where relevant, before the broader implications of these frameworks are drawn out to examine the wider contribution that anthropological theory can make to geographic citizen science.

The ethnographic approach deployed in the first stage of a project, involving building trust and seeking consent, is grounded on the premise that we do not know what trust building and ethical conduct look like in a different cultural context. People relate to each other in different ways (Schneider 1968; Strathern 1988; Astuti 1995) and organise their societies differently (Evans-Pritchard 1936; Clastres 1977; Lewis 2014) in accordance with different ethical principles (Fabian 1983). This means that unless we seek to inform ourselves about local norms for approaching and proposing projects to a given community, we risk unwittingly causing offence and creating suspicion and hostility, which pose challenges to project acceptance and implementation, and might even render the entire project impossible. The cultural specificity of ethical codes also holds broader implications, for it invites us to consider how the project will be evaluated in light of different concepts of desirable and undesirable behaviour. This may offer new perspectives on the acquisition of consent for scientific research in other areas of inquiry.

The ethnographic questions raised when community needs and local knowledge are addressed are of particular interest. They are grounded in the premise that people know the world in different ways (Latour 1993; Viveiros de Castro 1998; Ingold 2011; Kohn 2013) and know different things about it, which leads them to prioritise problems differently (Scheper-Hughes 1992) and propose different solutions. For example, when facilitating geographic citizen science projects which seek to collect environmental data for conservation monitoring, the existence of alternative ways of knowing the environment, and what is known of it, holds implications for what counts as environmental data, how it is classified and how it might inform broader conservation science. In Chapter 12 of this volume, Hoyte explains that Baka communities in Cameroon have 28 different words for elephant, depending on the animal's condition, its behaviour and its relation to others. Baka knowledge of their environment, their TEK, entails a level of precision and specificity articulated in those 28 different words. So, local strategies for environmental monitoring and responses will be understood through an epistemic framework which distinguishes 28 different conditions of being an elephant. This point is key, for it highlights the significance of, first, acknowledging the existence of alternative knowledge frameworks and, second, engaging with them as a potential source of credible knowledge. By placing scientific understanding as one form of knowledge among others, indigenous knowledge is positioned on an epistemic par – as a different, yet equally valid, approach to making sense of the world around us.

The ethnographic questions raised when designing suitable devices and interfaces are grounded on the premise that people relate to material objects in different ways (Latour 2005; Miller 2005; Ingold 2011), and have different ideas about the kinds of actions that particular objects can enable (Mauss [1936] 1979; Gibson [1977] 1986; Warnier 2009). The particular technical objects that populate a society shape the kinds of bodies that people acquire (Mauss [1936] 1979). So, the calloused hands of the Ashaninka in Acre, Brazil (Comandulli, Chapter 15 in this volume) are shaped by the instruments they use to cut, pound, dig and hunt in their environments, and they struggle to adapt to devices (touch screens) which embody assumptions about users that require new forms of dexterity and body technique (Mauss [1936] 1979; Foucault 1988).

Anthropological theories of technology go further still, and point to the ways in which human engagement with technical devices shape the body's capacity for action and therefore the subjectivities we develop of ourselves as agents (Warnier 2001; 2009). The cultural variation presented by human relationships to technical objects – and the things they choose to do with them – offers a counterpoint to arguments for technological determinism by showing the ways in which technology can be misused, subverted and otherwise 'hacked'. This point challenges concerns about the imposition of a specific modernity through the introduction of digital technologies, instead offering the possibility for these introductions to provide insights into how modernity may be thought of differently.

The ethnographic questions raised by participative evaluation are grounded on the premise that assessments of efficacy or usefulness are dependent on culturally specific notions of what is trying to be achieved (Coupaye 2009). As most extreme citizen science projects seek to meet the goals of two distinct cultural groups – the indigenous community and the institutional actors or researchers who wish to make use of the data – successful projects must simultaneously meet the evaluative criteria of both cultural frameworks, and perform an act of translation (Asad 1986) between one and the other. However, as the saying goes (Asad 1986; Viveiros de Castro 2014), 'translator, traitor' – translation is never a completely faithful replica, and always betrays one language in its attempt to grasp the sense of the other. Yet, as has been suggested by contemporary anthropological scholars (Viveiros de Castro 2014), if this 'betrayal' or transformation occurs in the new language (i.e. conservation science), instead of in the original one (i.e. TEK), then scientific discourse may gain by acknowledging the existence of 28 different conditions of being an elephant, instead of the Baka simplifying the precision of their ecological knowledge and reducing the elephant diversity they see and know into a single data point.

## 4. Conclusion

To conclude, then, in which ways can anthropological methods and theory support the design and implementation of geographic citizen science? Anthropological methods are key to informing the design and implementation of geographic citizen science projects which are attentive to local needs, sensitive to cultural specificity and consequently more successfully incorporated by local communities as part of their routines. Several of the case studies in this volume highlight the importance of the iterative participatory design process for the successful implementation of geographic citizen science projects because community participation in the project at every step makes it far more likely that it is appropriate, welcomed and used by local actors. This iterative process can be extended to geographic citizen science itself where case studies can help to inform the broader project of geographic citizen science and the disciplines supporting it. The Baka's TEK not only shapes conservation projects in the local area, but also can go on to inform global conservation science as a whole. Through this participatory and iterative approach, informed by local experiences of human–environment relations, we can build a body of knowledge better equipped to address the environmental challenges of our times. Moreover, the existence and use value of diverse knowledge systems presents professional scientific practice with an empirical variant that challenges claims that monopolise truth. Instead, the scientific project itself can grow in an iterative relation with the knowledge of others.

## References

Asad, Talal. 1986. 'The concept of cultural translation in British social anthropology'. In *Writing Culture*, edited by James Clifford and George Marcus, 141–64. Berkeley: University of California Press.

Astuti, Rita. 1995. '"The Vezo are not a kind of people". Identity, difference and "ethnicity" among a fishing people of Western Madagascar', *American Ethnologist* 22: 464–82.

Clastres, Pierre. 1977. *Society Against the State*. New York: Urizen Books.

Coupaye, Ludovic. 2009. 'What's the matter with technology? Long (and short) yams, materialisation and technology in Nyamikum Village, Maprik District, Papua New Guinea', *The Australian Journal of Anthropology* 20: 93–111.

Danielsen, Finn, Per M. Jensen, Neil D. Burgess, Ronald Altamirano, Philip A. Alviola, Herizo Andrianandrasana, Justin S. Brashares et al. 2014. 'A multicountry assessment of tropical resource monitoring by local communities', *Bioscience* 64: 236–51.

Descola, Philippe. 2005. 'On anthropological knowledge', *Social Anthropology* 13: 65–73.

Evans-Pritchard, Edward Evan. 1936. 'Customs and beliefs relating to twins among the Nilotic Nuer', *The Uganda Journal* 1936: 236.

Fabian, Johannes. 1983. *Time and the Other: How anthropology makes its object*. New York: Columbia University Press.

FAO. 2014. Respecting free, prior and informed consent: Practical guidance for governments, companies, NGOs, indigenous peoples and local communities in relation to land acquisition. Accessed 10 September 2020. http://www.fao.org/3/a-i3496e.pdf.

FAO. 2016. Free prior and informed consent: An indigenous peoples' right and a good practice for local communities. Accessed 10 September 2020. http://www.fao.org/3/a-i6190e.pdf.

Fernández-Llamazares, Álvaro, and Mar Cabeza. 2018. 'Rediscovering the potential of indigenous storytelling for conservation practice', *Conservation Letters: Journal of the Society for Conservation Biology* 11: e12398.

Foucault, Michel. 1988. 'Technologies of the self'. In *Technologies of the Self*, edited by Luther Martin, Hugh Gutman and Patrick Hutton, 16–49. Amherst: University of Massachusetts Press.

Gibson, James Jerome. (1977) 1986. 'The theory of affordances'. In *Perceiving, Acting and Knowing*, edited by Robert Shaw and John Bransford, 67–82. Hillsdale, NJ: Lawrence Erlbaum Associates.

Haklay, Mordechai (Muki). 2013. 'Citizen science and volunteered geographic information: Overview and typology of participation'. In *Crowdsourcing Geographic Knowledge: Volunteered geographic information (VGI) in theory and practice*, edited by Daniel Z. Sui, Sarah Elwood and Michael F. Goodchild, 105–22. New York: Springer.

Hockey, Jenny, and Martin Forsey. 2012. 'Ethnography is not participant observation: Reflections on the interview as participatory qualitative research'. In *The Interview: An ethnographic approach*, edited by Jonathan Skinner, 69–87. London: Berg.

Ingold, Tim. 2011. *Being Alive: Essays on movement, knowledge and description*. Abingdon, UK: Routledge.

Kohn, Eduardo. 2013. *How Forests Think: Toward an anthropology beyond the human*. Berkeley: University of California Press.

Krøijer, Stine. 2015. *Figurations of the Future: Forms and temporalities of left radical politics in Northern Europe*. New York: Berghahn Books.

Latour, Bruno. 1993. *We Have Never Been Modern*. Cambridge, MA: Harvard University Press.

Latour, Bruno. 2005. *Re-assembling the Social: An introduction to actor-network theory*. New York: Oxford University Press.

Lemonnier, Pierre. 1993. 'Introduction'. In *Technological Choices: Transformations in material cultures since the Neolithic*, edited by Pierre Lemonnier, 1–35. Abingdon, UK: Routledge.

Lewis, Jerome. 2012. 'How to implement free, prior, informed consent'. In Swiderska et al. (2012). *Participatory Learning and Action 65: Biodiversity and Culture - exploring community protocols, rights and consent* (15): 175–79. https://www.clpi.info/media/library/resources/biodiversity-and-culture-exploring-community-proto/1768iied.pdf#page=177

Lewis, Jerome. 2014. 'Pygmy hunter-gatherer egalitarian social organization: The case of the Mbendjele BaYaka'. In *Congo Basin Hunter-Gatherers*, edited by Barry S. Hewlett, 219–44. New Brunswick, NJ: Transaction Publishers.

Lewis, Jerome. 2019. *Flourishing Diversity: Learning from indigenous wisdom traditions*. London: Flourishing Diversity.

Lewis, Jerome, and Téodyl Nkuintchua. 2012. 'Accessible technologies and FPIC: Independent monitoring with forest communities in Cameroon', *Participatory Learning and Action* 65: 151–65.

Mauss, Marcel. (1936) 1979. *Sociology and Psychology: Essays*. Translated by Ben Brewster. Abingdon, UK: Routledge.

Miller, Daniel, ed. 2005. *Materiality*. Durham, NC: Duke University Press.

Oxfam. 2010. Guide to free, prior and informed consent. Accessed 16 September 2020. https://www.culturalsurvival.org/sites/default/files/guidetofreepriorinformedconsent_0.pdf.

Scheper-Hughes, Nancy. 1992. *Death Without Weeping: The violence of everyday life in Brazil*. Berkeley: University of California Press.

Schneider, David. 1968. *American Kinship: A cultural account*. Chicago: University of Chicago Press.

Strathern, Marilyn. 1988. *The Gender of the Gift: Problems with women and problems with society in Melanesia*. Berkeley: University of California Press.

Viveiros de Castro, Eduardo Batalha. 1998. 'Cosmological deixis and Amerindian perspectivism', *Journal of the Royal Anthropological Institute* 4: 469a88.

Viveiros de Castro, Eduardo Batalha. 2014. *Cannibal Metaphysics*. Minneapolis, MN: Univocal Publishing.

Wacquant, Loïc. 2004. *Body and Soul: Notebooks of an apprentice boxer.* New York: Oxford University Press.
Wagner, Roy. 1981. *The Invention of Culture.* Chicago: University of Chicago Press.
Warnier, Jean Pierre. 2001. 'A praxeological approach to subjectivation in a material world', *Journal of Material Culture* 6: 5–24.
Warnier, Jean Pierre. 2009. 'Technology as efficacious action on objects . . . and subjects', *Journal of Material Culture* 14: 459–70.

# Part II
# Interacting with geographic citizen science in the Global North

# Chapter 5
# Geographic expertise and citizen science: planning and co-design implications

Robert Feick and Colin Robertson

## Highlights

- Citizen science projects involve researchers and volunteers who can range widely in formal qualifications, skill sets and knowledge types.
- Understanding the types of expertise that participants have is critical for defining the design requirements for geographic citizen science activities and tools.
- A model of geographic expertise is applied to three citizen science projects to illustrate how activities and tools can be tailored to participants' geographic expertise.
- Fostering geographic and place-based expertise in tool and activity design may facilitate knowledge exchange between researchers and volunteers and lead to sustained participation.
- Creating educational pathways for different knowledge types may be a useful strategy for developing participants' geographic expertise and encourage long-term engagement.

## 1. Introduction

Geographic citizen science spans a range of natural and social science fields across a diverse array of community types and needs. This demands different approaches for developing tools to collect and add value to data, for recruiting and engaging participants and for mobilising different knowledge types. There is widespread enthusiasm for the potential for citizen and participatory science approaches to help bridge gaps between science and the lay public – a goal deemed especially critical in a time of rapid environmental change and declining trust in traditional knowledge authorities (i.e. experts; Collins and Evans 2002; Irwin 2018).

Citizen science initiatives are often geographic because they require participants to collect observations that contain data that include a locational component (Haklay 2013; Brown and Donovan 2014). Research questions that underpin citizen science projects are often geographic in nature and relate to, for example, changes in species distributions, spatial variations in environmental conditions (e.g. air and water quality) or other issues that require observations to be linked to explicit geographic locations. Collecting data to address these questions has been fuelled by widely available mobile and web-based information and communication technologies that make in situ recording of geographic data easier and, in many cases, a more social and collaborative process (Rotman et al. 2012).

Some research issues, such as neighbourhood liveability and sociocultural landscape valuing, require geographic data that are more contextualised or place based in nature (Brown and Donovan 2014; Haywood, Parrish and Dolliver 2016). Place is commonly used to describe the meaning space has to individuals or groups, or the emotional bonds people have to settings or areas (Lewicka 2011; Preece 2016). For many citizen science initiatives, place and locality can provide a focal point for integrating local knowledge and expertise that community members acquire through lived experiences, culture and tradition, and as a motivator for sustained participation (Haywood, Parrish and Dolliver 2016; Newman et al. 2017).

In this chapter, we consider how community members' geographic expertise – that is, their familiarity and knowledge of particular locales or with identifiable types of places (Robertson and Feick 2017) – may inform co-design of locally relevant citizen science tools and projects. Questions of expertise have become increasingly important to citizen science in recent years. Respect for scientific knowledge and authority has diminished, and some of the distinctions between amateurs and experts have been blurred as access to information and inexpensive technologies for collecting georeferenced data have broadened (Goodchild 2009; Haklay 2013; Irwin 2018). Expertise is traditionally conceptualised at the level of the individual as an accumulation of knowledge, qualifications and abilities which vary on a spectrum from amateur to expert (Collins 2013; Lave 2015). Citizen science delineates between those who are experts and those who are not, and often uses this to define how scientists and volunteers collaborate to complete tasks such as data collection, interpretation and study design (Haklay 2013; Bonney et al. 2016; Johnston et al. 2018).

However, new conceptualisations of expertise posit that it can differ in type and be widely distributed in a population, depending on the domain of interest (Carolan 2006; Collins 2013). In Section 2, a framework of geographic knowledge types is presented that draws from Collins's (2013) deconstruction of expertise across three core dimensions that recognise formal training and qualifications, tacit knowledge sharing and knowledge uniqueness. This framework is applied in Section 3 to three sample citizen science projects to illustrate how geographic expertise affected participants' involvement in each project and the design considerations that resulted. The chapter concludes with a broader discussion of how geographic expertise concepts may help to recast the binary expert–amateur divide to recognise a wider spectrum of knowledge in citizen science design.

## 2. Using geographic expertise to aid geographic citizen science tool design

There is growing acceptance that citizens, equipped with proper tools and training, can produce data similar in quality to that produced by experts (See et al. 2013; Simpson, de Loë and Andrey 2015). There is also recognition that community members can hold types of knowledge and expertise, rooted in local context, tradition or experience, which experts lack (Armitage et al. 2011). However, our understanding of how citizens' geographic knowledge relates to how and why people participate and, ultimately, informs tool design in citizen science is incomplete.

Geographic citizen science tool design is challenging in part because projects vary in thematic focus, application contexts (e.g. terrestrial vs. aquatic, urban vs. rural, etc.) and the roles participants have within individual projects. Bonney et al. (2009), for example, distinguish between contributory projects, where citizens are largely limited to collecting data using procedures that researchers design, and projects that are collaborative or co-designed, in which citizens have progressively more input, such as formulating research questions and defining how data are collected and analysed (Buytaert et al. 2014). Participation can also vary within a project, since volunteers differ in skills, reasons for participating, formal or informal training and the specific roles and tasks they assume (Preece 2016). Haklay (2013) illustrates this heterogeneity by framing a four-level typology of participation based on relationships between volunteers and professional scientists, which is extensively discussed in Chapter 1. This recognises that participation in a project can vary across activities, over time and from person to person. For example, most of the people in a project may be engaged in simple recording of observations (level 1 – crowdsourcing) and/or basic interpretation of data (level 2 – distributed intelligence), while smaller numbers of citizens are engaged more deeply with professional scientists in developing problem definitions or data-collection procedures (level 3 – participatory science, and level 4 – collaborative science). Participation becomes more knowledge and expertise based at higher levels of Haklay's typology, as volunteers rely more on their cognitive abilities, experience and training to complete tasks.

Among professional scientists, knowledge and expertise are typically bounded by disciplinary specialisations that define scientific content (e.g. grassland bird behaviour, habitats, etc.) and are underpinned by a foundational understanding of the nature of scientific research (e.g. role of theory, statistical testing, uncertainty, etc.; Jordan et al. 2011; Bonney et al. 2016). Volunteers' knowledge and expertise are more heterogeneous, with varying degrees of content and lay knowledge based in experience, culture or tradition (Leach and Fairhead 2002; See et al. 2013; Caley et al. 2014). Careful design of volunteers' training and experiences during a project can improve their scientific content knowledge and ensure that they collect data similar in quality to that collected by experts (Haywood, Parrish and Dolliver 2016; Johnston et al. 2018). Improving volunteers' understanding of the nature of science has been more difficult to achieve. However, more success has been noted in participatory or collaborative science projects where volunteers shape research questions and data-collection procedures (Jordan et al. 2011; Bonney et al. 2016).

Recognition that knowledge is socially produced, partial and often contested has contributed to efforts to design projects, research activities and tools that are more inclusive of different types of knowledge, expertise and ways of learning (Leach and Fairhead 2002; Bonney et al. 2016). For example, software user interfaces that are intuitive to volunteers with diverse backgrounds can reduce training requirements, help volunteers to complete tasks efficiently and improve retention (Sharp, Rogers and Preece 2019). In other contexts, software design may seek to embed expert knowledge in tools, such as digital filters that flag questionable species sightings based on a volunteer's location (Preece 2016). Similar opportunities exist to leverage participants' geographic knowledge and expertise in tool and project design.

Some types of knowledge and expertise – scientific as well as those based on experience or tradition – are linked to specific localities or to the understanding of how certain human or natural processes manifest differently from place to place (Golledge 2002; Caley et al. 2014). Others have noted that geographic context can also serve as a frame for integrating different types of knowledge and for anchoring the relevance of scientific research to the places that community members value (Haywood, Parrish and Dolliver 2016; Newman et al. 2017). Understanding volunteers' geographic knowledge better may help to design tools that can support efficient collection of relevant and high-quality data.

What is expertise? Expertise has traditionally been viewed, at least implicitly, in terms of knowledge and skills differentials (Ericsson 2018). People with high levels of knowledge and skills in a particular domain area, where such knowledge and skills are lower in the wider public, are generally considered experts (Caley et al. 2014). Recognition of expertise is often made on the basis of recognisable characteristics such as formal educational achievements and professional accreditations (Lave 2015). However, sociological studies of expertise have posited that expertise is as much social, with distinct social networks and language, as it is a matter of technical achievements and qualifications (Collins and Evans 2002). In this light, some types of expertise arise from sustained participation in a community of practice which provides exposure to a group's linguistic and social milieu, tacit knowledge and accepted behaviours (Collins 2013). Hence, an individual who may lack formal training can, through socialisation within a domain, develop experience-based knowledge and skills that are of an 'expert' level in many fields. This has important implications for citizen science and particularly the design of activities and tools that facilitate volunteers' participation in scientific activities.

To recognise different types of expertise, Collins and Evans (2002) distinguish between contributory and interactional expertise. Contributory expertise concerns the knowledge and skills that are needed to contribute to knowledge in a topic area, either in the form of more abstract and generalisable knowledge or from local or practical experience (Carolan 2006; Collins 2013). Interactional expertise is a form of expertise that depends on a person having contributory expertise as well as the ability to communicate and interact with others in that topic area. While the former is built through formal education and work experience within a domain, interactional expertise can be acquired by socialising with others in a field and being exposed to others' tacit knowledge (i.e. knowledge acquired by doing; Collins 2013). In Collins and Evans's (2002) studies of expertise and experience frameworks, domains can be wide or

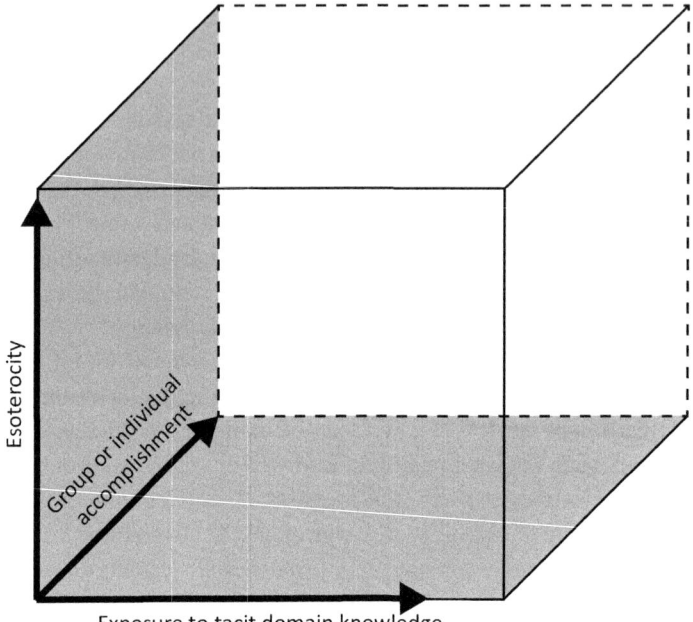

**Fig. 5.1** Expertise Space Diagram (ESD; after Collins 2013). Reprinted by permission from Springer Nature: 'Three dimensions of expertise' by Harry Collins 2013.

narrow (e.g. natural language speaking vs. computer programming), resulting in ubiquitous or specialised (esoteric) forms of expertise. They developed a three-dimensional model of expertise shown in Figure 5.1 by arraying contributory expertise (accomplishment) on the $x$-axis, interactional expertise (exposure to tacit knowledge) on the $y$-axis and esoterocity on the $z$-axis.

By visualising this model as an Expertise Space Diagram (ESD), the nature of individuals' or groups' expertise can be compared across the three domains (i.e. contributory, interactional and tacit knowledge, and esoterocity) rather than along a single scale that measures formal qualifications (Collins 2013). For example, scientists in a project focused on arctic climate-change impacts would have high levels of contributory expertise within their field (e.g. water chemistry, climate modelling, permafrost, etc.), some of which may be highly specialised or esoteric (e.g. permafrost thawing and shoreline erosion). Similarly, local residents who have extensive experience on the land or a rich knowledge base passed on from elders may also have contributory expertise accumulated from their observations of changes in animal migration patterns, timing of sea ice melt and snowpack.

Scientists would also need enough interactional expertise outside of their specialised subfields to develop a project-wide understanding of the terminology and methods. Note that interactional expertise is not confined to scientific domains. A scientist needs enough interactional expertise in the language of community members' domains (e.g. hunting and travelling on the land) to access their tacit knowledge effectively and vice versa (Collins 2013). We believe that this concept has important implications for geographic citizen science co-design.

In previous work, a geographic ESD (GESD) was proposed to represent the place-specific nature of geographic expertise more explicitly (Robertson and Feick 2017). One change in the GESD is that the esoterocity dimension is reoriented to 'thematic specificity' to capture differences in knowledge depth in a geographic context (Figure 5.2). For example, knowledge of major landmarks in London is more ubiquitous than highly detailed knowledge of travel routes, traffic conditions and which specific route to take on any given day. In the first case, any regular visitor to London might become an expert in where major landmarks are located, whereas in the second case, only those who drive throughout the city regularly (e.g. taxi drivers) would acquire such specialist knowledge. In this example, both the tourist and taxi driver could be considered 'experts', but the domain of knowledge differs by its level of thematic specificity.

A more significant change from Collins's (2013) model is the recasting of tacit knowledge from a topic or specialist orientation towards two related types of place-based expertise: locale familiarity (LF) and place type (PT; Figure 5.2). LF refers to knowledge that individuals or groups accumulate from experience within a specific place and is broadly analogous to local knowledge. It is more likely to be relational and fuzzy in nature rather than metric; it is often rich in detail and narrative, and can include verifiable facts as well as beliefs and community practices that are tied, in some fashion, to a specific place. PT expertise pertains to geographic archetypes, such as urban parks, alpine meadows or tidal pool ecosystems, which share fundamental properties and underlying processes of change. Since PT expertise centres on a general class of place, it is transferable across locations. In this way, an expert on coral ecosystems could contribute to a marine park management plan, even if they had little direct experience with an island's specific reef systems. PT expertise is often the product of formal education and training, but it can also be developed through experience (Collins and Evans 2002).

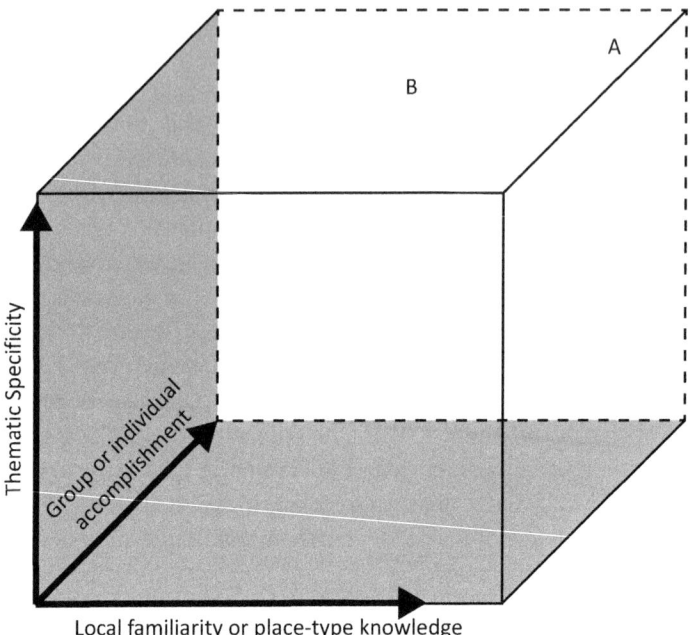

**Fig. 5.2** Geographic ESD (GESD).
Source: author.

Conceptualising geographic expertise in this way may contribute to more community-relevant and scientifically valuable tools and project designs. Application requirements, for example, may be defined more accurately if the multifaceted nature of geographic expertise within a user community is better understood. Liebenberg et al. (2017) illustrate this by describing how indigenous trackers shed new light on animal behaviour and health in Africa by using CyberTracker in concert with their holistic understanding of place and highly specialised tracking skills. Relative to Figure 5.2, trackers would rank highly in terms of locale familiarity, thematic specificity and contributory expertise that are rooted in their traditional knowledge and experience (see A in Figure 5.2). A researcher may have similar levels of contributory expertise due to their professional experience and training. Yet, they are not likely to have the same understanding of local ecosystems and animal behaviour (see B in Figure 5.2). Iterative and user-centred approaches to eliciting requirements and co-designing applications (e.g. see Stevens et al. 2014) are key to uncovering representative types of geographic expertise within a community and could also be pivotal to identifying the expertise that volunteers require to complete specific tasks or activities (Haywood, Parrish and Dolliver 2016).

Considering geographic expertise in design is of value in two other ways. First, it may help researchers to be cognisant of the social context that technologies and scientific methods are used and developed within. Technology is usually viewed in terms of software and hardware elements and the task-based capabilities they offer. Franklin (1999) presents a more human-centred view by casting technology as formalised cultures of practice that are based on socially accepted ways of completing tasks, defining content and control. In this sense, there is a 'technology' of fly-fishing based on a fisher's assessment of weather and river conditions, choice of flies and long-practiced casting techniques. Situating tool design and use within cultures of practice provides direct linkages to community members' tacit knowledge as well as related place-based knowledge such as vernacular place names, beliefs and values. As Simpson, de Loë and Andrey (2015), Preece (2016) and Skarlatidou et al. (2019) note, these linkages are important to technology acceptance and adoption, ensuring that tools are appropriate to social contexts and issues of concern (e.g. personal privacy protection), and in terms of embedding project activities within routine activities of specific communities or the lives of individuals.

In addition, considering geographic expertise may help researchers to understand the local uniqueness and conditions of using geographic spaces and the needs that this creates within an ever-increasing array of citizen science projects and applications. PT similarity may reveal opportunities for geographically dispersed groups investigating the same issue to network, share resources and learn from each other (Preece 2016; Newman et al. 2017). This may also counter what Haklay (2018) termed the 'not developed here' problem, where researchers create new applications instead of adopting an existing tool with similar functionality that others have developed.

## 3. Case studies: contextual characteristics

This section provides context for the subsequent discussion of geographic expertise through three citizen science projects that the authors have been jointly or individually involved in over the past five years. These projects span urban and rural contexts, had participant pools that ranged from very specific and small (i.e. dozens of volunteers) to general and moderate (i.e. thousands of volunteers) in size, and varied in geographic scope from city and county levels to a national scale. All these projects were developed in a university research environment with the design and deployment of two applications (GrassLander and Wildlife Health Tracker)

**Table 5.1** Case-study characteristics

| Project | Target community | Purpose | End-user application components |
|---|---|---|---|
| RinkWatch | People with backyard rinks – Canada (primary), also United States | Engage community members about local-scale climate-change impacts through measurements of winter temperature and ice conditions variability | Web map for viewing patterns of 'skateability', web forms for recording objective and experiential observations of ice conditions, user forums for sharing experiences and advice on ice-making, weather, etc. |
| GrassLander | Farmers (by crop, location, environment) – Southern Ontario | Gather population and nesting data on threatened grassland bird species and explore relationships with farm field practices | Web map and linked forms for farmers to record endangered bird sightings and nests, map-based drawing tools for delineating fields and recording crops and farm activities (e.g. planting, haying) |
| Wildlife Health Tracker | Hunters and biologists – Southern Ontario | Engage hunters and ecologists in rural areas to submit observations of dead or diseased wildlife to aid disease surveillance and modelling | Web map and linked forms for recording sightings of dead or diseased animals, species type and observations of animal health |

involving collaboration with non-profit/government domain intermediaries. Table 5.1 provides a high-level comparison of the user communities, project purposes and the primary application environment that participants were exposed to.

All projects used a map-centric design for users to view their own data and, with some exceptions, others' observations, and to enter new observations through linked web dialogs and forms. All the tools were web based and followed responsive design principles that allow application windows and controls to resize and rearrange dynamically for use on mobile phones, tablets and desktop computers (Turner-McGrievy et al. 2016). Using mobile-friendly web applications rather than native iOS or Android apps simplified development, since only one code base needed to be supported, and participants could choose the platform they felt was most suitable for specific tasks. For example, a farmer in the GrassLander project could draw their fields on a desktop or laptop computer and later record an eastern meadowlark nesting site using their mobile phone.

## 3.1 RinkWatch: where skating meets environmental science

RinkWatch is a web-based geographic citizen science project that aims to get people who are involved in outdoor skating on backyard and community ice rinks across North America to report skating conditions. The goals of RinkWatch are to obtain data on weather-related impacts on outdoor skating and to engage citizens and communities in climate-change research and related ecosystem services. The project started in the 2012–13 winter season. In the first two years of operation, almost eleven thousand observations from more than nine hundred and fifty rinks were accumulated, and by 2018, more than two thousand rinks were registered (Robertson, McLeman and Lawrence 2015). Volunteer 'rink watchers' are hockey and skating enthusiasts who maintain ice rinks and are highly attuned to how sensitive ice characteristics (e.g. smoothness, brittleness, friction) are to weather variations. 'Skateability', then, is a useful proxy for daily, seasonal and longer-term changes in local weather conditions that rink watchers have a vested interest in gathering and sharing.

Since the initial 2011 prototype, RinkWatch has evolved through three redesigns in response to users' requests, project scaling and maintenance needs. Across all versions, registered users associate an email address with a specific rink and then submit periodic assessments of 'skateability' through a simple web form (Figure 5.3). Initially, users only reported if a rink was 'skateable' or 'not skateable' on a given day. However, user feedback led to additional controls being added for recording assessments of ice quality, ranging from 'barely skateable' through to 'fantastic' (see Figure 5.3), and for optionally adding text comments or photos.

User forums and multilingual functionality were also added as incremental updates in response to participants' requests following the second season of operation. For the 2014–15 season, the initial Google Maps application was replaced with a new design based on ArcGIS Server (spatial data – rink locations) and coupled with a PostgreSQL database (user data, observations) that provided enhanced visualisation tools, multi-language support and updated user forums (Figure 5.4). The most frequently used functionality was centralised in a panel that integrated instructions for contributing data, user forums and easy-to-use controls to map patterns of skateability by category, date ranges or heat maps.

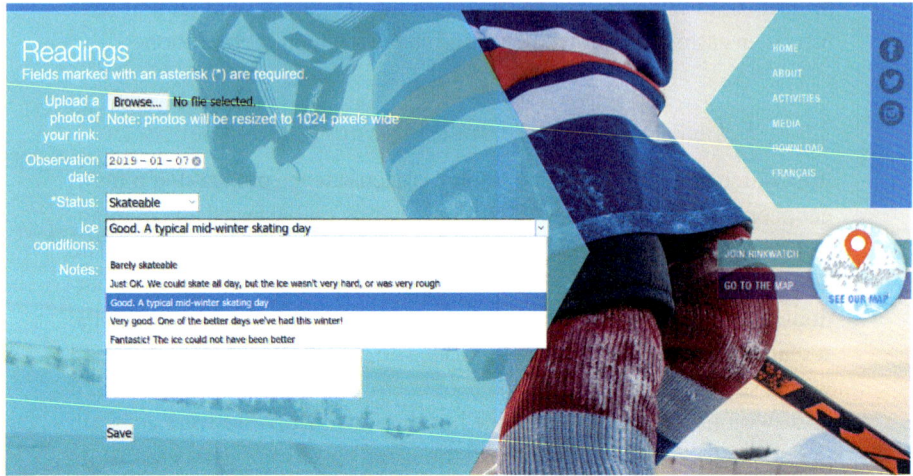

**Fig. 5.3** RinkWatch: rink conditions data entry form.
Source: RinkWatch.org.

## 3.2 GrassLander: grassland birds, citizen science and farming communities

GrassLander is a prototype geographic citizen science project based in Ontario that aims to recruit farmers to report on grassland bird sightings on their farms and agricultural management practices related to bird habitat. In many parts of Canada and the United States, agricultural landscapes serve as critical habitats for wildlife. However, since these landscapes are privately owned, they are largely missing from official wildlife surveys and broad-based citizen science initiatives. GrassLander was designed to meet the unique needs of farmers and engage them in conservation efforts for two threatened grassland bird species: the eastern meadowlark and the bobolink. Both species have substituted hayfields and pastures as nesting sites in response to losing native grassland habitats. This is problematic, since their nests are difficult to see in fields and can easily be disturbed when hay is cut.

GrassLander was built using the EsriLeaflet application programming interface (API) and a university-hosted ArcGIS Server and PostgreSQL database installation. Farmer feedback and design requirements were conveyed to the researchers through a non-profit intermediary: the

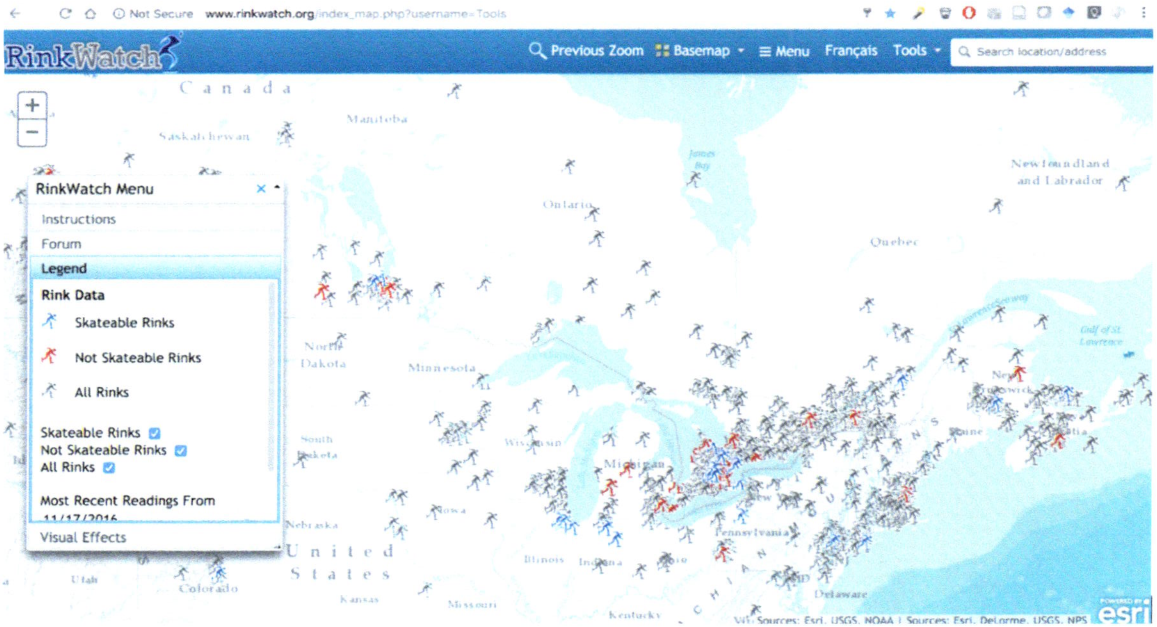

**Fig. 5.4** RinkWatch: visualisation of rink 'skateability' across North America. Source: RinkWatch.org. Basemap © Esri.

**Fig. 5.5** GrassLander: sample farm and field boundaries, and bird observations.
Source: Grasslander.org. Basemap © Esri.

Ontario Soil and Crop Improvement Association (OSCIA). Privacy was a key design requirement, and farmers could only view their own data following a secure login. No data, including who was participating, were shared publicly. Feedback on prototypes identified a desire to group functionality based on frequency of use. Tools for the initial set-up of a farm, for example, are consolidated under the 'Your Farm' menu. This includes map-based editing tools for selecting parcels that define a farm's boundaries and onscreen digitising of individual fields from aerial imagery (see Figure 5.5) as well as web forms for documenting field use (e.g. pasture, hay, etc.; Figure 5.6). Although GrassLander is mobile friendly, these set-

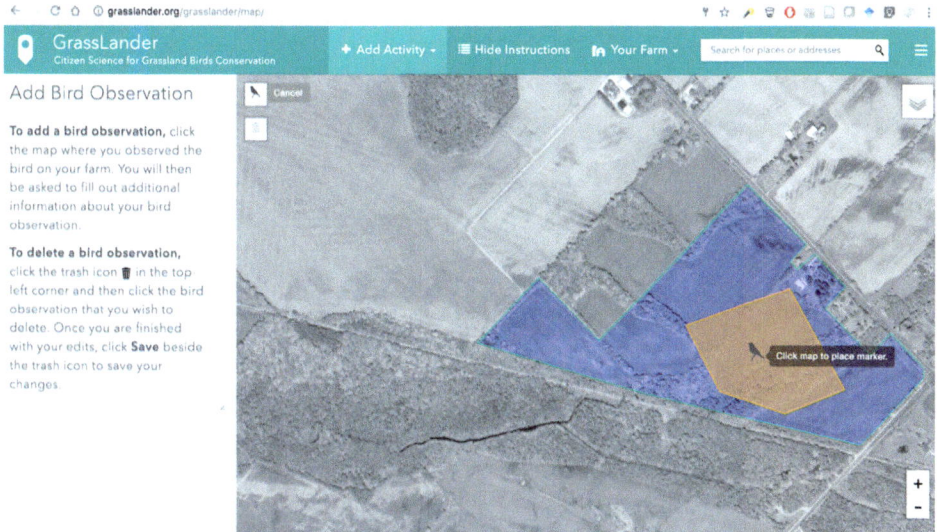

**Fig. 5.6** GrassLander: adding a bird observation (desktop computer view).
Source: Grasslander.org. Basemap © Esri.

up tasks are typically completed with a desktop or laptop computer that offers a larger screen and a mouse to digitise fields.

Throughout the spring to autumn season, farmers record activities for each field (e.g. timing of planting or haying) and locations of adult bird sightings and nests. GrassLander permits farmers to record observations with mobile devices, computers or in combination, as preferred. For example, a farmer could use their mobile device to add an observation by tapping on the map where the bird or nest was seen (Add Bird | Nest Observation) and then add required data (e.g. number of birds, species if known, date) on the following form. They could also capture observations in the field as geotagged photos and then use the photos to situate nest or bird markers on the map using a desktop computer (see Figure 5.6). Context-dependent help and instructions are available in panels that can be hidden as required.

### 3.3 Wildlife Health Tracker: enhancing wildlife disease surveillance in Ontario

Wildlife Health Tracker was designed as a pilot project to fill key data gaps in wildlife disease surveillance activities in Ontario, and to provide

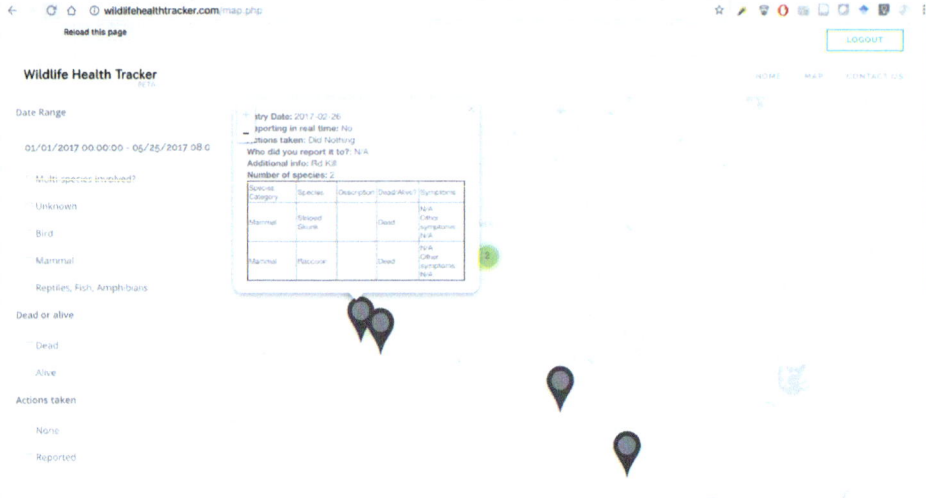

**Fig. 5.7** Wildlife Health Tracker.
Source: Wildlifehealthtracker.com. Basemap © OpenStreetMap.

more information about wildlife population morbidity and mortality events. The project recruited from two participant pools in Ontario – hunters and biologists/ecologists – in order to gather and interpret observations of diseased and dead wildlife, respectively. Hunters used an EsriLeaflet API-based interface to submit observations of dead or diseased animals. Data collected include notes on species, disease symptoms (behaviour and appearance), location, time of observation, photos if available and whether they reported the animal to the authorities (Figure 5.7). A goal of the project was to examine how hunters' understanding of local environments and wildlife behaviour could identify previously unknown patterns of disease diffusion and how these perceptions compared to an 'expert' group of biologist/ecologists. Visualisation of reported observations was available to project participants (but not to the public).

## 4. Interactions

Each of these projects had different levels of participant engagement in project and tool design (Table 5.2). This influences how volunteers used the tools, project evolution and sustainability, and the types of geographic knowledge and expertise required to contribute data (Figure 5.8). For example, RinkWatch was developed initially under the project leaders'

Table 5.2  Geographic expertise and participant involvement in design

| Project | Participants – types of geographic expertise | Participant engagement in design (after Haklay 2013) |
|---|---|---|
| RinkWatch | Rink maintainers: high locale familiarity and thematic specificity (i.e. their rink), low place type (nearby rinks with similar site characteristics) | Medium/high (level 2/3): researchers drove design; participants provided input on user interface, functionality (e.g. forums, visualizations) and information tracked |
| GrassLander | Farmers: high locale familiarity (their fields, farming activities, bird sightings); moderate to high place type and thematic specificity (farm landscapes and activities over dispersed fields) | High (level 3): research team, intermediary and participants |
| Wildlife Health Tracker | Hunters: moderate locale familiarity Biologists: low place type – derived from hunters' data | Low (level 2): domain experts and research team; no participant involvement |

own initiative, without participant input. The first user interface design was built using the Google Maps API linked to a PostgreSQL database and provided simple capabilities for adding a rink by address geocoding and submitting 'skateable' versus 'not skateable' ice quality observations. Local and national media attention in the first season of the project led to an explosive growth in participants and email requests for feature upgrades (e.g. the ability to add retrospective observations, gradations of skateability, visualise own observations, discuss with other users, etc.). Email was the main communication channel, as participants were distributed over a large area. Subsequent years of the project continued this style of participant-led design refinement, with engagement moving from bulletin-board style forums to a Facebook page where users frequently interact with project developers about website functionality. Technology platforms shifted to ArcGIS Server/PostgreSQL, then to EsriLeaflet/ArcGIS Online and now back to Google Maps. In the end, a simpler user interface and underlying architecture was required to ease maintenance for frequently changing student interns and project staff.

In terms of geographic expertise, RinkWatch requires submission of localised ice rink skating conditions, which we characterise as highly thematically specific (i.e. a small, well-defined community engaged in building outdoor rinks who know about rink conditions) and localised (i.e. people generally only report on their own backyard conditions). In contrast, the actual domain knowledge required is minimal (e.g. rating skating conditions). As such, participants were motivated to provide feedback about this new initiative, we believe, since this application was tapping into a previously unrecognised aspect of their social life that required significant investment of resources and leisure time. It may be that more thematically specific (or esoteric) domains have an easier time in engaging participants as a result of the selective nature of inclusion.

GrassLander was built using the EsriLeaflet mapping API that draws upon map services and a PostgreSQL database that is hosted within an ArcGIS Server website. This technology platform eased development of a responsive (mobile and desktop friendly) application and made it simple to support user-level access to observations. The design stage featured high engagement with participants facilitated through the non-profit OSCIA's existing relationships with farmers across the province. This partnership was vital in identifying participants for testing and gaining early feedback on the web digitising tools, documentation and reporting functions, and the 'My Farm' and 'Add Activity' user interface design that grouped tools by frequency of use (see Figure 5.6). Direct engagement with a test group of participants was done by phone, since testing coincided with the busy spring planting which precluded group meetings. One design feature that resulted from pilot testing was a completely secure application where participants could only see their own data, and data were not viewable on the public website. Farmers in Ontario are traditionally guarded about reporting information to authorities, and this can be exacerbated for potentially controversial issues such as conservation. Working with a trusted intermediary was vital to soliciting meaningful feedback on user interface design in this context.

Contributors were farmers with interest and knowledge of local grassland birds (high thematic specificity) who contributed information that was moderately localised (e.g. bird sightings and farm activities on their properties). Interestingly, farmers were found to be very competent at digitising farm boundaries through the web application, most likely due to frequent computer use for business and their detailed knowledge of their properties (high locale familiarity) that helped them to interpret the aerial imagery (see Figure 5.8). Although we could not directly observe how farmers used GrassLander, there is some anecdotal evidence that

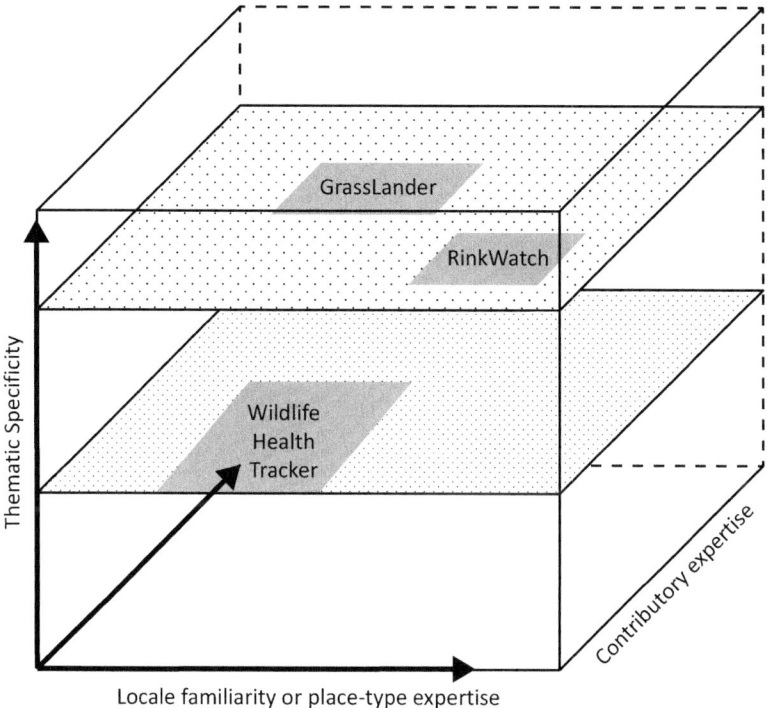

**Fig. 5.8** Dimensions of expertise for reviewed geographical citizen science projects (after Collins 2013).

farmers used the software on mobile and desktop devices for different tasks. As mentioned in Section 3.2, initial digitising of field boundaries was exclusively a desktop task, while recording bird and nest locations was often done in the field on mobile devices. In some cases, observation locations made in the field (e.g. geotagged photos) were adjusted later using a desktop computer (see Figure 5.6) to compensate for the offset between the location where a geotagged photo was taken and the actual location of a nest or an adult bird, which were often on or near fence lines.

Wildlife Health Tracker requires knowledge from volunteers that is less specific in theme, as many people, especially in rural communities, who report diseased or dead wildlife are hunters. Participation in Wildlife Health Tracker is opportunistic and thus benefits from the potential scaling up that geographic citizen science can provide (Loss et al. 2015). Only moderate geographic expertise and minimal contributory knowledge are required to contribute high-quality submissions of wildlife behaviour, morbidity and mortality (see Figure 5.8). As a result, less participant engagement in design and involvement resulted. Even though the research plan had originally called for workshops with hunters, feedback from hunter groups obtained by phone suggested these would be unsuccessful unless the application addressed issues that were highly localised (e.g. reporting on habitat quality or changes in abundance over time). Geographically, the scope was quite extensive across south central Ontario, which detracted from the locality motivator of participants. For higher levels of engagement and co-design, locale-specific versions of the app may need to be developed for individual watersheds or natural areas.

## 5. Leveraging and developing geographic expertise

Geographic citizen science projects can take a variety of approaches to user engagement and co-design. While deep engagement and co-design of projects are often desirable, these may not always be feasible or even preferred. One way to frame co-design in geographic citizen science is through the lens of expertise – who has what kinds of expertise, how does expertise map on to activities and how can we engage with communities of expertise in appropriate ways. In the three projects reviewed here, we found that the projects that required higher levels of place-based knowledge from participants (RinkWatch and GrassLander) benefited from higher levels of co-design engagement. Likely this related to participants feeling more personally invested in the issues and projects that intersected with their local/place-based interests (Newman et al. 2017). In contrast, while observations in Wildlife Health Tracker are geographic in nature, the substance of the observations (i.e. sick or dead animals) is less locally unique.

One of the interesting implications is that there is a trade-off between projects that are generic enough to solicit wide participation and enable pooling of observations over space and time, and the degree of local knowledge required to participate. If projects are highly localised and dependent on locale familiarity or place type knowledge, the community to draw from for participation can be limited. This may be offset by more committed volunteers who are, at least potentially, more vested in producing quality data, as described by Haywood, Parrish and Dolliver (2016). Conversely, if volunteers perceive a project to be place agnostic in focus and collected data, there may be greater difficulties in engaging users to find participation meaningful and useful. This tension between local and generic knowledge, in a geographic sense, arises from the deconstruction of expertise provided by Collins (2013). As such, there may be ways to utilise these concepts in project planning to leverage community participation and design strategies.

One way to consider both participation and tool design in geographic citizen science is as an educational pathway. Rather than focusing solely on the levels of knowledge sought or the tool functionality required at the outset, we might instead focus on how knowledge and expertise develop over time through project participation. For example, participants with deep place-based knowledge might, through interacting with experts and other participants, gain broader domain knowledge that aligns their local observations with established theory and increases their interactional expertise. Since our interaction with participants was largely limited to digital communication (e.g. email, social media) or through an intermediary, it was not possible to test if this form of expertise exchange and growth was realised. However, we see enough promise in the concept that it will be built into future project designs. These dialogues between experts and participants can also benefit domain experts by providing empirical anecdotes that either confirm or contradict expectations. Moreover, software applications tend to have short upgrade cycles that are often driven by technological reasons, and these dialogues may also foster redevelopment that, following Franklin (1999), is expertise and knowledge led and leverages existing cultures of practice. Ultimately, considering place and place-based expertise in tool and activity design may help foster joint knowledge development and exchange between researchers and volunteers, lead to sustained participation and increase the legitimacy of geographic citizen science outputs in decision making and the broader community (Buytaert et al. 2014).

## 6. Lessons learned

- Place and geographic context are important concepts for citizen science. Projects can benefit from more engaged participation by developing tools and design processes (recruitment, events, etc.) that highlight local issues and encourage local knowledge exchange among participants and researchers. Such exchanges can inform project design components across the participation/expertise spectrum.
- How projects are scoped – geographically and thematically – determines the potential pool of participants and their contributions, and informs how tools should be developed and/or co-designed. As projects become more generic in theme and over larger geographic extents, engaging place-based expertise becomes more challenging.

- Participant expertise varies and evolves through project activities. Developing pathways for different knowledge types may be useful for engaging participants in a sustainable and long-term way.
- Improving interactional expertise among participants is key and can be developed through field activities with scientists and participants, social events, digital communication tools and networks, and so on.

## References

Armitage, Derek, Fikret Berkes, Aaron Dale, Erik Kocho-Schellenberg and Eva Patton. 2011. 'Co-management and the co-production of knowledge: Learning to adapt in Canada's Arctic', *Global Environmental Change* 21: 995–1004.

Bonney, Rick, Heidi Ballard, Rebecca Jordan, Ellen McCallie, Tina Phillips, Jennifer Shirk and Candie C. Wilderman. 2009. Public participation in scientific research: Defining the field and assessing its potential for informal science education. A CAISE Inquiry Group report. Accessed 19 March 2020. https://eric.ed.gov/?id=ED519688.

Bonney, Rick, Tina B. Phillips, Heidi L. Ballard and Jody W. Enck. 2016. 'Can citizen science enhance public understanding of science?', *Public Understanding of Science* 25: 2–16.

Brown, Greg, and Shannon Donovan. 2014. 'Measuring change in place values for environmental and natural resource planning using public participation GIS (PPGIS): Results and challenges for longitudinal research', *Society and Natural Resources* 27: 36–54.

Buytaert, Wouter, Zed Zulkafli, Sam Grainger, Luis Acosta, Tilashwork C. Alemie, Johan Bastiaensen, Bert De Bièvre et al. 2014. 'Citizen science in hydrology and water resources: Opportunities for knowledge generation, ecosystem service management, and sustainable development', *Frontiers in Earth Science* 2: 26.

Caley, Michael Julian, Rebecca A. O'Leary, Rebecca Fisher, Samantha Low-Choy, Sandra Johnson and Kerrie Mengersen. 2014. 'What is an expert? A systems perspective on expertise', *Ecology and Evolution* 4: 231–42.

Carolan, Michael S. 2006. 'Science, expertise, and the democratization of the decision-making process', *Society and Natural Resources* 19: 661–8.

Collins, Harry. 2013. 'Three dimensions of expertise', *Phenomenology and the Cognitive Sciences* 12: 253–73.

Collins, Harry, and Robert Evans. 2002. 'The third wave of science studies: Studies of expertise and experience', *Social Studies of Science* 32: 235–96.

Ericsson, K. Anders. 2018. 'An introduction to the second edition of *The Cambridge Handbook of Expertise and Expert Performance*: Its development, organization, and content', In *The Cambridge Handbook of Expertise and Expert Performance*, 2nd ed., edited by K. Anders Ericsson, Robert R. Hoffman, Aaron Kozbelt and A. Mark Williams, 3–20. Cambridge: Cambridge University Press.

Franklin, Ursula. 1999. *The Real World of Technology*. Toronto, Canada: House of Anansi.

Golledge, Reginald G. 2002. 'The nature of geographic knowledge', *Annals of the Association of American Geographers* 92: 1–14.

Goodchild, Michael. 2009. 'NeoGeography and the nature of geographic expertise', *Journal of Location Based Services* 3: 82–96.

Haklay, Mordechai (Muki). 2013. 'Citizen science and volunteered geographic information: Overview and typology of participation'. In *Crowdsourcing Geographic Knowledge: Volunteered geographic information (VGI) in theory and practice*, edited by Daniel Z. Sui, Sarah Elwood and Michael F. Goodchild, 105–22. New York: Springer.

Haklay, Muki. 2018. Panel discussion at the Lessons Learned from Volunteers' Interactions with Geographic Citizen Science' Workshop, University College London, London, 27 April 2018.

Haywood, Benjamin K., Julia K. Parrish and Jane Dolliver. 2016. 'Place-based and data-rich citizen science as a precursor for conservation action', *Conservation Biology* 30: 476–86.

Irwin, Aisling. 2018. 'No PhDs needed: How citizen science is transforming research', *Nature* 562: 480–2.

Johnston, Alison, Daniel Fink, Wesley M. Hochachka and Steve Kelling. 2018. 'Estimates of observer expertise improve species distributions from citizen science data', *Methods in Ecology and Evolution* 9: 88–97.

Jordan, Rebecca C., Steven A. Gray, David V. Howe, Wesley R. Brooks and Joan G. Ehrenfeld. 2011. 'Knowledge gain and behavioral change in citizen-science programs', *Conservation Biology* 25: 1148–54.

Lave, Rebecca. 2015. 'The future of environmental expertise', *Annals of the Association of American Geographers* 105: 244–52.

Leach, Melissa, and James Fairhead. 2002. 'Manners of contestation: "Citizen science" and "indigenous knowledge" in West Africa and the Caribbean', *International Social Science Journal* 54: 299–311.

Lewicka, Maria. 2011. 'Place attachment: How far have we come in the last 40 years?', *Journal of Environmental Psychology* 31: 207–30.

Liebenberg, Louis, Justin Steventon, Nate Brahman, Karel Benadie, James Minye and Horekhwe Karoha Langwane. 2017. 'Smartphone icon user interface design for non-literate trackers and its implications for an inclusive citizen science', *Biological Conservation* 208: 155–62.

Loss, Scott R., Sara S. Loss, Tom Will and Peter P. Marra. 2015. 'Linking place-based citizen science with large-scale conservation research: A case study of bird-building collisions and the role of professional scientists', *Biological Conservation* 184: 439–45.

Newman, Greg, Mark Chandler, Malin Clyde, Bridie McGreavy, Mordechai (Muki) Haklay, Heidi Ballard, Steven Gray et al. 2017. 'Leveraging the power of place in citizen science for effective conservation decision making', *Biological Conservation* 208: 55–64.

Preece, Jennifer. 2016. 'Citizen science: New research challenges for human–computer interaction', *International Journal of Human–Computer Interaction* 32: 585–612.

Robertson, Colin, and Rob Feick. 2017. 'Defining local experts: Geographical expertise as a basis for geographic information quality'. In *13th International Conference on Spatial Information Theory (COSIT 2017)*, 22:1–22:14. Wadern, Germany: Schloss Dagstuhl – Leibniz-Zentrum für Informatik.

Robertson, Colin, Robert McLeman and Haydn Lawrence. 2015. 'Winters too warm to skate? Citizen-science reported variability in availability of outdoor skating in Canada', *The Canadian Geographer/Le Géographe Canadien* 59: 383–90.

Rotman, Dana, Jenny Preece, Jen Hammock, Kezee Procita, Derek Hansen, Cynthia Parr, Darcy Lewis and David Jacobs. 2012. 'Dynamic changes in motivation in collaborative citizen-science projects'. In *Proceedings of the ACM 2012 Conference on Computer Supported Cooperative Work*, 217–26. New York: ACM.

See, Linda, Alexis Comber, Carl Salk, Steffen Fritz, Marijn Van Der Velde, Christoph Perger, Christian Schill, Ian McCallum, Florian Kraxner and Michael Obersteiner. 2013. 'Comparing the quality of crowdsourced data contributed by expert and non-experts', *PLoS One* 8: e69958.

Sharp, Helen, Yvonne Rogers and Jennifer Preece. 2019. *Interaction Design: Beyond human–computer interaction*, 5th ed. Indianapolis, IN: John Wiley.

Simpson, Hugh, Rob de Loë and Jean Andrey. 2015. 'Vernacular knowledge and water management – Towards the integration of expert science and local knowledge in Ontario, Canada', *Water Alternatives* 8: 352–72.

Skarlatidou, Artemis, Alexandra Hamilton, Michalis Vitos and Muki Haklay. 2019. 'What do volunteers want from citizen science technologies? A systematic literature review and best practice guidelines', *Journal of Science Communication* 18: 1–23.

Stevens, Matthias, Michalis Vitos, Julia Altenbuchner, Gillian Conquest, Jerome Lewis and Muki Haklay. 2014. 'Taking participatory citizen science to extremes', *IEEE Pervasive Computing* 13: 20–9.

Turner-McGrievy, Gabrielle M., Sarah B. Hales, Danielle E. Schoffman, Homay Valafar, Keith Brazendale, R. Glenn Weaver, Michael W. Beets et al. 2016. 'Choosing between responsive-design websites versus mobile apps for your mobile behavioral intervention: Presenting four case studies', *Translational Behavioral Medicine* 7: 224–32.

# Chapter 6
# Citizen science mobile apps for soundscape research and public spaces studies: lessons from the Hush City project

Antonella Radicchi

## Highlights

- Mobile apps have been increasingly developed as participatory tools within the context of citizen science projects on environmental noise. However, fewer apps for the combined identification and assessment of quiet areas have been developed.
- Public quiet areas can be essential for healthy cities, being key to counterbalancing the detrimental effects of noise pollution on human health, biodiversity and the environment.
- The free citizen science Hush City app, released in 2017, enables users to create an open access map of quiet areas, with the potential of orientating plans and policies for healthier living.
- Drawing on the experience of the Hush City app, 15 people-centred recommendations are proposed potentially to inform the design, build and use of citizen science mobile apps in soundscape research and public spaces studies, aimed at generating a greater health-related quality of life.

## 1. Introduction

According to the latest trends in European urbanisation, most European cities are expected to grow and cover greater areas than in the past, and they will likely have to deal with an increase in global environmental issues, such as noise pollution (Vandecasteele et al. 2019). Noise constitutes the second most harmful environmental stressor in Europe, affecting more than 125 million people every year (EEA 2014). Long-term exposure to noise can affect environmental biodiversity, have detrimental effects on health (WHO 2018) and have a high cost for society (WHO 2011).

In 2002, the Environmental Noise Directive (END; EC 2002) was released with the aim of establishing a common methodology among member states to reduce noise pollution. One of the noise reduction measures introduced by the END is the creation of a plan for quiet areas in open country and agglomerations. The importance of protecting quiet areas in cities has also been recently suggested by the World Health Organization (WHO 2018). Indeed, access to quiet areas can provide benefits to health and well-being by facilitating restoration, improving concentration, favouring good sleep quality and boosting mental health (Öhrström et al. 2006; Gidlöf-Gunnarsson, Öhrström and Öhrgren 2007). Furthermore, as Rowcroft et al. (2011) suggest, access to quiet areas brings direct and indirect economic benefits, for example by saving on health costs and increasing worker productivity. But how can quiet areas be identified so as to be protected?

As sound is both a subjective and objective phenomenon, the literature recommends the adoption of qualitative criteria, in line with the soundscape concept (Schafer 1977; ISO 2014), to compensate the limits which emerged from the application of quantitative criteria to identify quiet areas in urban contexts (EEA 2014). The application of the soundscape approach to the identification of urban quiet areas implies studying the way the acoustic environment in context is perceived, experienced and/or understood (ISO 2014). However, applying the soundscape approach to identify urban quiet areas opens up further questions. How can we involve people in the identification and evaluation of urban quiet areas? How can we access and share people's knowledge about finding quietness in cities?

Against this backdrop, this chapter presents the Hush City app, which was developed as a citizen science tool to address these open questions, within the context of a more comprehensive framework (Radicchi 2017b; 2019). First, Hush City's rationale (Section 2) and its mapping interface (Section 3) are described. Then, benefits and barriers experienced by Hush City users are illustrated, followed by how they can be exploited in future development of the app (Section 4). In conclusion, an original framework of 15 people-centred recommendations is proposed to inform the design, build and use of citizen science mobile apps for soundscape research and public spaces studies aimed at generating a greater health-related quality of life (Section 5).

## 2. The Hush City app

### 2.1 Rationale for the development of a citizen science app to map and assess quiet areas

The Hush City citizen science mobile app was envisioned to address an issue framed at the European environmental policy level (EC 2002; WHO 2018): how to identify and map urban quiet areas properly so as to protect them by applying the soundscape approach.

The idea of developing a citizen science mobile app emerged from studying the literature. Trends in citizen science were observed regarding the practice of involving citizens in addressing open questions in science by exploiting mobile apps (Haklay 2016; Hecker et al. 2018; Luna et al. 2018). First, a review of the existing mobile apps developed for environmental noise assessment was conducted. Since 2008, 28 mobile apps have been developed, but none of them could be used by people to map and assess quiet areas specifically by collecting mixed data (Radicchi 2017a; 2017c; 2018). Therefore, Hush City was developed with the aim of addressing the following goals on multiple levels:

- Participation: exploiting mobile technology to favour citizen engagement in the planning and policy process;
- Science: helping scientists understand what people value when they search for quietness in cities;
- Policy: validating a participatory methodology to identify and map quiet areas in cities so as to protect them;
- Health and well-being: helping people find places to recover from sensory overload, by creating an open access web-based map of urban quiet areas; and
- Education and civic awareness: inducing self-reflection on the impact of noise on health and biodiversity and the importance of protecting quietness.

### 2.2 Identifying Hush City users

When Hush City was under development in 2016, there was no target core group of users with whom to co-design and test the app. To overcome this potential weakness, colleagues and friends were invited to test the app and provide initial feedback. The design principles learned by the author during participation in the workshop organised by the European Citizen Science Association in Berlin in the autumn of 2016 were also used as a reference (Sturm et al. 2018).

A user analysis was then performed during the first two years after the launch of the Hush City mobile app. From this user analysis, six types of Hush City users were identified. The general public constitutes the core group of users. Activist groups concerned with ecological and environmental issues use the app to evaluate and monitor the quality of public spaces in their communities. Local municipalities in charge of developing and updating their Plan of Quiet Areas every five years (according to the END) are also users of Hush City; they are also potential partners for the project, as in the case of the municipality of Berlin within the context of the Berlin Plan of Quiet Areas 2019–23. Researchers and academics also constitute a core group of users. They can use the data collected with the Hush City mobile app to understand better what people value when they search for quietness in cities and to investigate similarities and differences related to context variation (Radicchi and Vida Manzano 2018). Finally, the media and journalists represent an important group of actors who help raise and retain the participation of the public, and build awareness about the importance of living in places of high (acoustic) quality.

## 3. The Hush City app: concept, interface and technology

Hush City is a novel and free citizen science mobile app, launched in April 2017 to enable people to map, evaluate and discover public quiet areas. A second version of the Hush City mobile app, available in four languages (English, German, Spanish and Italian), was released in June 2018, along with the web-based version of the app. The mobile version of Hush City is available for both Android and iOS, and both the mobile and the web-based version of the app are free to use.

The mobile version of the Hush City app was developed to allow for the in situ mapping of quiet areas and the collection of data related to these quiet areas. The web-based version of Hush City was also developed to make the data accessible to those who might not own a smartphone but are nevertheless interested in exploring the quiet areas crowdsourced via the Hush City mobile app.

Innovative aspects of the Hush City app relate primarily to data collection. By using the Hush City mobile app, both qualitative and quantitative data related to the quiet areas can be crowdsourced through a data-collection process articulated in four sequential steps. First, users record a 30-second audio recording, and then they calculate its sound pressure levels. Next, they take a picture of the quiet area where they are, and lastly, they reply to a questionnaire (Figures 6.1 and 6.2). After completion of the four data-collection steps, users can submit the data, which is then linked in real time to the open access web-based map. The questionnaire is composed of 20 predefined questions (Table 6.1), which relate to the multifaceted factors influencing the environmental experience (Herranz-Pascual, Aspuru and García 2010). The questionnaire and the reply options were designed in 2016, prior to the release of the 2018 ISO norm on data-collection methods for soundscape (ISO 2018), referring to established questionnaires used in previous soundscape and quiet areas studies (e.g. Carfagni et al. 2014).

After the submissions of the data sets through the mobile app, data are geo-referenced and time stamped in real time to the Hush City Map, which is available via both the mobile and the web-based version of the app (Figures 6.1–6.3).

### 3.1 Hush City mobile app: interface design concept

The interface design concept of the Hush City mobile app has been designed to favour a user-friendly experience. After accessing the home page, users are offered two options. They can start mapping and evaluating the quiet areas by clicking on the button 'Map the quietness around you', and they can discover quiet areas crowdsourced by other users in their city or worldwide by clicking on the button 'Quiet Areas' (see Figure 6.1).

On the home page, there is also a menu which allows users to return to the home page at any point, consult and eventually delete their own data submissions, access the list of the monthly 'Hush City Ambassadors' and manage their account settings (e.g. they can change their password, select the language, provide feedback on the app, close their account and so on).

In detail, the 'Hush City Ambassadors' feature was introduced in the second version of the Hush City app, in 2018, to set up a rewarding mechanism to motivate users and retain participation. When users map and share quiet areas, they enter a list of 'Hush City Ambassadors' which is updated monthly. At the end of the month, users get a pop-up message, notifying them that they have been nominated 'Hush City Ambassador of [name of the city]', and they can choose whether they want to have their name featured in the Hush City's monthly newsletters and on the Hush City Ambassador web page.

Additional home-page features include a 'Localizer icon', which indicates the user's position on the map while using the app. By clicking on it, users can refresh and double-check their geographic position before starting the mapping and data-collection process. A search button ena-

**Table 6.1** Reporting the questionnaire in English embedded in the Hush City mobile app

| Q no. | Questions | Reply options |
|---|---|---|
| 1 | What prompted you to record this sound? | *Multiple choice and open entry*<br>Pleasure \|-\| Comfort \|-\| Irritation \|-\| Distraction \|-\| Happiness \|-\| Sadness \|-\| Calm \|-\| Anger \|-\| Nostalgia \|-\| Anxiety \|-\| Surprise \|-\| Shame \|-\| Fun \|-\| Disgust \|-\| Boredom \|-\| Interest \|-\| Other |
| 2 | In which category would you place this sound? | *Multiple choice and open entry*<br>Human voices \|-\| Human movement \|-\| Natural elements \|-\| Animals \|-\| Vegetation \|-\| Construction \|-\| Ventilation and electronics \|-\| Motorised Transport \|-\| Non-motorised transport \|-\| Social/signals \|-\| Music \|-\| Other |
| 3 | Using the words given below, please describe the sound you recorded. | *Multiple choice and open entry*<br>Lively \|-\| Boring \|-\| Familiar \|-\| Unfamiliar \|-\| Stressing \|-\| Relaxing \|-\| Meaningful \|-\| Meaningless \|-\| Pleasant \|-\| Unpleasant \|-\| Informative \|-\| Uninformative \|-\| Preferred \|-\| Unpreferred \|-\| Natural \|-\| Artificial \|-\| Friendly \|-\| Unfriendly \|-\| Beautiful \|-\| Ugly \|-\| Other |
| 4 | Rate how quiet the soundscape is in this location. | *Five-point linear scale*<br>Not quiet–very quiet |
| 5 | Enter the sounds that contribute in a positive way to your sense of quietness in this location. | *Open entry*<br>Free text |
| 6 | Enter the sounds that disturb your sense of quietness in this location. | *Open entry*<br>Free text |
| 7 | To what extent do the sounds in this location promote social interaction? | *Five-point linear scale*<br>Not much–very much |
| 8 | To what extent do the sounds in this location encourage you to have conversations here? | *Five-point linear scale*<br>Not much–very much |

(*continued*)

**Table 6.1** (continued)

| Q no. | Questions | Reply options |
|---|---|---|
| 9 | Can you hear other people's conversations around you? | *Dual scale*<br>Yes/no |
| 10 | Enter the sounds that contribute to the identity of this place. | *Open entry*<br>Free text |
| 11 | Are there people around? | *Multiple choice*<br>No one \|-\| A few \|-\| Many |
| 12 | What are people doing here? | *Multiple choice and open entry*<br>Passing through \|-\| Working \|-\| Relaxing \|-\| Recreationing \|-\| Waiting \|-\| Reading \|-\| Talking \|-\| Listening to music \|-\| Playing \|-\| Other |
| 13 | Personal information regarding where the user lives. | *Multiple choice and open entry*<br>I live in this area \|-\| I work in this area \|-\| I live in this city, but not in this area \|-\| I am a tourist \|-\| Other |
| 14 | How is the weather? | *Multiple choice and open entry*<br>Windy \|-\| Snow \|-\| Rainy \|-\| Humid \|-\| Foggy \|-\| Sunny \|-\| Cloudy \|-\| Stormy \|-\| Dry \|-\| Icy \|-\| Warm \|-\| Cold \|-\| Clear \|-\| Hot \|-\| Calm \|-\| Other |
| 15 | Rate the overall quality of this location. | *Five-point linear scale*<br>Not good–very good |
| 16 | Rate the overall cleanliness of this location. | *Five-point linear scale*<br>Not good–very good |
| 17 | Rate the overall maintenance of this location. | *Five-point linear scale*<br>Not good–very good |
| 18 | Rate the feeling of security in this location. | *Five-point linear scale*<br>Not good–very good |
| 19 | Rate the overall accessibility to this location. | *Five-point linear scale*<br>Not good–very good |
| 20 | Please add your additional comments and thoughts in the blank space below. | *Open entry*<br>Free text |

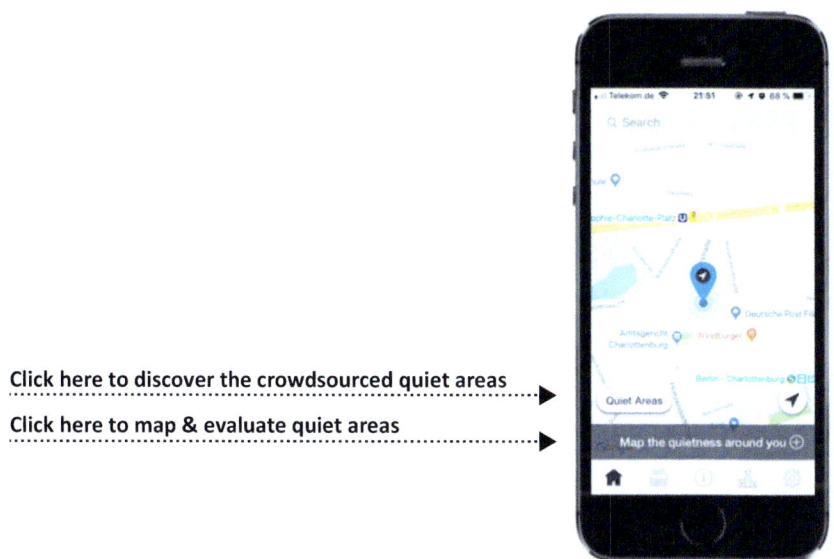

**Fig. 6.1** Interface of the Hush City mobile app. © Antonella Radicchi 2017. Basemap © Google Maps.

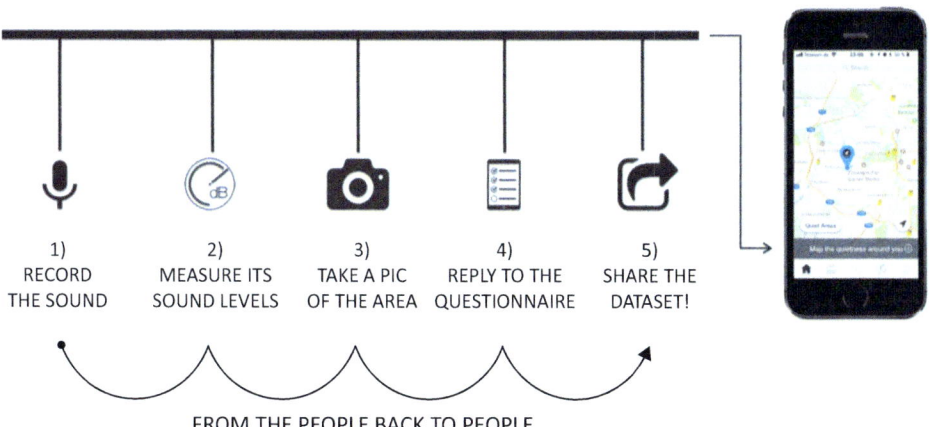

**Fig. 6.2** Concept of the Hush City mobile app data-collection process. © Antonella Radicchi 2019. Basemap © Google Maps.

bles users to search for geographic areas and be directed to the city where they are located or to other cities around the world. In the 'Quiet Areas' mode, which can be activated by clicking the button, the background map turns to black, and users can access three additional features: 'Legend', 'Filter' and 'List View'.

**Fig. 6.3** Hush City Map as displayed on the web-based version of the Hush City app. © Antonella Radicchi 2020. Basemap © Google Maps.

'Legend' informs users about the meaning of the markers' colours and the numbers embedded in the markers displayed on the map. The colours refer to the sound pressure levels measured by the Hush City app, whereas the numbers displayed on the markers refer to the numbers of data sets collected on the same quiet area. This feature was provided with the aim of giving users the possibility of collecting more data sets in the same quiet area and monitoring the status of the quiet area over a certain period of time.

'Filter' allows users to filter the quiet areas according to several parameters, such as sound levels, semantic descriptors, perceived quietness, visual quality and accessibility. These parameters can be selected individually or in combination. For example, users can find those quiet areas which are perceived as very quiet and accessible by setting the respective parameters and activating the Filter feature. The quiet areas can also be sorted by 'Number', 'City' and 'Noise Levels', and displayed in ascending or descending order by accessing the 'List View' feature. This feature was implemented to facilitate the screening and browsing of quiet areas via smartphones.

## 3.2 The Hush City mobile app: data-collection interface concept

To initiate the mapping and assessing of quiet areas, users in a quiet area can access the mobile app and click on the button 'Map the quietness around you' displayed on the home page (see Figure 6.2).

Users are then invited to take an audio recording of the quiet area, which after 30 seconds automatically stops. They then measure the sound levels by clicking on the button 'Analyse'. Afterwards, the app displays the weighted sound pressure levels of the sound recorded as: $L_{eq}$ (equivalent continuous sound level), $L_{min}$ (minimum sound level) and $L_{max}$ (maximum sound level). After that, users take a picture of the quiet area where they are, and then they evaluate the quiet area by replying to the questionnaire (see Table 6.1). After the completion of the questionnaire, users can review the data set and choose whether to submit the data set.

## 3.3 Hush City web-based application: interface design concept

The web-based version of the Hush City app displays the quiet areas crowdsourced worldwide via the Hush City mobile app; it does not allow for mapping and assessing quiet areas. Users can access the home-page features to filter the quiet areas and visualise them through the 'List View' mode. Users can also select the language they prefer from Italian, English, German, Spanish and Portuguese. They can also obtain information on the project and provide feedback on the Hush City app.

As of July 2020, the Hush City Map, accessible through the mobile and the web-based version of the app, contains more than four thousand quiet areas crowdsourced by more than five hundred users in different cities around the world (see Figure 6.3).

## 4. Interacting with the Hush City app: benefits and barriers

Data collection through mobile apps is usually unsupervised in nature, which does not necessarily affect the effectiveness and capability of such apps to produce reliable, consistent and accessible data (Theunis, Stevens and Botteldooren 2017). Nevertheless, how the apps are built and designed is key to addressing a mobile app's data reliability. Consequently, in the process of designing and building the Hush City mobile app, recommendations discussed in the citizen science literature (Luna et al. 2018) were taken into account, especially regarding interoperability, participant-centred design and agile development, user interface and experience design, and users' motivational factors.

As a result, was Hush City able to meet the needs and fulfil the expectations of its users? What benefits and barriers did Hush City users encounter using the app? Which features would they recommend to improve it? How do Hush City users interact with the app?

To answer these questions, Hush City users' feedback received over the past two years via the mobile and web-based versions of the app, paper forms and a survey conducted via MonkeySurvey tool was analysed using a qualitative approach to data synthesis.

For the purpose of this chapter, users' feedback will be presented so as to highlight whether and to what extent the questionnaires which are used to enable data input are helpful for interaction design and, if so, how these can inform the future of geographic citizen science app design. The chapter also explains how the lessons learnt from Hush City users' feedback will inform potential future improvements of the app.

Regarding the app rating – provided by users through the web-based app, iTunes and Google Play stores, and the survey – the results show that 51 out of 81 users rated the app with four or five stars. Although software developers are usually interested in this kind of rating, in my role as Hush City inventor and principal investigator, I find such ratings informative yet not explicative of the reasons which lead users to rate the app positively or negatively.

In terms of user behaviour, the results show that there are two distinct group of users: 'one-session' users who installed the mobile app only to use it once and/or to test it, and 'long-term' users who use the mobile app on a regular basis, for example several times a week, several times a month, monthly and/or when the occasion arises. This pattern is in line with trends that emerged in mainstream citizen science projects according to Seymour and Haklay (2017). The majority of Hush City users almost never share the data collected with Hush City app via social media, although this function is embedded in the mobile and web-based versions of the app. This result can be used to implement new features to encourage data sharing via social media, for example by embedding in the app 'pop-up messages' and/or a short video explaining the feature and its rationale.

From the analysis of the results, it also emerged that the majority of user-experience problems are related to unstable and unreliable Internet access, Internet cost and app crashes (the latter especially occurring on smartphones running on Android OS). This limitation can be addressed, giving the users the possibility to collect data offline and upload them when Internet access is established.

Other barriers highlighted by the Hush City users refer to the questionnaire embedded in the mobile app (Table 6.1). Some users find the questionnaire too long, while others find it annoying not having the option to skip questions. This is a critical point, which was extensively discussed when the first version of the mobile app was under development in 2016. The discussion indeed revolved around whether to provide users with the option to skip the questions. Ultimately, it was decided to make replying to the questionnaire mandatory, so as to collect consistent data which can be used for research purposes.

Some users also suggested creating a webspace for engagement and discussion (i.e. a forum). A few others recommended improving the mobile app's readability by increasing the font size of the text and implementing a kind of a 'sunscreen feature', so that the mobile app is more easily accessible when used outdoors in extreme light conditions. One user suggested making the mobile app more interactive by implementing 'filters that can block the noise of traffic, just to see how quiet a place can be without these factors in real time'. All these comments and recommendations are valuable, and they can be implemented in future versions of the app, depending on the financial budget. Among them, creating a webspace for discussion and improving the readability by enlarging the font will be the easiest to implement.

Along with user recommendations for future improvements, it is equally important to account for the features the users like the most and to consider them in future steps. The majority of users most like the mobile app's functionality and the way it looks and feels. The mobile app's 'user-friendliness' was appreciated by the users in the New York soundwalks (Radicchi 2019), with one user commenting that 'the app and the approach [were] useful and helpful in understanding the importance of subjective environmental acoustic awareness'. The Hush City mobile app was also defined as a tool 'that could be used to be part of a social change' and 'a good [one] for children too'. Positive feedback also related to the research idea behind the app, that is, giving people the ability to find quiet places and map them. This feature is particularly appreciated by one user who is the parent of an autistic child, who 'find[s] the app a great tool to find quiet places if [their] son is having a meltdown'. This comment can be used to implement pop-up features which recommend to users the quiet areas crowdsourced by other users in the city where they are. For example, pop-up notifications can be implemented to notify the users when they are in the proximity of a quiet area. This feature can only be operationalised if users consent to share their position with the app.

Since its launch in April 2017, public interest in the Hush City app has grown, and the crowdsourcing process, initiated by the author in 2017 within the context of a pilot study in a Berlin neighbourhood, has spontaneously scaled up the app to the worldwide level. Today, Hush City can count on an international community of engaged citizen scientists who have crowdsourced quiet areas from different countries, spanning from Europe to America to Asia.

## 5. People-centred recommendations for soundscape and public spaces studies

This chapter has presented Hush City, a novel free citizen science mobile app, launched in April 2017 to address an open issue in European environmental policy: how to enable people to map, evaluate and discover public quiet areas. First, Hush City's rationale along with the mobile and web version of the app were introduced. Then, benefits and barriers experienced by the Hush City users were illustrated to explain how users' feedback can be exploited in the future development of the app. In conclusion, an original framework of 15 people-centred recommendations (Table 6.2) for the design, build and use of mobile apps for soundscape research and public spaces studies is presented. In detail, recommendations 1–4 are drawn from a framework developed in citizen science by Luna et al. (2018), whereas recommendations 5–15 are an original contribution by the author (Table 6.2).

First, it is of paramount importance to ensure that citizen science mobile apps are accepted among society at large by favouring their discussion and negotiation in complex decision-making processes among different actors and the public (Königstorfer and Gröppel-Jlein 2012).

To place the user at the centre of the design process is also recommended, by involving the participants in the design of the citizen science mobile app, possibly in each step of the project (from start to end). Considering and incorporating motivational factors is then relevant to elicit and retain participation in order to ensure successful utilisation of the apps and also sustainability. In the case of the Hush City project, for example, the most active participants are publicly acknowledged and nominated as 'Hush City Ambassadors' of the cities where they have been active in crowdsourcing quiet areas.

Other key recommendations include strategies for favouring knowledge dissemination generated by the use of citizen science mobile apps. To this end, communication and data representation are key factors to account for, which can be achieved by building user-friendly digital dashboards and maps which enable interaction with users. Knowledge sharing can also be favoured by creating open access web-based repositories where data collected via citizen science mobile apps can be accessed by stakeholders and society at large and used within the context of bottom-up integrated urban planning. Also, ensuring that data collected via the mobile apps are open access and linked to open web-based platforms can contribute to data democratisation (Morozov and Bria 2018) and to novel forms of multilevel governance.

These 15 recommendations are informed by the citizen science literature and the author's experience gained through the Hush City app development and implementation over the past three years. In sharing them, it is hoped that they can inform and orientate the design, build and use of citizen science mobile apps for the collection of reliable and consistent data which can be used within the context of soundscape research and public spaces studies, aimed at generating a greater health-related quality of life.

**Table 6.2** Framework of 15 people-centred recommendations for the design, build and use of citizen science mobile apps for soundscape and public spaces studies

1. **Ensure interoperability**: e.g. ensuring data quality, data sharing with the participants, data reuse though CC licenses, data privacy.
2. **Place the user at the centre of the design process**: e.g. involving the participants in the design of mobile apps at each step of the projects (start to end).
3. **Follow a user experience design**: e.g. using 'effective and efficient' design elements (e.g. icons, arrows, etc.) to guide participants through the data-collection process.
4. **Consider and incorporate motivational factors**: e.g. bringing relevant motivations for participation to the foreground, addressing the 'six motivational categories'.
5. **Ensure the app is accepted among society at large**: e.g. favouring the negotiation of technological innovation, such as mobile apps, in society and complex decision-making processes among different actors (Königstorfer and Gröppel-Jlein 2012).
6. **Curate data representation to facilitate communication and user interaction**: e.g. exploiting interactive data-representation techniques to build user-friendly digital dashboards and maps.
7. **Enable communication and interaction with the users**: e.g. training academics and professional in innovative dissemination and communication techniques, referring to trends in citizen science and media and communication studies.
8. **Favour in situ and context-based assessment**: e.g. exploiting mobile apps to allow for in situ perceptual evaluation, fulfilling the definition of the landscape (EC ETS 2000) and soundscape (ISO 2014) concepts.
9. **Encourage intimate sensing–related practices**: e.g. implementing mobile apps as a means to favour the return to an intimate sensing of places, counterbalancing place detachment inducted by trends in remote sensing (Porteous 1990).
10. **Boost extreme participation**: e.g. designing citizen science mobile apps so as to help move from an information/consulting level of participation to a new one where citizens can control and act from the beginning of the process (Arnstein 1969; Haklay 2016).
11. **Raise environmental awareness**: e.g. including information about the projects in the mobile apps to raise awareness of the importance of living in healthy environments.

**Table 6.2** (continued)

12. **Make data open access**: e.g. ensuring that data collected via the mobile apps is open access and linked to open, web-based platforms, contributing to data democratisation (Morozov and Bria 2018) and to novel forms of multilevel governance.
13. **Enhance data collection and mapping**: e.g. exploiting data collection and mapping via mobile apps to complement traditional mapping methods.
14. **Favour bottom-up, integrated urban planning:** e.g. creating open access, web-based repositories where data collected via mobile apps can be accessed by stakeholders and society at large.
15. **Develop comparative scientific interdisciplinary studies and inter-sectoral projects**: e.g. designing data-collection processes via mobile apps so as to obtain reliable and consistent data which can then be used for comparative scientific studies and inter-sectoral projects.

## 6. Lessons learned

- The involvement and retainment of participation is key to the successful implementation of citizen science projects. It is recommended that training courses in communication and social media management become part of academic and professional curricula, and communication skills are acknowledged as added values for scientists.
- The use of mobile apps as participatory tools for data collection, mapping and sharing has not yet been fully accepted, especially among academics and public officials, although the majority of the mobile apps for environmental noise assessment released between 2008 and 2018 have been developed within the context of academic and inter-sectoral projects. It is recommended that the reliability of technological innovations such as citizen science mobile apps is discussed further within society at large and with different stakeholders.
- Empirical evidence gained through Hush City project shows the potential and benefits of exploiting mobile apps as participatory tools within the context of soundscape research, public spaces studies and planning for healthy cities. It is advisable that critical issues

regarding the design, building and implementation of mobile apps are addressed by ISO norms and further investigated in future research.

## Acknowledgements

Hush City was invented by Dr Arch. Antonella Radicchi in 2015 and then developed at the Technical University of Berlin in the framework of the projects: Beyond the Noise: Open Source Soundscapes (2016–18, first version of the Hush City mobile app) and Hush City Mobile Lab (2018–20, second version of the Hush City mobile app and first version of the web-based app).

Both the aforementioned projects have been envisioned and developed by Dr Arch. Antonella Radicchi (Technical University of Berlin). Project supervisors: Prof. Dr D. Henckel (Technical University of Berlin) and Dipl. Ing. J. Kaptain (Berlin Senate, Senate Department for the Environment, Transport and Climate Protection). Soundscape adviser: Prof. Dr B. Schulte-Fortkamp (Technical University of Berlin). Acoustic advisers: Dipl. Ing. M. Jäcker-Cüppers (ALD, Technical University of Berlin), M.A. M. Frost (Berlin Senate, Senate Department for the Environment, Transport and Climate Protection) and Dipl. Ing. M. Cobianchi (Bowers & Wilkins, UK).

The project Beyond the Noise: Open Source Soundscapes (2016–18) received funding from the TU Berlin IPODI-Marie Curie Program.

The project Hush City Mobile Lab (2018–20) received funding from the HEAD-Genuit Foundation (P-17/08-W). The support of the Foundation is gratefully acknowledged.

The author would like to thank Prof. Dr B. Schulte-Fortkamp for having drawn her attention to Königstorfer's Acceptance Diagram.

## References

Arnstein, Sherry R. 1969. 'A ladder of citizen participation', *Journal of the American Planning Association* 35: 216–24.

Carfagni, Monica, Chiara Bartalucci, Francesco Borchi and Lapo Governi. 2014. 'LIFE+ 2010 QUADMAP project (Quiet Areas Definition and Management in Action Plans): The new methodology obtained after applying the optimization procedures'. In *Proceedings of 21st International Congress on Sound and Vibration*, 2576–83, Beijing, China, 13–17 July 2014.

EC (European Parliament and Council). 2002. 'Directive 2002/49/EC of 25 June 2002 relating to the assessment and management of environmental noise', *Official Journal of the European Communities L* 189: 12–26.

EC ETS. 2000. The European Landscape Convention of the Council of Europe, Florence. Accessed 17 September 2020. https://www.coe.int/en/web/conventions/full-list/-/conventions/treaty/176.

EEA (European Environmental Agency). 2014. *Good Practice Guide on Quiet Areas, Technical report n.4*. Luxembourg: Publications Office of the European Union.

Gidlöf-Gunnarsson, Anita, Evy Öhrström and Mikael Öhrgren. 2007. 'Noise annoyance and restoration in different courtyard settings: Laboratory experiments on audio-visual interactions'. In *INTER-NOISE and NOISE-CON Congress and Conference Proceedings*, 1040–9, Istanbul, Turkey, 28–31 August 2007.

Haklay, Mordechai (Muki). 2016. 'The three eras of environmental information: The roles of experts and the public'. In *Participatory Sensing, Opinions and Collective Awareness*, edited by Vittorio Loreto, Andreas Hotho, Jan Theunis, Mordechai Haklay, Francesca Tria, Gerd Stumme and Vito D. P. Servedio, 163–79. Cham, Switzerland: Springer.

Hecker, Susanne, Muki Haklay, Anne Bowser, Zen Makuch, Johannes Vogel and Aletta Bonn. 2018. *Citizen Science. Innovation in open science, society and policy*. London: UCL Press.

Herranz-Pascual, Karmele, Itziar Aspuru and Igone García. 2010. 'Proposed conceptual model of environmental experience as framework to study the soundscape'. In *INTER-NOISE and NOISE-CON Congress and Conference Proceedings*, Lisbon, Portugal, 13–16 June 2010.

ISO/DIS 12913-1. 2014. *Acoustics. Soundscape – Part 1: Definition and conceptual framework*. Geneva: International Standardization Organization.

ISO/DIS 12913-2. 2018. *Acoustics. Soundscape – Part 2: Data collection and reporting requirements*. Geneva: International Standardization Organization.

Königstorfer, Jörg, and Andrea Gröppel-Jlein. 2012. 'Consumer acceptance of the mobile Internet', *Marketing Letters* 23: 917–28.

Luna, Soledad, Margaret Gold, Alexandra Albert, Luigi Ceccaroni, Bernat Claramunt, Olha Danylo, Muki Haklay et al. 2018. 'Developing mobile applications for environmental and biodiversity citizen science: Considerations and recommendations'. In *Multimedia Tools and Applications for Environmental and Biodiversity Informatics*, edited by Alexis Joly, Stefanos Vrochidis, Kostas Karatzas, Ari Karppinen and Pierre Bonnet, 9–30. Cham, Switzerland: Springer.

Morozov, Evgeny, and Francesca Bria. 2018. *Rethinking the Smart City: Democratizing urban technology*. New York: Rosa Luxemburg Stiftung.

Öhrström, Evy, Annbritt Skånberg, Helena Svensson and Anita Gidlöf-Gunnarsson. 2006. 'Effects of road traffic noise and the benefit of access to quietness'. *Journal of Sound and Vibration* 295: 40–59.

Porteous, J. Douglas. 1990. *Landscapes of the Mind*. Toronto, Canada: University of Toronto Press.

Radicchi, Antonella. 2017a. 'A pocket guide to soundwalking. Some introductory notes on its origin, established methods and four experimental variations'. In *Perspectives on Urban Economics*, edited by Anja Besecke, Josiane Meier, Ricarda Pätzold and Susanne Thomaier, 70–3. Berlin: Universitätsverlag der TU Berlin.

Radicchi, Antonella. 2017b. 'Beyond the noise: Open source soundscapes. A mixed methodology to analyse, evaluate and plan "everyday" quiet areas', *Proceedings of Meetings on Acoustics* 30: 040005.

Radicchi, Antonella. 2017c. 'The HUSH CITY App. A new mobile application to crowdsource and assess "everyday quiet areas" in cities'. In *Invisible Places. Proceedings of the International Conference on Sound, Urbanism and the Sense of Place*, 511–28, São Miguel Island, Azores, Portugal, 7–9 April 2017.

Radicchi, Antonella. 2018. 'The use of mobile applications in soundscape research: Open questions in standardization'. In *Proceedings of EURONOISE 2018*, Crete, Greece, 27–31 May 2018.

Radicchi, Antonella. 2019. 'A soundscape study in New York. Reflections on the application of standardized methods to study everyday quiet areas'. In *Proceedings of ICA 2019*, Aachen, Germany, 9–13 September 2019.

Radicchi, Antonella, and Jeronimo Vida Manzano. 2018. 'Soundscape evaluation of urban social spaces. A comparative study: Berlin–Granada', *The Journal of the Acoustical Society of America* 144: 1660.

Rowcroft, Petrina, Paul Stuart Shields, Cody Skinner, Stuart Woodin and Abigail L. Bristow. 2011. 'Is quiet the new loud? Towards the development of a methodology for estimating the economic value of quiet areas'. In *Proceedings of INTERNOISE 2011*, Osaka, Japan, 4–10 September 2011.

Schafer, R. Murray. 1977. *The Soundscape. Our sonic environment and the tuning of the world*. New York: A. Knopf.

Seymour, Valentine, and Mordechai (Muki) Haklay. 2017. 'Exploring engagement characteristics and behaviours of environmental volunteers', *Citizen Science: Theory and Practice* 2: 5.

Sturm, Ulrike, Sven Schade, Luigi Ceccaroni, Margaret Gold, Christopher Kyba, Bernat Claramunt, Mordechai (Muki) Haklay et al. 2018. 'Defining principles for mobile apps and platforms development in citizen science', *Research Ideas and Outcomes* 4: e23394.

Theunis, Jan, Matthias Stevens and Dick Botteldooren. 2017. 'Sensing the environment'. In *Participatory Sensing, Opinions and Collective Awareness*, edited by Vittorio Loreto, Andreas Hotho, Jan Theunis, Mordechai Haklay, Francesca Tria, Gerd Stumme and Vito D. P. Servedio, 21–46. Cham, Switzerland: Springer.

Vandecasteele, Ine, Claudia Baranzelli, Alice Siragusa, Jean Philippe Aurambout, Valentina Albertis, Maria Alonson Raposo, Carmela Attardo et al. 2019. *The Future of Cities*. Luxembourg: Publications Office of the European Union.

World Health Organization (WHO). 2011. *Burden of Disease from Environmental Noise. Quantification of healthy life years lost in Europe*. Brussels: Regional Office for Europe/European Commission Joint Research Centre.

World Health Organization (WHO). 2018. *Environmental Noise Guidelines for the European Region*. Geneva: World Health Organization.

# Chapter 7
# Using mixed methods to enhance user experience: developing Global Forest Watch

Jamie Gibson

## Highlights

- Managing and protecting forests is critical to the continuing survival of every living thing on Earth.
- The Global Forest Watch platform, launched in 2014, was built to address a gap in the data about how much the world's forests are changing.
- A mixed-methods approach can be used to research the needs of users, their behaviour on the platform and the barriers they face.
- This approach has helped to find ways to increase the amount of time spent on the site continually, and it may be useful for others aiming to visualise data for sustainable development topics.

## 1. Introduction

Forests are an essential part of the global ecosystem. They provide the clean air, water, shelter, food and medicine every species on this planet depends on for survival. For this reason, the protection of forests has been incorporated into the United Nations (UN) Sustainable Development Goals, with the target to promote 'sustainable management of all types of forests, halt deforestation, restore degraded forests and substantially increase afforestation and reforestation globally' (UNDESA n.d.).

The Global Forest Watch (GFW) platform aims to supply a global community of actors with the data they need to monitor and manage the world's forests sustainably. Within a few minutes, the platform can show you a global picture of how forests are changing in near real time, based on free and reliable data (Hansen et al. 2016). Whether you are managing a protected area, researching the impacts of dams on your surroundings or evaluating national forest policy, the data available on the platform can help identify where forest area is being lost or gained and some of the characteristics of that area that may explain why that change is happening.

Reducing the rate of global forest loss is no small task. In 2017, the world's tropical forests lost a football field's worth of trees every minute – in sum, that's 15.8 million hectares (Weisse and Goldman 2018). Tropical forest loss has increased year-on-year over the last decade: in 2001, only six million hectares of forest area was lost (Weisse and Goldman 2018). There are clear environmental costs to this, as well as risks to the indigenous communities that inhabit and rely on these areas (Mendes 2018). While increased data access alone cannot solve this challenge, the platform aims to remove 'access to data' as a barrier to action and raise awareness so that the issue is harder to ignore.

The GFW team has spent much of the last five years reaching out and listening to people who use the platform. Through this engagement, the team has amassed a large amount of information about user needs, patterns of behaviour and barriers to using this information for action. This chapter introduces some of the user research that I have completed (in collaboration with the GFW team) in order to inform the development of the platform and its companion applications.

## 2. Introducing the Global Forest Watchers

Before explaining the functionality of the GFW applications, I want to begin by outlining why forests are being cut down around the world and the plethora of personas who need to be engaged for effective forest management.

In recent work conducted by Curtis et al. (2018), the authors built a model to attribute the main drivers of global forest loss detected between 2001 and 2015. They identified a number of key drivers, summarised by Harris et al. (2018):

(1) Commodity-driven deforestation: where a forest is cut down in order to use the land for other means, such as mining, agriculture, oil or gas.

(2) Forestry: forest areas lost due to the forestry industry.
(3) Shifting agriculture: communities practising a specific form of small-scale agriculture where forest land is cleared for crops and then left to recover after harvest.
(4) Urbanisation: where forest area is cleared for the expansion of settlements.
(5) Wildfire: forest areas that are lost due to fires (either caused naturally or by human action).

This complements previous work looking at the difference between natural drivers of deforestation, such as drought, fire, storms and disease, and the multiple anthropogenic drivers, such as agribusiness, mining and energy (Keenan et al. 2015). The model results indicate that commodity-driven deforestation, forestry, shifting agriculture and wildfire are each associated with about 25 per cent of global forest loss, while urbanisation is responsible for just 0.6 per cent of forest loss.

Combating each of the deforestation drivers requires different combinations of actors working together to make a difference. Most of those actors have been characterised by the GFW team using the method of user personas. User personas are character descriptions that represent the characteristics and expectations of a large group of similar people, usually deducted from interviews (US HHS n.d.; see also Chapter 3). When designing websites, software applications or products, the personas help evaluate whether an application is likely to satisfy or be useful to the target audience (Harley 2015). When considering how the GFW platform may be used to manage forests, the team considered the following user personas:

- National governments and their agencies. This group has responsibility for setting, implementing or evaluating forestry policy or other policies related to forestry. As national points of authority, they are naturally linked to all deforestation drivers. They want data about deforestation trends in their country and the contexts in which deforestation is happening. They often have their own national sources of forest data and information, but they use GFW's independent data sets to verify information received from line agencies and local offices.
- Businesses. There are a number of businesses connected with forest change around the world, from businesses that own and operate in forest areas (often called 'concessions') to the companies that buy the final products extracted from those areas (minerals, soy, palm

oil, timber, etc.). People in businesses often want to know if there has been any deforestation in or near areas related to their operations, and may want to view satellite images to see for themselves what is happening.
- Forest monitors. Across the globe, there are managers, guards, guides and researchers with a mandate to monitor protected areas, indigenous and community lands, and certified concessions and other specific areas (IUCN n.d.). This group uses deforestation and fire alerts paired with satellite imagery to prioritise areas for follow-up and field investigation, and to include in reports.
- Researchers and academics. These are people with specific analytical skills who are responsible for performing and delivering research about the world's forests. They will likely have a particular geographic area of interest, and will want to browse the data available for that area, perform some simple analyses and download any interesting data for further analysis.
- Non-governmental organisations (NGOs). Among the GFW partners are a number of international and local NGOs, with goals related to advocacy, community outreach, research or direct forest management. These users likely have existing sources of information, but GFW serves as an independent data set to verify their own data, as well as allowing them to perform other simple analyses. These users often include GFW data and images to document deforestation in reports or advocacy materials.
- UN institutions. GFW is used by a number of UN organisations. Depending on the agency, they may be more interested in forestry or commodity drivers. We believe their use case is more about data exploration, viewing trends across different countries and validating country-reported data.
- The general public and the media. While there are specific actors who can take action, it is also important to build global awareness and public pressure on institutions to reduce deforestation. GFW aims to reach concerned citizens and science or environment writers in news organisations in order to build this baseline level of awareness. For the general public, the need is around engaging stories and attractive visuals which help increase awareness about the topic. For journalists, the need is around data exploration and being able to build graphics from the data, and quickly access key statistics and insights about where and why forests are changing.

GFW was developed following the ideas of user-centred design (see also Chapter 3) that 'aims to make systems usable and useful by focusing on the users, their needs and requirements, and by applying human factors/ergonomics, and usability knowledge and techniques' (ISO 2010). This means the team has tried to involve potential users in the development process as much as possible. To gather the necessary insights to deliver a global-scale platform, a mixed-methods approach was needed to derive a set of broad global behaviour trends backed up with a detailed understanding of key users. Five main techniques have been used to gather information about GFW's users over the last few years:

(1) Interviews: talking to potential or existing users to learn specific details about their roles, responsibilities, capabilities, preferences and needs. Most interviews have taken place via videoconferencing software, though many in-person interactions have taken place.
(2) Google Analytics: metrics related to how people interact with a website. Each metric can be explored further by filtering via groups of users with specific characteristics (e.g. country, device being used). This is our main source of quantitative data about GFW's audience and their behaviour. It has two main limitations: (1) it can only tell you what happened not 'why', which is where the interviews come in; and (2) there is very limited demographic and ethnographic data to allow us to analyse or compare specific groups of people (e.g. analysts in governments vs. analysts in NGOs).
(3) Usability testing sessions: a more structured discussion where someone is asked to complete a series of tasks on the platform. The testing sessions help reveal the ease with which people can complete basic tasks on the site. Due to the global nature of the user base, many of the testing sessions have been conducted via web conference, but a number of field trips over the last two years have allowed the team to gather more detailed feedback from users in person.
(4) Feedback on prototypes: at many points during the development cycle, the team has presented prototypes for users to view and give feedback on. The prototypes also work as props to stimulate discussion about the aims of the site and the needs of the interviewee. Prototypes can take the form of static images, interactive designs (using Invision software; https://www.invisionapp.com/) or simple versions of the page in question that are ready to use on a website.
(5) Surveys: one major survey was conducted in 2016, receiving about 350 responses. It was used to understand needs at that time and has had limited usefulness since it was completed.

For a more detailed analysis on the methods, the reader may refer to Chapter 3. All methods have their strengths and weaknesses, and can reveal different parts of the user experience. A mixed-methods approach, balancing broad audience-level and deeper individual engagement, helps to build the most complete picture. In this chapter, I will focus mainly on the insights derived from Google Analytics reviews and interviews, as these are the most important methods for understanding our audience and how they use the whole site. Feedback on prototypes and usability testing have, so far, just focused on specific parts of the site and not the whole experience, while the survey has less relevance as the site has changed since the survey was conducted.

## 3. The GFW and Forest Watcher applications

### 3.1 The main GFW platform

GFW is a web application that is publicly accessible. Anyone with an Internet connection can access the data and learn something about the state of the world's forests. It is in many ways a classic geographic citizen science platform: using technology to simplify and speed up a workflow that used to take many skilled geographic information system (GIS) analysts a lot of time to create, and present the results in an open, accessible and usable way.

This chapter will mostly refer to the mapping component of the platform. So, it is useful to list some of the main features within this component (see also https://www.globalforestwatch.org/map/). When a user arrives at the map, they see three data layers: global forest cover, global forest loss and global forest gain. Using the menu, users can add more layers to the map to find out how forests are changing, how they are currently being used and how forests relate to climate change and biodiversity. Data layers are grouped by thematic categories (see Figure 7.1), which link to different concerns of the personas (e.g. protecting forests for conservation purposes, sustainable management and use).

Once the user has found an area of interest to them, they can run an analysis on that area. This could be a predefined administrative boundary, a shape denoting a particular land use (i.e. a mining concession, a protected area), a watershed or a user-drawn shape. The forest change layers that are active on the map at that time will be analysed as a result.

The core data sets on GFW – the annual loss, gain since 2001 and Global Land Analysis and Discovery (GLAD) layers – are generated by the University of Maryland. The University of Maryland team has created an algorithm to analyse satellite images, compare the amount of forest in each one and then assign a 'loss' or 'gain' label where there is less or more

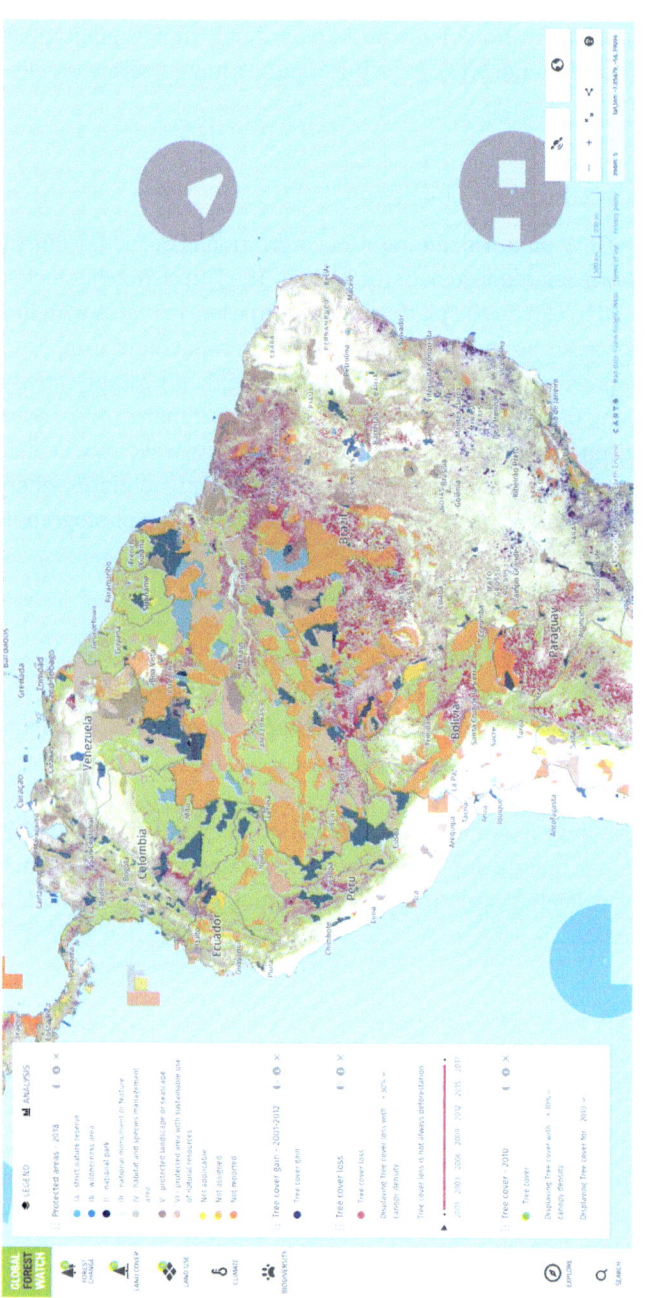

**Fig. 7.1** Screenshot of the Global Forest Watch map, showing protected areas in Brazil in relation to forest cover and tree-cover loss and gain. Basemap © Mapbox, © OpenStreetMap contributors. Source: globalforestwatch.org.

tree cover, respectively. For more information on the methodology, see Hansen et al. (2016). These data are complemented by hundreds of other layers, such as the location of key bird areas, dams, timber plantations, mining areas and mangroves, as well as options to add recent satellite imagery.

## 3.2 The Forest Watcher mobile app

To access the GFW platform, an Internet connection is needed. Unfortunately, forests are not renowned for their Wi-Fi. The Forest Watcher mobile app is designed to extend the platform's reach to users in places with limited access to desktops and with inconsistent Internet connection, especially for users who go out into forests and investigate recent activity. Instead of using paper maps and a rough list of coordinates, or wandering along a given set of trails, forest guardians can now walk into the field with the latest data at their fingertips and head directly to places where trees are being felled. When they get there, they can take photographic evidence to support their reports.

The application contains a few key screens:

- Setting up areas that users want to 'watch'.
- A map screen to view and navigate towards forest loss alerts.
- A report screen so that users can submit photos and feedback about deforestation.

Vizzuality also built a desktop web application allowing administrators to customise the mobile app experience for the specific needs of their team, such as collecting specific types of data and viewing different data layers in the map.

## 3.3 Other GFW-related applications

In addition to the main GFW platform and the Forest Watcher app, there are four thematic applications showcasing a small subsection of the GFW data set with a few custom analytical capabilities in order to meet specific needs. The four applications cover climate, commodities, fires and water. There is also a set of national Forest Atlases, which are managed by national government agencies and combine GFW data with national data sets, to meet the needs of specific national partners. The four thematic applications and the Forest Atlases are not discussed explicitly in this case study, but some of the findings from user research conducted for these projects have been included in the wider platform research presented here.

## 4. How do people interact with GFW?

In this section, I outline some of the key findings gathered about what users need from GFW, how they behave on the site and some of the barriers they have encountered.

### 4.1 What do the Global Forest Watchers need?

*Basic needs*

There are a few key needs that almost all users of GFW have. As with anyone coming to a web-based informational resource, they want to see a clear story or insight very quickly. Furthermore, their first experience with the platform should anchor their further interactions with the information. Much of the user feedback received by the team over the course of 2017 and early 2018 repeated the demand for a simpler introduction to the topic. So, simplicity and curation were guiding principles when Vizzuality redesigned and rebuilt the map interface (launched in November 2018).

The second basic need is related to the idea of intrigue and curiosity: people want to learn something new, or confirm something they already knew. As a user's path through the site is likely to be driven by that need, if they do not feel intrigued enough to continue exploring, or if they do not get that 'reward' from learning something early in the experience, they may leave the site. Meeting this need in an interface requires a mixture of performant technology (removing frustration from slow loading), pleasant design, high-quality micro-interactions (hovers, animations and so on that surprise users) and an understanding of the questions people want to answer.

*Intermediate needs*

After fulfilling these needs, there is a second tier of needs the platform aims to satisfy. In many interactions with users, the idea of a 'dashboard' or 'alerts' comes up time and again. Digging a bit deeper, and matching these requests up against the user personas, exposes the underlying needs (which can be solved with a dashboard):

- I need to know whenever something 'negative' is happening in the area I manage, so that I can take action as soon as possible to limit the damage and make sure I meet my goals.
- Every few months when I have to submit a report to someone or when I want to get some overview statistics about a place before making a decision, I want to see a page with simple summary statistics, so that I can quickly finish the report or quickly see the need for action (if any).

There are also regular conversations with users about getting data or visualisations out of the platform and into other places. This kind of need is consistent with many other citizen science platforms: release data to the public, allow them to create new knowledge and then share that with a wider community.

Some users, especially researchers and analysts, want the raw data rather than a visualisation in order to analyse the data further themselves. Some are looking for simple tables, some for spatial data. Others are asking for direct access to application programming interfaces so that they can access the raw data live from the database.

## Advanced needs

The first advanced need is around the ability to analyse data and obtain key objective information to support decision-making processes. In a variety of different contexts, many users wish to learn more about specific parts of the planet than the basic interface can provide. The core need being expressed when users say, 'I need to be able to analyse' is usually 'I need to see specific statistics about an area I have to manage/am concerned about to help me make choices about what to do in that area'. This can be seen as an extension of the basic need to learn more.

The second advanced need users express is around verification and validation. Especially since the launch of the GLAD alert system, which shows weekly forest loss at a much finer spatial resolution, users have been asking for access to high-resolution satellite imagery. The core need isn't to view satellite images; the need is to be able to see what is happening on the ground and verify that the alert is correct.

Finally, a very advanced need, expressed by a small group of people, relates to reusing and customising the GFW technology stack for specific purposes. This need is mostly expressed in regards to:

- the Forest Watcher application, where users want to customise the application to their specific circumstances. Some users want to

gather custom pieces of data in reports, while others want to show specific custom base maps and data sets to people in their team, using the application to help them monitor, patrol and navigate their area.
- the Forest Atlases, where government users want to share confidential information, or provide a set of specific analytical and monitoring features to make it easier for their team to do their job of managing the forests in that country.

I see this need consistently in the open-source, citizen science community: the ability to copy, reuse or refine pieces of technology that are built for generic, global purposes for specific, more local aims. These two examples – with customisation at the core – show how it is possible to meet that need.

## 4.2 How is GFW used to watch forests?

In this section, I outline some of the key insights about user behaviour on the GFW platform over the last two years, mostly derived from analysis of Google Analytics data, but also embellished or verified by testimonies gathered during user interactions.

The first thing to note is how popular the map part of the website is. In 2017, around 66 per cent of all time on the website was spent looking at the map, or 18,000 hours of time. In 2018, this figure had leapt to more than 25,000 hours – an increase of nearly 40 per cent compared to 2017. Additionally, the average time spent looking at the map in a session is almost 10 minutes. In terms of meeting a core need – helping people learn something and explore their curiosity – I feel confident that the platform delivers on this. User testimony from a feedback survey on the new map backs this up ('The new looks are quite nice!', 'It is excellent'.).

The second interesting observation is that around 15 per cent of time spent looking at the GFW data is not spent on the GFW platform. Users can embed a visualisation from GFW on to another website. In 2017 and 2018, people spent 9,800 hours of time looking at GFW maps on other websites compared to 54,770 hours looking at GFW maps on the platform itself. I believe this shows GFW has succeeded in increasing access to data and helping people around the world create and share knowledge.

When evaluating how users interact with the map, in addition to the three layers that are shown automatically (which answer many basic and popular user requirements), many people look at how the land is being used. Just over 20 per cent of all the events recorded in Google

Analytics are people clicking to view data about land use or conservation (e.g. dams, mines, concessions, protected areas or tiger conservation).

Some of the data that can be viewed on GFW is country specific. People have submitted layers showing concession boundaries, protected areas and other characteristics of the land for certain countries. The aim was to satisfy demand for additional contextual data for a few priority countries. In our review of platform use at the end of 2016, it was disappointing to see that most of the users of country data were based in the United States and Europe. These new data were being used as a tool for international monitoring, not in those countries submitting the data. Our 2017 and 2018 reviews indicated that this was changing, with an increased proportion of the users viewing a country's data coming from that country (particularly Indonesia and Peru).

Once I had investigated what data were being viewed, I then looked at how many people used the analysis tool and what data were being analysed. In 2017, just 8 per cent of all users completed an analysis, and most of the entities being analysed were countries in South America or South East Asia. Having identified this low rate of usage, the team made some changes to the interface to increase the visibility and ease of use of the analysis tool, which, when combined with outreach and promotion of the feature, appear to have improved its utilisation. The new map was launched in November 2018. Since then, 16.43 per cent of users have completed a data analysis.

One final part of these annual reports is an exploration of how the satellite imagery feature is being used. In particular, the team wants to know which country a user is from and what part of the world they are looking at. In 2016, the satellite imagery was used 21,000 times by 4 per cent of all users. Mostly, it was being used by people in North America, South America and Europe to look at Peru, Brazil and South East Asia. During 2017, it was used more than 19,000 times, and again it was being used by people from the same geographic regions. However, there was increased use across the whole of Latin America and Africa and South East Asia. The two most popular areas for people to look at were in Norway and Mali. This suggests that quite a lot of people are using the satellite images as expected – people verifying forest loss in their country – but that (as with the country-specific data) this tool is still serving a more 'Western' audience.

## 4.3 What barriers do Global Forest Watchers face?

At the same time as learning about the needs of users and the behaviour they exhibit on the site, the team has also heard their frustrations. To meet the wider goal of forest protection and sustainable management, there are a number of barriers that users face. I want to close this chapter by mentioning two of the biggest platform-specific barriers that the team has encountered in the last few years.

The first, and perhaps the biggest, is stable and reliable access to the Internet. This is a multifaceted barrier. At its simplest level, it is about having access. Even Forest Watcher, the mobile app where you can take GFW offline into the field, still relies on the device having periods of connectivity to update the data. The second element is Internet speed and reliability, which can cause frustration when some parts load slowly or not at all. This can be solved to some extent in the way the code is written (to make it faster to load on slower connections). However, bearing in mind GFW is a large and fairly complex web-based application, there is only so much that can be done to deliver a good experience for all connection strengths.

Second is a user's familiarity with the data sets available and with modern web mapping platforms. In general, Vizzuality follows a principle called progressive disclosure to make sure a user's first interactions on a site explain the data and features available in a stepwise manner in order not to overwhelm them with everything all at once. This can mean that some features are slightly hidden. Many of our priority users are quite technically minded, such as the users in businesses and governments analysing deforestation rates and risk, and the researchers doing the same to keep them to account. When observing them in usability tests, they tend to be happy to click around an interface and discover everything on their own. Other users – often those who are less skilled in data science and analysis – are less keen to do so and often express a feeling of not knowing what to do next. In order to keep the more advanced users satisfied, but also open up the data to as many people as possible to interrogate, the redesigned map interface has a series of topic descriptions, providing additional introductory steps that can pull people deeper into the interface.

## 5. Conclusion

In this case study, I have explained a few of the things that I have learned about users of the GFW platform from years of research. From some of the findings presented – showing how much time is spent on the map and how often different features are being used – I have concluded that the GFW platform delivers a valued and usable experience to people all over the world. The research findings also help point to areas where the platform can be improved, for example the low use of the analysis and satellite imagery functions. However, it is the consultative and iterative development process by which these findings are acted upon – where users are invited to comment on and help improve early versions of our work – which helps us move towards a more usable, understandable and useful platform.

The findings presented in the chapter have been gleaned from a mixed-methods research strategy – a strategy which helps us attain the level of breadth and depth necessary to make reasonable insights. Using analytics alone risks missing 'why' some actions are performed and other nuances of user interaction observable when you watch someone use an application. Interviews alone risk making judgements on a limited and potentially non-representative sample and also risk 'extracting' from users rather than engaging with them. As alluded to above, I consider the consultation and feedback process a key method in our research, not as a means to generate primary data but as a way to verify our interpretations and bring a different perspective to the analysis of the problem to be solved.

When thinking about the future of geographic citizen science, this participation in the development process is essential for making sure the end product is a global public good. GFW aims to overcome a complex usability challenge: to make something that can be used as easily by a relative novice as by the highly skilled people who have been coming back every month. Many platforms do not address this challenge – they build for an audience without fully soliciting their advice (leading to poor usability of and uptake of the solution), or they build for a niche and closed audience (so only a handful of people are able to use the platform). I think GFW provides a framework that the sustainable development community can use to continue to produce innovative (and much-needed) citizen GIS platforms. Forest loss is a major global problem that needs many people reviewing the data, finding insights and sharing knowledge in order to make sure pro-forest decisions are taken. The same can be said for many of the big sustainable development challenges. Building platforms that reach millions is more likely with constant user consultation and the triangulation of different data sets to produce insights that are sectorally broad and individually deep.

## 6. Lessons learned

There are five key lessons learned that I want to emphasise to conclude this chapter:

- Despite the increasing use of mobile phones around the world and greater access to the Internet (via physical network expansion and reduction of cost), access to the Internet still remains a factor that limits the use of Internet-based citizen science platforms.
- With the GFW platform, there are still some occasions where users report slow load speed hindering their experience. With the Forest Watcher app, the need to visit larger towns regularly (with slightly better Internet access) can hinder data updates and transmission across teams.
- Curation is always welcomed, even by 'expert users' who ultimately want to access a lot of data and advanced functionality. The progressive disclosure principle is one way to balance the need for simple introductions and access to complexity. Preliminary analysis suggests a more curated approach is leading to greater engagement with the map.
- Building partnerships around a platform is essential to reach many users and bring their thinking into the design process. These more formalised relationships, with regular interaction points, help improve the quantity and quality of the feedback received, and the likelihood of securing a response to requests to interview, test a site or complete a survey.
- Inserting more user-focused steps – not just extractive research during project initiation but constant engagement and discussion – into a project can help increase its use and usability, and ultimately the impact of the project. This requires projects to be developed in more stepwise agile ways, so that users can be brought in regularly to test assumptions, validate findings and improve the end product.

## References

Curtis, Philip G., Christy M. Slay, Nancy L. Harris, Alexandra Tyukavina and Matthew C. Hansen. 2018. 'Classifying drivers of global forest loss', *Science* 361: 1108–11.

Hansen, Matthew C., Alexander Krylov, Alexandra Tyukavina, Peter V. Potapov, Svetlana Turubanova, Bryan Zutta, Suspense Ifo, Belinda Margono, Fred Stolle and Rebecca Moore. 2016. 'Humid tropical forest disturbance alerts using Landsat data', *Environmental Research Letters* 11: 034008.

Harley, Aurora. 2015. Personas make users memorable for product team members. Accessed 13 March 2020. https://www.nngroup.com/articles/persona/.

Harris, Nancy, L., Elizabeth D. Goldman, Mikaela Weisse and Alyssa Barrett. 2018. When a tree falls, is it deforestation? Accessed 11 January 2020. https://blog.globalforestwatch.org/data/when-a-tree-falls-is-it-deforestation.

ISO. 2010. ISO 9241-210:2010(en) Ergonomics of human–system interaction – Part 210: Human-centred design for interactive systems. Accessed 21 September 2020. https://www.iso.org/obp/ui/#iso:std:iso:9241:-210:ed-1:v1:en.

IUCN. n.d. Protected area categories. Accessed 11 February 2019. https://www.iucn.org/theme/protected-areas/about/protected-area-categories.

Keenan, Rodny J., Gregory A. Reams, Frédéric A. Achard, Joberto V. de Freitas, Alan Grainger and Erik Lindquist. 2015. 'Dynamics of global forest area: Results from the FAO Global Forest Resources Assessment', *Forest Ecology and Management* 352: 9–20.

Mendes, Karla. 2018. Deforestation in the Amazon is putting uncontacted tribes at risk. Accessed 21 September 2020. https://www.weforum.org/agenda/2018/09/deforestation-in-the-amazon-is-putting-uncontacted-tribes-at-risk/.

UNDESA. n.d. Goal 15: Sustainable development knowledge platform. Accessed 11 February 2019. https://sustainabledevelopment.un.org/sdg15.

US HHS. n.d. Personas. Accessed 11 February 2019. https://www.usability.gov/how-to-and-tools/methods/personas.html.

Weisse, Mikaela, Elizabeth D. and Goldman 2018. 2017 was the second-worst year on record for tropical tree cover loss. Accessed 18 September 2020. https://www.wri.org/blog/2018/06/2017-was-second-worst-year-record-tropical-tree-cover-loss.

# Chapter 8
# Path of least resistance: using geo-games and crowdsourced data to map cycling frictions

Diego Pajarito Grajales, Suzanne Maas,
Maria Attard and Michael Gould

## Highlights

- Cycling provides a low-cost, low-pollution, active mode of transport. Cycling can provide a solution to urban mobility issues such as congestion, traffic accidents, transport poverty, inequality and public health issues such as inactivity and obesity. Policies and programmes promoting cycling have increased cycling dramatically in cities worldwide.
- The Cyclist Geo-C app has been developed to involve citizens in open cycling data collection by tracking cyclists' movements and collecting feedback on their journeys through the selection of tags.
- The Cyclist Geo-C app allows for the identification of areas where cycling takes place, where areas of friction are located and where infrastructural or design improvements are needed to address the identified frictions.
- Volunteers confirm the suitability of collaboration-based gamification to engage bicycle commuters in collecting open cycling data, in contrast with the more commonly used competition-based strategies used in sports cycling apps.

## 1. Introduction

Urban cycling is an alternative mode of transport promoted by cities worldwide to reduce congestion and pollution and to increase citizens' physical activity (Oldenziel et al. 2015). Cycling data, such as information about the cycling modal share, preferred routes and the main constraints or frictions faced during cycling, can be used as an evidence base for urban planning, cycling infrastructure design, cycling advocacy campaigns, promotion of alternative commuting and the assessment of impacts and benefits of cycling planning and promotion (Gössling 2018). The same data also have wider applicability in planning cycling policies, for instance to evaluate the impact of cycling on individuals and public health and environmental quality (Wiggins and Crowston 2011), to reduce congestion and pollution and to create comfortable commuting routes. The data sets can also serve to evaluate changes in physical activity, reduction of stress on the road and reduction of time and money spent on driving and parking (Pooley et al. 2011; Martinez Tabares 2017). Unfortunately, both city managers and citizens have limited data available about the amount, location and conditions of cycling in their cities.

Different strategies to produce better cycling data have emerged in recent years (Braun et al. 2016). With the current availability of mobile devices and location-based services (LBS) – that is, services using the user's location to provide particular functionalities such as the closest bus station, taxi, store or restaurant – many software tools have been developed for collecting urban mobility data (Grant-Muller et al. 2015; Boss et al. 2018). In tandem, there has been an increase in the capacity and willingness of citizens to collect and share data using their smartphones, increasing the possibilities for crowdsourcing information. These developments have enabled the growth of geographic citizen science, which provides the means and tools to enable people to collect data and generate information with a specific reference to places and locations. In the context of mobility and transport, volunteered geographic information has become a valuable resource for decision makers who are aiming to understand people's mobility choices and to support a shift towards sustainable mobility (Goodchild 2007; Attard, Haklay and Capineri 2016).

In this chapter, we share our experience with mapping cyclists' routes through the volunteered collection of open cycling data, and we discuss the role of mobile apps in encouraging bicycle commuting and the use of geographic citizen science in this context. This chapter continues with a description of the case study, the Cyclist Geo-C mobile app and volunteers' interaction with the app. We then present how our analysis found that volunteers were generally willing to share their location data and trip movements to support cycling research by providing scientific evidence for campaigning or lobbying efforts demanding better cycling conditions. Thereafter, we explain the exploratory analysis of the geographic citizen science data and its potential use in urban planning. Finally, we discuss the results and lessons learned during the study. These results focus on volunteers' feedback and suggestions, as well as the future applications for geographic citizen science involving cyclists, LBS and crowdsourced data collection.

## 2. Geographic citizen science applications used in urban cycling

Despite the current hyper-connected status of cities and citizens, in part due to apps and services from companies such as Amazon, Alphabet and Apple (Weigend 2017), urban cyclists, cycling advocacy groups and city planners struggle to access data describing urban cycling behaviour for policymaking (Gössling 2018). On one hand, cities trying to promote urban cycling often lack data in order to plan infrastructure development and to promote sustainable mobility. On the other hand, cities that already have a bicycle culture often lack data to understand cycling patterns and frictions commonly faced during cycling.

Urban cyclists appear to be motivated by different factors than sports cyclists; for example, comfortable commuting routes and reducing environmental pollution (Pooley et al. 2011; Martinez Tabares 2017). Health and environmental research projects increasingly involve urban cyclists as experimental subjects (Wiggins and Crowston 2011), while app developers tend to tailor user interfaces for their particular needs. After reviewing 10 cycling apps having more than ten thousand downloads in mobile stores, we found that only Bike Citizen and Biko specifically target urban cyclists. Other mobile cycling apps, such as the popular Strava, are characterised by their focus on sports or extra-urban cycling, and mainly provide feedback on physical performance and promote social interaction through challenges or competition between users (Pajarito Grajales and Gould 2018). Four of the reviewed apps – Strava, Human, Bike Citizens and Biko – explicitly offer advisory and data services for mobility planning in cities as part of their business (Barratt 2017; Boss et al. 2018). However, none of these provides free access to the collected data.

Existing open-source data-collection platforms, such as BikeMaps.org and OpenCycleMap.org, which is OpenStreetMap's initiative for cycling, are limited to collecting data on infrastructure and do not provide functionality to record bicycle trips or to describe cycling conditions. Despite being the most accessible data source, these platforms do not adequately serve city councils, planners and researchers when it comes to the analysis and design of public policy for urban cycling (Yeboah and Alvanides 2015; Sultan et al. 2017).

To date, many cities rely only on traditional analysis strategies for measuring bicycle commuting, that is, household travel surveys, bicycle counts, travel diaries or cycling promotion (e.g. posters, newspapers, magazines, TV ads, social media and so on; Handy, van Wee and Kroesen 2014). Having better cycling data, including data collected through mobile apps, has the potential to provide better diagnosis of, and solutions to, current cycling issues in cities.

Our geographic citizen science application – the Cyclist Geo-C mobile app – was designed to address shortcomings identified in urban cycling data collection. Our research provides a mobile app for urban cyclists to collect open data on their bicycle trips, that is, geo-spatial route data which are collected by constantly taking Global Positioning System–based point measurements, to construct the path followed between origin and destination. Additionally, cyclists can use the app to collect information about their trip through semantic tags describing their experience. Users can choose up to three predefined tags to describe each trip, and by analysing these tags, researchers can understand how users perceive the cycling environment.

We tested the Cyclist Geo-C app at three sites, all of which were European cities with different cycling cultures and social and geographic contexts. Groups of 20 volunteers per city used the Cyclist Geo-C app to record their bicycle trips. Volunteers were blindly assigned one of two interfaces which implemented either a competition- or collaboration-based gamification strategy. The comparison aimed to identify the most successful strategy for crowdsourcing open cycling data collection with urban commuters. We used the recorded location points of all bicycle trips to identify cycling routes and potential frictions inhibiting urban cycling (Pajarito Grajales and Gould 2018).

## 2.1 Cultural and social context of testing and using the Cyclist Geo-C app

To test the Cyclist Geo-C app in different real-life contexts, we chose the European cities of Münster in Germany, Castelló in Spain and the urban area around Valletta in Malta because of their contrasting cycling environments (see Table 8.1). Volunteers from these cities represented different levels of cycling experience and involvement in bicycle advocacy groups. Volunteers also had varied perspectives on the implications of sharing personal data through mobile apps for cyclists. In particular, we

Table 8.1 General context of the cities where the app was tested

| City | Population | Location | Topography | Cycling network | Cycling modal share |
|---|---|---|---|---|---|
| Münster | 300,000 | North Rhine–Westphalia region in North-West Germany | Mostly flat | An extended and high-quality network of dedicated bicycle lanes | 39% – high |
| Castelló | 180,000 | Mediterranean coast of Spain, 60 km north of Valencia city | Mostly flat | A growing network of bicycle lanes covering the main urban corridors and connecting with the surrounding towns | 3% – low |
| Valletta | 210,000 (Valletta conurbation) | A conurbation of small towns, located on the eastern shore of Malta | Multiple bays and low hills | Almost no cycling infrastructure apart from a small number of dedicated bus lanes shared with cyclists | <1% – very low |

found volunteers from Germany had a more critical perception of their movements being tracked, sharing their location or allowing automatic recording functionalities outside of experimental conditions. Finally, most of them agreed to participate, as they identified with the research goal and the planned use of the collected data. Volunteers from Spain and Malta were generally less critical about the location features, and in some cases even suggested the use of the very same automated functionalities as new app features.

Münster city council proudly presents its bicycle-friendly character as a key element of its marketing strategy. In contrast, in the two other cities, the promotion of cycling as a solution that enables a healthier lifestyle and a response to social and environmental issues (i.e. traffic congestion or air pollution) is not currently high on the political agenda. However, in Castelló, for example, the expansion of a public bicycle-sharing system and the improvement of the bicycle lanes network are part of the city's sustainable transport strategy. In Malta, two bicycle-sharing schemes have started running in recent years, and in late 2018, the country published a draft National Cycling Strategy, although only limited construction of dedicated cycling infrastructure is foreseen.

The three cities differ in their geography, orography and climate – all factors that impact the popularity of cycling (Heinen, van Wee and Maat 2010). A Mediterranean climate with warm summers, mild winters and relatively little precipitation facilitates cycling in Castelló and, to a certain extent, in Valletta, although high summer temperatures can also hamper cycling (Médard de Chardon, Caruso and Thomas 2017). On the other hand, colder average temperatures and a higher incidence of rainy and windy weather do not seem to hamper cyclists in Münster, which has a high cycling modal share, even compared to other cities in Germany.

The observed differences in cycling modal share are perhaps better explained by the historical, social and cultural environment rather than by their geographic conditions. The positive perception of bicycle commuting, a dense and connected bicycle lane network and a strong cycling culture in Münster are the result of a society that embraces cycling as a way of life and of political commitment to the promotion of cycling.

## 3. The Cyclist Geo-C mobile app

The Cyclist Geo-C app is an Android location-based mobile app for tracking cycling trips. The app uses two features provided by Google (i.e. 'Location'[1] and 'Fit'[2]) to control the recording of the cycling trips. It uses a dedicated server to store location information and a set of computer routines to turn location information into paths compatible with geographic information systems (GIS). The app provides options such as a user profile, a list of 30 predefined tags to describe the journey upon arrival and a dashboard to compare one's activity to that of other cyclists in the city.

The app's source code, as well as the anonymised geographic data collected (e.g. aggregated bicycle track information without personal details to minimise the risk of identifying individual behaviour patterns in order to comply with General Data Protection Regulation (GDPR) guidelines, and to fulfil the requirements of ethical committees from the University of Münster and University Jaume I) are available for anyone to use, from citizens, developers and local authorities to advocacy and research groups. The Cyclist Geo-C app is a tool developed to empower citizen participation and enable open cities as part of the Open City Toolkit (see Degbelo, Bhattacharya et al. 2016; Degbelo, Granell et al. 2016). The app currently supports four languages: English, European Spanish, German and Catalan. Four researchers tested the app in Münster, and identified all the security, communication and storage features needed to publish location-based apps.

## 3.1 Data collected through the app

Using the app, 57 volunteers (19 in Münster, 20 in Castelló and 18 in Malta) recorded 793 trips in the three cities: 343 trips in Castelló, 335 trips in Malta and 115 trips in Münster. Volunteers were aged between 15 and 58 years (mean 33.4 years, median 32.5 years), were mainly single (23 single and 12 in a relationship but not living together) and included 24 female volunteers (42 per cent). Despite the different geographic and sociocultural environments, most volunteers frequently commuted by bicycle and, in some cases, reported the bicycle as their only means of transport. Each volunteer who completed the experiment received a nominal €10 reward, which was part of research funding from the Association of Geographic Information Laboratories in Europe.

The collected data not only allowed us to identify volunteers' preferred streets for cycling, but also provided insights into the use of cycling infrastructure. Volunteers in Germany mostly recorded urban trips during workdays, while in Malta and Spain, volunteers also recorded their weekend leisure trips venturing into the surrounding countryside. On average, each volunteer recorded 13 trips during the experiment or about 9.3 trips per week. The mean trip length was 5.6 km, with an average duration of 30.2 minutes.

The data collected during the experiment revealed differences between the cities, especially related to the cities' geographic context and cycling infrastructure. The maps in Figure 8.1 summarise the recorded bicycle trips (green), the bicycle infrastructure (blue) and the identified frictions (red).

Münster's road network consists of a series of concentric ring roads with segregated unidirectional bicycle lanes in most of the streets. To access or go around downtown, riders tend to take the main promenade ring – an exclusive and spacious bicycle lane circling the city centre (see Figure 8.1). The identified frictions were mainly related to road intersections, especially when crossing highways, or traffic lights (Figure 8.2).

In Castelló, the road network is a mixture of narrow streets in the city centre and a few north–south and east–west axes to the surrounding towns. The cycling infrastructure is a mix of painted lanes on sidewalks, shared streets with speed limits and segregated bicycle paths. The recorded trips were mainly from the city centre to the university campus on the western border of the city, the port area in the east and the northern towns, as well as the north-west mountain area used for leisure and sports cycling (see Figure 8.1). The identified frictions were mostly related to roundabouts, intersections or pedestrian areas (see Figure 8.2).

The road network in Malta is a mixture of narrow residential roads and a few larger arterial roads connecting the different towns and cities making up the conurbation. Urban cyclists in Malta lack dedicated bicycle infrastructure and must deal with a hilly topography, busy roads and a strong prevalence of one-way streets. The recorded trips follow the streets with cyclable slopes, which usually connect through secondary roads and shortcuts chosen by cyclists trying to avoid hills – sometimes they are forced to go against the traffic flow (see Figure 8.1). The trips were mainly on flat streets along the coastal promenade, while the identified frictions were along the streets with steep slopes or at intersections where cyclists must dismount to use pedestrian crossings (see Figure 8.2).

Complementing the bicycle trip data, volunteers recorded 791 tags that described their cycling experience, as well as their perception of either urban cycling benefits or environmental conditions faced during the trip. We classified these tags or words according to their meaning into positive ($n = 273$; i.e. a word expressing a positive feature of the trip such as 'fast' or 'enjoyable'), neutral ($n = 285$; e.g. 'normal') and negative ($n = 192$; e.g. 'boring', 'noisy'). The tags provided usually referred to trip features such as cycling speed (e.g. fast, quick, moving, slow), comfort (e.g. relaxed, crowded, intensive) or the environment (e.g. efficient, secure, inspiring, risky, safe).

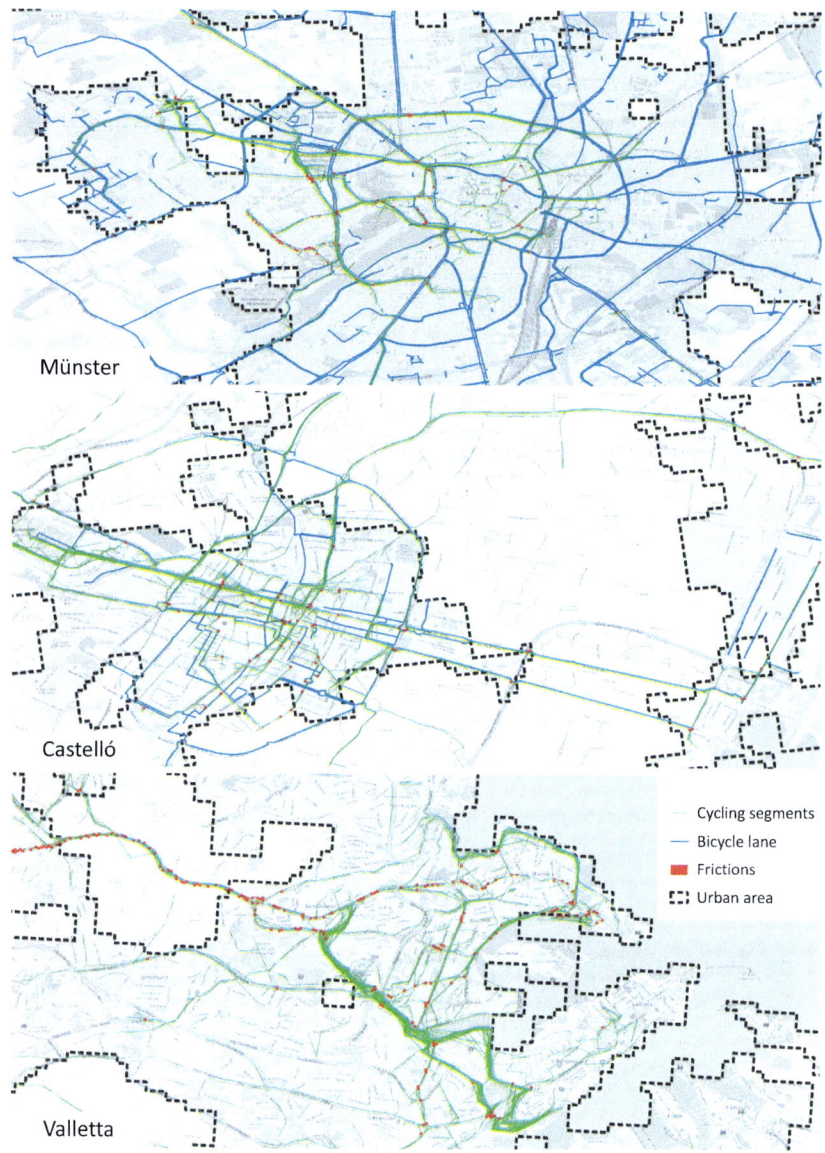

**Fig. 8.1** Maps of the distribution of bicycle trips recorded by volunteers using the Cyclist Geo-C app. Basemap © OpenStreetMap contributors. Bicycle trip data: research data from Geotec research group – Universitat Jaume I. Approximated scale 1:50,000.

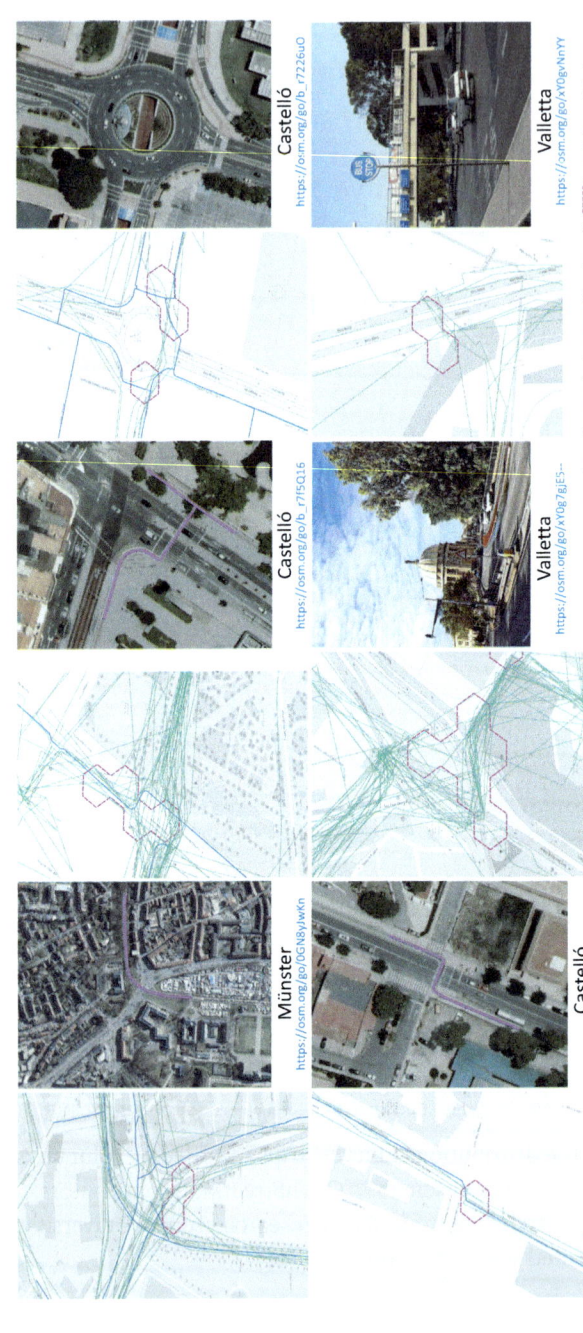

**Fig. 8.2** Examples of frictions and their cartographic representation.
Source: Cyclist Geo-C app. Basemap © OpenStreetMap. Orthophoto: Castelló from Infraestructura Valenciana de datos espaciales (IDEV), Münster from Geoportal Münsterland, Valletta pictures by Diego Pajarito Grajales. Bicycle trip data: research data from Geotec research group – Universitat Jaume I. Approximated scale 1:10,000.

## 3.2 Collaboration- and competition-based rewards

Cyclist Geo-C uses two types of rewards via its user interface (see Figure 8.3): collaboration-based rewards, expressed as the contribution to the city's total number of trips as a percentage; and competition-based rewards, expressed as the position on the city's leader board of volunteers with the most trips and tags. We included a collaboration-based reward system in the app to test whether such a strategy, in addition to existing competition-based rewards that are present in sport cycling apps, might lead to increased intention and engagement with urban cycling, as suggested by Halko and Kientz (2010). To evaluate volunteers' experience with the two types of rewards, we tested the feature in an experiment in which volunteers were randomly assigned either the collaboration- or competition-based interface, and asked for their opinions after interacting with the reward system.

## 3.3 Analysing collected data and cycling frictions

For the analysis of cycling patterns, crowdsourced bicycle trip segments (i.e. the lines connecting two consecutive pairs of coordinates recorded using the app) were classified on the basis of speed into either walking segments (i.e. speed lower than 5 km/h) or cycling segments (i.e. speed between 5 and 50 km/h). Thereafter, using a 30-metre-wide hexagonal grid to overlay the segments, we counted and compared walking versus cycling segments in each cell to identify the potential frictions inhibiting bicycle commuting. We defined the 'friction intensity' as the ratio between the number of walking and cycling segments (see Figure 8.4; Pajarito Grajales and Gould 2018).

Aggregating and mapping the crowdsourced trips using GIS tools allows for the visualisation of the preferred streets and most intensely used cycling routes in a city. The maps also show extra features such as trip origin, destination and the bicycle lane network, as well as the urban area limits surrounding city infrastructure and landmarks. This analysis is intended to be a useful tool for both urban planners and cycling advocacy groups to understand the city's cycling environment. Combining the recorded trips, the identified frictions and the bicycle lane network, it should be possible to prioritise interventions in the bicycle infrastructure or urban cycling promotions. We consider this capability the most relevant outcome of our geographic citizen science application due to its relative simplicity and high relevance for local authorities and advocacy groups.

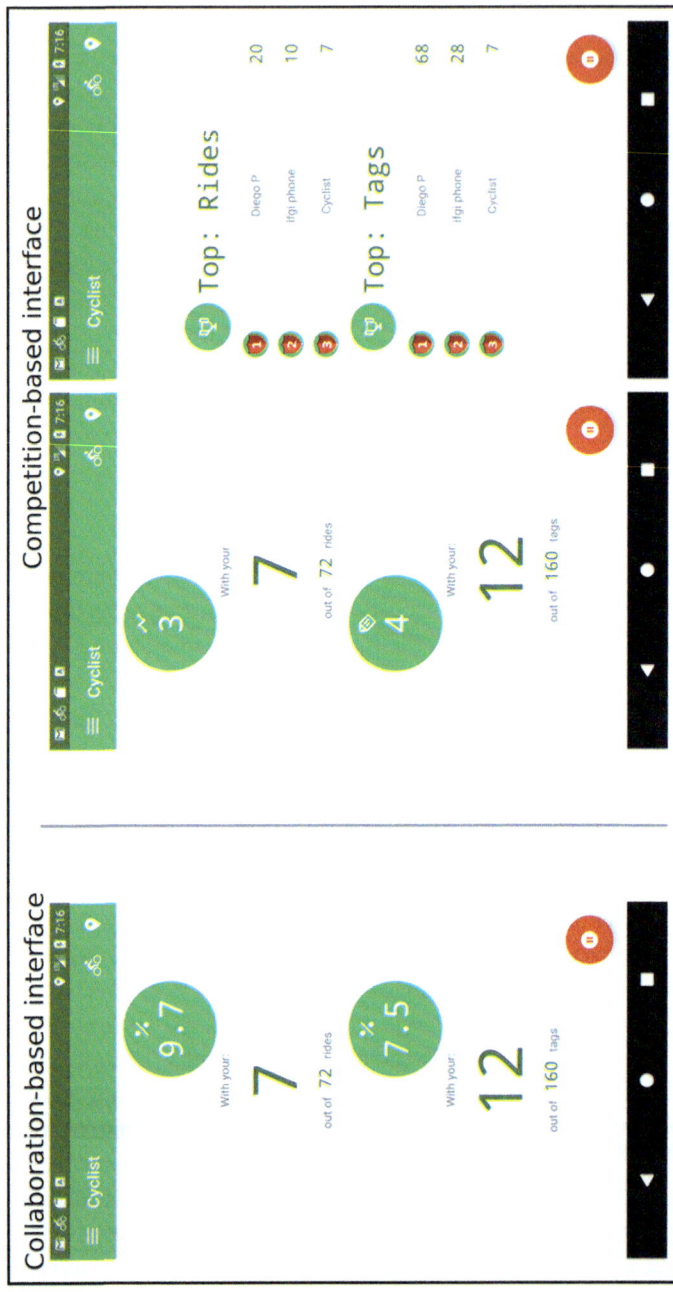

**Fig. 8.3** Cyclist Geo-C app interface. Left: participant's trips as a percentage of the city's total contributions. Right: participant's position on city leader board based on the number of trips.
Source: Cyclist Geo-C app.

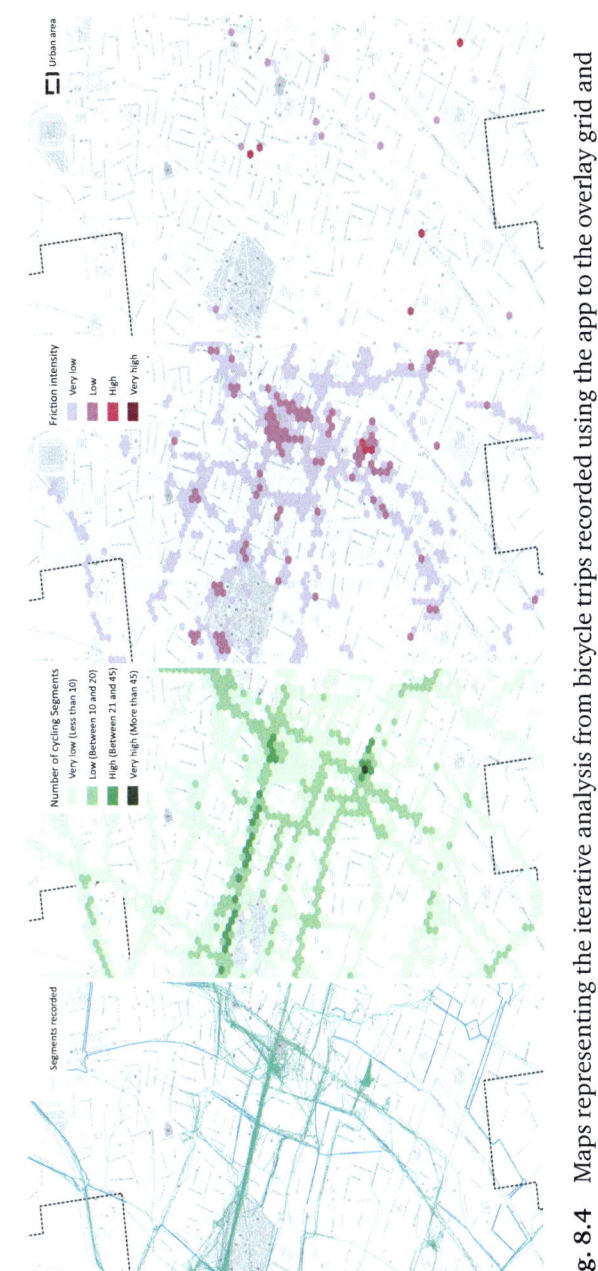

**Fig. 8.4** Maps representing the iterative analysis from bicycle trips recorded using the app to the overlay grid and identified frictions. Basemap © OpenStreetMap. Bicycle trip data: research data from Geotec research group – Universitat Jaume I. Approximated scale 1:20,000.

## 4. User experience perspectives: testing in three cycling cities

Testing the Cyclist Geo-C app with volunteers in three different European cities aimed to understand users' needs and barriers better when interacting with the app.

### 4.1 Understanding users' needs

Although the market offers a vast variety of sensors, devices and apps to track physical activities such as cycling, few volunteers in our experiment used mobile cycling apps (46 per cent), and only 15 volunteers (26 per cent) used wearable devices. Future research might explore how urban cyclists use apps and wearable devices. Volunteers who used mobile apps mostly referred to using mapping functionalities and shortest path visualisations. Volunteers thought of mobile apps for tracking bicycle trips mostly as a sport-related instrument and not relevant when commuting. However, volunteers responded positively to the Cyclist Geo-C app. They saw the potential for its personal use and benefits for their cycling organisation, especially having records of the most convenient cycling routes (e.g. 'I see the value of using the app as a motivational tool, especially for beginners' and 'I will continue using the app so that I can help with the future of cycling in Malta').

The digital versions of friction maps and data analysis performed received multiple reactions on social media and strong interest from the bicycle advocacy groups in Castelló and Malta. In general, they agreed with the identified frictions, intersections, roundabouts and pedestrianised spaces which were difficult to navigate by bicycle (see Figure 8.2), as well as on the need for expanding the bicycle lane network. A newspaper article in the *Times of Malta* in October 2018 cited our study as a scientific resource to consider when promoting urban cycling in Malta. All identified actions serve as evidence of the need to involve citizens in other research and analysis tasks.

Volunteers suggested two sets of improvements. First, they requested features such as reminders for recording trips based on personal routines (e.g. 'Perhaps you could implement the possibility to turn on a reminder'), better control over trip recordings with manual options to edit or delete (e.g. 'It would be good if you could edit your former journeys' and 'Maybe an option to start the count when you wish rather than when opening the app') and more intuitive ways to add new tags. Second, they suggested app extensions such as using the smartphone to map infrastructure quality using location and vibration data sensed while cycling. Volunteers also suggested additional game rewards based on topography (e.g. points for 'climbing hills'), automatic trip tracking and integration with existing sports cycling apps (e.g. 'I think the app should start automatically, maybe by using a third-party app such as Google Fit').

## 4.2 Barriers experienced by users

Volunteer feedback reported barriers with regards to both the interaction with mobile cycling apps and participation in our experiment. For instance, the volunteers found it cumbersome to start and stop the app manually for every trip. Despite being given instructions, it was not easy for all the volunteers to remember to open the app for every single trip.

Another main frustration voiced by users was the lack of a cycling reference map in the app interface, as the app did not provide a way to visualise the collected routes or the identified frictions. Volunteers foresaw the use of maps as a tool to share popular cycling routes and commonly faced difficulties, which could be particularly useful for new cyclists not familiar with these specific routes. The mapping features could therefore evolve into a personalised cycling coach when it comes to choosing the most convenient route (Wiggins and Crowston 2011). Additionally, some volunteers suggested creative map-based functionalities such as underlining alternative bicycle-friendly routes, selecting meeting points for commuting groups with neighbours and workmates, and an interface which highlights flat, green or calming routes. Other suggestions included extending the scope of the app to include other active modes of transport such as walking and creating special functionalities for local or time-bound events, such as conferences and festivals. Adding or testing these geo-visualisation functionalities in the future would serve to improve future citizen science applications.

Prior familiarity with cycling and fitness apps such as Strava, or lack thereof, created divergent experiences for the volunteers. Cyclists who had never used such apps before needed more guidance and assistance to use the app properly and to remember to use it for their rides, while those already familiar with cycling apps needed to adjust to the fact that the Cyclist Geo-C app had limited capabilities and a simpler interface. However, through participating in our experiment, both types of users got to experience collecting open-source cycling data as well as interacting with an interface and a gamified strategy developed specifically with cycling commuters in mind.

## 4.3 Analysing user feedback on the gamified reward system

The results of the experiment provided feedback on the gamified strategy in three main directions. First, volunteers agreed that they enjoyed cycling and that they intended to continue using their bicycles for commuting purposes. Second, the experiment served to validate volunteers' willingness to contribute to research related to urban cycling in their city. We see these two directions as an opportunity to develop further data-collection tools for geographic citizen science projects with urban cyclists. Third, we found that volunteers enjoy collaborating with other cyclists more than competing against them (Pajarito Grajales, Degbelo and Gould 2020; e.g. 'The app should have an option to interact or send messages to cyclists in my city'). Therefore, app designers and developers might explore new types of user interfaces aligned with collaboration schemas rather than competition.

Volunteers' continuous use of the app demonstrated positive feelings towards cycling. Some volunteers kept recording trips and tags after the experimental period was over, while the recorded trips showed how volunteers consistently used the bicycle during and after the experiment. In these findings, we saw the potential of mobile apps to enable cycling data collection to serve urban planning in cities in the long term. However, when it comes to usability, our app user interface needs to work better with commuters' routines (e.g. 'In using the app I had issues with the ability of the phone in determining my location and recording any data, which led to me having a generally poor experience with the app').

## 4.4 Considerations of privacy for geographic citizen science

Volunteers expressed divergent stances when it came to collaboration and trust regarding sharing their personal data. On one hand, a volunteer from Münster reported a strong concern for sharing these data (i.e. personal trips and locations), agreeing to do it only for the experiment. On the other hand, volunteers from Castelló and Valletta expressed their intentions to share their trips recorded through third-party cycling apps either to cover a more extended period or to provide additional insights for the analysis. Some volunteers also suggested an app feature for automatically recording users' activity instead of manually starting and stopping recording every trip – something that other users might consider intrusive.

Other privacy issues and personal data-sharing concerns cropped up at different stages of our pilot: when notifying the user while recording location data and when providing an option to stop such recording at any given time or after the publication of anonymised geographic data collected (e.g. aggregated bicycle tracks without personal information to minimise the risk of identifying individual behaviour patterns). These issues led us to multiple considerations when it came to user privacy during the experiment. However, these considerations give rise to the need for a strict protocol to manage the information coming from LBS in geographic citizen science studies. Such a protocol needs to adapt to contrasting social and cultural contexts from those most reluctant to support any monitoring system, such as in Germany, to those more open to the idea of tracking activities and sharing information at different levels, such as Mediterranean, North American or Asian countries.

## 5. Discussion

As found in the cycling literature (Heinen, van Wee and Maat 2010) and reflected in the data collected by volunteers' feedback, a well-designed, connected and direct bicycle lane network enables multiple bicycle trips and complements the promotion of urban cycling. In our case, the differences observed in the cycling modal share of overall transport uses and the characteristics of the bicycle lane networks in the three cities match the differences in cultural adoption of bicycle commuting, private car dependence, the use of multimodal transport systems and volunteers' cycling profiles.

Our pilot experiment showed that participants preferred collaboration-based gamification strategies (Pajarito Grajales, Degbelo and Gould 2020). This finding can aid the improvement of mobile app design for urban cycling data collection. Our results potentially enhance the research outcomes from educational games and transport studies adopting gamified tools to boost trip data collection and to encourage behavioural changes towards sustainable modes of transport (Berri and Daziano 2015). Our results also complement research related to purpose-oriented (Wunsch et al. 2016) and user-specific engagement campaigns (Schrammel et al. 2015) while offering an alternative to prevent the negative impact of competition-based gamification in extremely competitive users (Barratt 2017).

A change towards collaboration-based gamification strategies would encourage engagement of urban cyclists and strengthen the relevance of both mobile cycling apps and crowdsourced cycling data collection for transport analysis. Additionally, the open architecture of the Cyclist Geo-C app and the open cycling data set fits in with the more general promotion of smart and open cities (Degbelo, Bhattacharya et al. 2016; Degbelo, Granell et al. 2016).

In addition to contributing to data collection, volunteers could use the data to analyse cycling patterns in their city, thus taking a role in data analysis as well as collection, as put forward by Haklay in his typology of citizen science strategies (Haklay 2013). However, upscaling the experiment of crowdsourcing cycling data collection to a long-term geographic citizen science strategy would require several steps. First, the app would need to be managed based on a structured and sustainable business model with defined responsibilities, financial resources and the involvement of stakeholders. Second, such a strategy would demand validation for identified frictions, the registering of minor or non-visible frictions omitted by the algorithmic analysis and a joint interpretation between researchers and cyclists. Lastly, a geographic citizen science application for urban cycling should address the barriers identified by volunteers in the experiment and consider incorporating the suggestions provided to improve interaction with the app. The research also pointed to alternative scenarios to deploy the app such as conferences or festivals in which automatic enrolment and tracking of trips would provide for the analysis of time-specific conditions.

Developers and scientists need to consider the impact of tracking and surveillance technologies as well as the GDPR guidelines (in Europe) when collecting private data. These issues reflect the need for establishing stronger connections between geographic citizen science and privacy regulation.

## 6. Lessons learned

- Testing the Cyclist Geo-C app in different cycling environments in three cities, each with a distinctive user base, geography, cycling culture and facilities, demonstrated the potential of mobile technology as a tool to enhance urban cycling promotion and analysis in a variety of urban contexts.
- This approach would be particularly beneficial for transport research. The data collected enabled friction analysis which can aid in the visualisation of cycling patterns and in identifying the elements potentially inhibiting the use of bicycles.
- Experiment volunteers enjoyed using gamified tools and showed a preference for collaboration-based rewards over competition-based

ones – a fact that might help to improve usability and engagement with mobile apps in similar contexts.
- There are at least five ways to improve this geographic citizen science application:
    (1) Involving citizens in results validation, framing further research questions and doing analysis together with researchers;
    (2) Adding functionalities such as geo-visualisation or analysis tools for different cultural environments;
    (3) Extending the data-collection task to other sustainable modes of transport such as walking;
    (4) Adapting functionalities to support local or time-specific events such as conferences or festivals; and
    (5) Streamlining and automating the process of enrolment and tracking of trips to promote broader usability.

## Acknowledgements

The authors of this document gratefully acknowledge funding from the European Union through the GEO-C project (H2020-MSCA-ITN-2014, Grant Agreement No. 642332, http://www.geo-c.eu/); from the Citizen Science COST Action CA15212 http://www.cs-eu.net; and from the small bursary provided by the Association of Geographic Information Laboratories in Europe (AGILE) for gathering research data. The experiment was approved by the Committee of Ethics at Universitat Jaume I and the Committee of Ethics at the Institute for Geoinformatics, University of Münster.

## Notes

1 Location is a feature offered by Google to record the device location with a defined frequency.
2 Fit is a software component offered by Google which uses mobile phone sensors to identify the type of activity a user performs (i.e. walking, jogging, cycling or driving a car).

## References

Attard, Maria, Muki Haklay and Cristina Capineri. 2016. 'The potential of volunteered geographic information (VGI) in future transport systems', *Urban Planning* 1: 6.
Barratt, Paul. 2017. 'Healthy competition: A qualitative study investigating persuasive technologies and the gamification of Cycling', *Health and Place* 46: 328–36.
Berri, Akli, and Ricardo Daziano. 2015. 'Workshop synthesis: Caring for the environment', *Transportation Research Procedia* 11: 413–21.
Boss, Darren, Trisalyn Nelson, Meghan Winters and Colin J. Ferster. 2018. 'Using crowdsourced data to monitor change in spatial patterns of bicycle ridership', *Journal of Transport and Health* 9: 226–33.

Braun, Lindsay M., Daniel A. Rodriguez, Tom Cole-Hunter, Albert Ambros, David Donaire-Gonzalez, Michael Jerrett, Michelle A. Mendez, Mark J. Nieuwenhuijsen and Audrey de Nazelle. 2016. 'Short-term planning and policy interventions to promote cycling in urban centers: Findings from a commute mode choice analysis in Barcelona, Spain', *Transportation Research Part A: Policy and Practice* 89: 164–83.

Degbelo, Auriol, Devanjan Bhattacharya, Carlos Granell and Sergio Trilles. 2016. 'Toolkits for Smarter Cities: A Brief Assessment'. In *Ubiquitous Computing and Ambient Intelligence. IWAAL 2016, AmIHEALTH 2016, UCAmI 2016*, edited by Carmelo R. García, Pino Caballero-Gil, Mike Burmester and Alexis Quesada-Arencibia. Lecture Notes in Computer Science Vol. 10070, 431–6. Cham, Switzerland: Springer.

Degbelo, Auriol, Carlos Granell, Sergio Trilles, Devanjan Bhattacharya, Sven Casteleyn and Christian Kray. 2016. 'Opening up smart cities: Citizen-centric challenges and opportunities from GIScience', *ISPRS International Journal of Geo-Information* 5: 16.

Goodchild, Michael F. 2007. 'Citizens as sensors: The world of volunteered geography', *GeoJournal* 69: 211–21.

Gössling, Stefan. 2018. 'ICT and transport behavior: A conceptual review', *International Journal of Sustainable Transportation* 12: 153–64.

Grant-Muller, Susan, Ayelet Gal-Tzur, Einat Minkov, Tsvi Kuflik, Silvio Nocera and Itay Shoo. 2015. 'Transport policy: Social media and user-generated content in a changing information paradigm'. In *Social Media for Government Services*, edited by Surya Nepal, Cécile Paris and Dimitrios Georgakopoulos, 325–61. Cham, Switzerland: Springer.

Haklay, Mordechai (Muki). 2013. 'Citizen science and volunteered geographic information: Overview and typology of participation'. In *Crowdsourcing Geographic Knowledge: Volunteered geographic information (VGI) in theory and practice*, edited by Daniel Z. Sui, Sarah Elwood and Michael F. Goodchild, 105–22. New York: Springer.

Halko, Sajanee, and Julie A Kientz. 2010. 'Personality and persuasive technology: An exploratory study on health-promoting mobile applications'. In *Persuasive Technology*, edited by Thomas Ploug, Per Hasle and Harri Oinas-Kukkonen. Lecture Notes in Computer Science Vol. 6137, 150–61. Berlin: Springer.

Handy, Susan, Bert van Wee and Maarten Kroesen. 2014. 'Promoting cycling for transport: Research needs and challenges', *Transport Reviews* 34: 4–24.

Heinen, Eva, Bert van Wee and Kees Maat. 2010. 'Commuting by bicycle: An overview of the literature', *Transport Reviews* 30: 59–96.

Martinez Tabares, Carolina. 2017. 'Individual factors related to utilitarian urban cycling: Representations, motivations and perceived aggression'. PhD diss., Université Paris VIII – Vincennes-Saint-Denis.

Médard de Chardon, Cyrille, Geoffrey Caruso and Isabelle Thomas. 2017. 'Bicycle sharing system "success" determinants', *Transportation Research Part A: Policy and Practice* 100: 202–14.

Oldenziel, Ruth, Martin Emanuel, Adri A. Albert de la Bruheze and Frank Veraart. 2015. *Cycling Cities: The European experience*. Eindhoven, The Netherlands: Foundation for the History of Technology and Rachel Carson Center for Environment and Society.

Pajarito Grajales, Diego Fabian and Michael Gould. 2018. 'Mapping frictions inhibiting bicycle commuting', *ISPRS International Journal of Geo-Information* 7: 396.

Pajarito Grajales, Diego Fabian, Auriol Degbelo and Michael Gould. 2020. 'Collaboration or competition: The impact of incentive types on urban cycling', *International Journal of Sustainable Transportation* 14: 761–76.

Pooley, Colin G, Tim Jones, Dave Horton, Ann Jopson, Caroline Mullen, Alison Chisholm and Sheila Constantine. 2011. Understanding walking and cycling: Summary of key findings and recommendations. Accessed 19 September 2020. http://eprints.lancs.ac.uk/50409/1/Understanding_Walking_Cycling_Report.pdf.

Schrammel, Johann, Sebastian Prost, Elke Mattheiss, Efthimios Bothos and Manfred Tscheligi. 2015. 'Using individual and collaborative challenges in behavior change support systems: Findings from a two-month field trial of a trip planner application'. In *Persuasive Technology. PERSUASIVE 2015*, edited by Thomas MacTavish and Santosh Basapur. Lecture Notes in Computer Science, Vol. 9072, 160–71. Cham, Switzerland: Springer.

Sultan, Jody, Gev Ben-Haim, Jan Henrik Haunert and Sagi Dalyot. 2017. 'Extracting spatial patterns in bicycle routes from crowdsourced data', *Transactions in GIS* 21: 1321–40.

*Times of Malta*. 2018. Mass transit issues. Accessed 21 September 2020. https://timesofmalta.com/articles/view/mass-transit-issues.691708.

Weigend, Andreas. 2017. *Data for the People: How to make our post-privacy economy work for you*. London: Hachette UK.

Wiggins, Andrea, and Kevin Crowston. 2011. 'From conservation to crowdsourcing: A typology of citizen science'. In *2011 44th Hawaii International Conference on System Sciences*, 1–10. New York: IEEE.

Wunsch, Matthias, Agnis Stibe, Alexandra Millonig, Stefan Seer, Ryan C. C. Chin and Katja Schechtner. 2016. 'Gamification and social dynamics: Insights from a corporate cycling campaign'. In *Distributed, Ambient and Pervasive Interactions. DAPI 2016*, edited by Norbert Streitz and Panos Markopoulos. Lecture Notes in Computer Science, Vol. 9749, 494–503. Cham, Switzerland: Springer.

Yeboah, Godwin, and Seraphim Alvanides. 2015. 'Route choice analysis of urban cycling behaviors using OpenStreetMap: Evidence from a British urban environment'. In *OpenStreetMap in GIScience*, edited by Jamal Jokar Arsanjani, Alexander Zipf, Peter Mooney and Marco Helbich, 189–210. Cham, Switzerland: Springer.

# Chapter 9
# Geographic citizen science in citizen–government communication and collaboration: lessons from the ImproveMyCity application

Ioannis Tsampoulatidis, Spiros Nikolopoulos, Ioannis Kompatsiaris and Nicos Komninos

## Highlights

- The voluntary engagement of citizens in the collection of digital geographic information opens up new forms of interaction between citizens and the government.
- Citizens can become the living sensors of the city by contributing data about georeferenced non-emergency issues, which strengthens the sense of community, increases the responsiveness of local authorities, optimises budget and resource allocation, strengthens trust in government and promotes transparency.
- Creating an interactive space for exchange and communication can help citizens to feel that their voice is being heard by local authorities. Even if the response is not the expected one, momentum is built, promoting and encouraging citizen involvement in voluntary activities.
- Discovering hidden patterns through geographic data aggregation and visualisation, and translating these into knowledgeable insights, is feasible through interactive and customised dashboards such as those supported by the ImproveMyCity application.
- Integrating traditional communication channels in a modern, unified software platform provides local authorities with a more flexible, transparent and efficient system for receiving and managing non-emergency issues.

## 1. Introduction

Even though local authorities need to listen to and engage with their citizens, few channels exist for meaningful and modern direct communication and collaboration. The application ImproveMyCity (IMC) aims to promote a participatory culture in local communities and to act as an instrument for the concerned citizen whose quality of life can be improved by utilising their smartphone. IMC is an open-source scalable software solution, initially launched in 2012, in the context of the EU's 'People' research programme (FP7). The aim was to enable citizens to report non-emergency local issues, such as potholes, blocked bike lanes, street-light outages, broken sidewalks, discarded trash bins and other deficiencies in their community. The reported issues are automatically routed to the appropriate local authority department which monitors, manages and schedules remedial action. This voluntary engagement of citizens and the collection of digital geographic information data open up new forms of interaction between citizens and their government. The adoption of a non-emergency reporting tool such as IMC allows governments to improve their service delivery and accountability. It also encourages citizens to be more engaged and to play a more explicit role in becoming 'the eyes and ears' of their local authorities.

Citizen–government communication and collaboration tools such as IMC aim to provide cities with the means to advance governance from traditional practices towards modern, analytically driven and decision-making oriented means, under the 'smart everything' paradigm, as stated by Komninos (2018), which converges digital technologies, user engagement and collaboration networks. In the field of citizen participation (GeoParticipation), IMC is considered a local project which uses geospatial tools in order to support citizen participation, and belongs to the fifth wave of changes in the understanding and implementation of public participation where 'the role of citizens has changed from being objects of geographical research to becoming the creators of the agenda and decision-makers within their community' (Pánek 2016, 304). According to Haklay's (2013) taxonomy (see also Chapter 1 in this book), IMC is a geographic citizen science application which supports geographic data collection and reporting by citizens, and which is based on scientific tools and sensors provided by the most recent mobile devices. It in turn supports various other purposes, including improving local knowledge and promoting advocacy as well as more environmentally friendly and sustainable attitudes to mention just a few. IMC aims to collect geospatial data which are then made available to urban planners, decision makers, local administrations and communities to support evidence-based decision making. Inviting citizens to collaborate directly with local authorities strengthens the sense of community service, increases the responsiveness of authorities to people's real needs, optimises budget and resource allocation, strengthens trust in government, promotes transparency and allows the future of the city to be planned collectively. Furthermore, by using platforms such as IMC, citizens become aware of the sometimes hidden work and effort of government entities.

Nevertheless, the adoption of an open and transparent geographic citizen science application that allows citizens to report issues and track the actions of local authorities brings some challenges for governments and comes with some political concerns. During personal communication with the authors during the early stages of IMC adoption in 2015, local authority politicians expressed concerns about opening up the process to public scrutiny in case they were not able to satisfy citizen requests. They were worried that their popularity might be reduced and that this could influence election outcomes. This, however, has already been disproved. According to Buell, Porter and Norton (2018), there is significant evidence that citizens' perceptions about their governments and their willingness to engage can be reshaped and enhanced if the government's operational transparency is promoted.

This chapter describes the implementation of the IMC platform in the case of the municipality of Thessaloniki, Greece. The territorial extent of the city and its population growth, combined with continuous funding cuts, led the municipality to turn to more effective solutions to support its operational needs. The goal was to increase the city's efficiency and to achieve better results with fewer resources. Applying modern information and communications technology (ICT) solutions and engaging the citizens in this process led to the adoption of IMC, which has been gradually embedded in the daily operational capacity of the municipality.

## 2. Prioritising citizen engagement as best practice for local authorities: the case of Thessaloniki

The municipality of Thessaloniki is the second largest in Greece after Athens. According to the 2011 census, it has a permanent population of 325,182 residents. It comprises the central metropolitan area with its historical centre and surrounding areas, as well as districts that extend to the east of the city. Thessaloniki is divided into five local public administration areas. As an organisation, the municipality of Thessaloniki employs about 2,600 employees and consists of 2 directorates-general, 22 directorates and 8 independent departments as of early 2019.

## 2.1 Managing non-emergency issues pre IMC

According to Vasilopoulos (2017), prior to the introduction of IMC, the municipality of Thessaloniki relied on traditional administrative practices for reporting non-emergency issues. Citizens could report issues and complaints through the following methods:

- Completing a paper form along with optional relevant evidence such as printed photographs (this method required citizens to visit the municipal premises);
- Telephoning the Call Centre which forwarded the incoming request to the appropriate offices; or
- Completing an online form, which required including personal details with each submission.

These different channels for citizen–government communication were not linked, causing extra delays and difficulties in the overall management. A more detailed description of each method, as explained to the authors by the head of the e-Gov Department for the municipality of Thessaloniki and by the supervisor of the Citizen Transparency and Service Directorate, is provided below.

### *Paper form*

Every incoming document addressed to local authorities in Greece is handled based on specific procedures. In the case of the municipality of Thessaloniki, the steps include: (1) registration of the document and assignment of a reference number; (2) forwarding of the document to the relevant authority involved in the processing and resolution of the request; and (3) information to the appropriate overseeing services or departments and their supervisors, such as the Office of the Deputy Mayor, based on the content of the request. The document is registered in each of the respective directorates, with the Head Officer assigning the issue to the corresponding departments. The head of each department must undertake the necessary actions based on the current availability of resources in terms of personnel, logistics, active subcontracts and materials, and communicate back to the citizen, either orally or in writing, the actions taken. When the issue is resolved, the competent service closes the request by informing the relevant directorates and the citizen in writing about any outcomes. During the processing of requests, citizens are not aware of their progress unless they manage to locate the department and contact the civil servants managing the task, either orally or in writing.

### Call Centre

In Greece, there is no dedicated phone line at a national level for reporting non-emergency issues such as the 311 phone line in the United States that was created by the US Federal Communication Commission in February 1997 with the goal of relieving congestion on the emergency line 911 (Chatfield and Reddick 2017). In Greece, citizens call the relevant local authority offices directly or contact the municipal Call Centre if they are not aware of the appropriate department to contact. The Call Centre forwards the call to the relevant office, where a civil servant records the request and the contact details of the citizen, and the problem is recorded on paper. The report is communicated to the head of the respective department. The process follows the same workflow as with paper-form applications. Again, citizens need to contact the relevant local authority office directly to track the progress of their report.

### Online form

An alternative communication channel offered by the municipality is an online web form that allows citizens to submit their issues. Personal details such as identity card number, name, phone, email, home address, problem description and location are compulsory fields. In the online form submission, the data are sent via email to a member of the Call Centre who then forwards it to the relevant department. Depending on the content of the reporting issue, it might also be necessary to print the form and receive an official reference number from the Registrar. Upon submitting the online form, the only feedback to citizens is a confirmation message that the issue has been successfully reported. Citizens are not able to track their issues or be informed about their status unless they contact the responsible local authority office again. Because citizens are not familiar with the municipality's internal structure, identifying the correct office from which to obtain feedback is a cumbersome procedure.

Besides the bureaucracy and the lack of transparency, these approaches have some other significant drawbacks: (1) the requirement of a citizen's physical presence in the municipal premises (for the paper form); (2) filling in documents and forms which are extremely time-consuming; (3) delays caused by posting the hard copy documents and paper forms to the relevant departments; (4) the lack of feedback to citizens about the progress of their reported issues; (5) the lack of in-house knowledge concerning the amount and type of submitted issues; and (6) the lack of a centralised unified management system, which leads to delays and difficulties in setting up an optimal management strategy for resolving the reported problems. In addition, traditional approaches have a negative impact not only on the processing speed but also in terms of ineffective allocation of human resources, equipment, facilities and budget. This has also affected the overall medium-term operational planning of the municipality. These deficiencies made it obvious that the convergence and the integration of the existing communication channels under a unified platform was a necessity, and the need for a more flexible and efficient system for receiving and managing incoming non-emergency issues, such as IMC, was raised by the mayor and the 2015 City Council of the municipality.

## 2.2 Managing non-emergency issues post IMC

The municipality of Thessaloniki launched the web-based IMC application in June 2015. In February 2016, the IMC mobile app for Android and iOS was released and offered for free via the Google Play Store and Apple App Store, respectively. The introduction of the IMC app for smartphones was aimed at further promoting citizen involvement in local government, since it was expected that locating and reporting local issues whilst on the move, using a mobile's Global Positioning System (GPS) sensor and camera, would be more efficient for citizens. As of the first quarter of 2019, almost fifty thousand issues had been reported by approximately twelve thousand registered citizens in the municipality. Reporting has steadily increased each year (Table 9.1).

Approximately 63 per cent of the reported issues have been resolved, with their status declared closed, and more than 750,000 notifications have been pushed to citizens and civil servants, significantly advancing interaction between citizens and local government. Based on heat maps and geohash grid maps (see Figure 9.1), the spatial distribution shows that citizens from all neighbourhoods participate, with some areas being more active during specific periods or events. For example, every September, more issues are reported around the location of the Thessaloniki International Fair (TIF) due to more residents visiting the surrounding area.

**Table 9.1** Annual number of reports submitted through ImproveMyCity by registered citizens in the municipality of Thessaloniki, Greece

| Year | Yearly reported issues | Percentage change | Total number of reported issues |
|---|---|---|---|
| 2015 (2nd semester) | 1,880 | – | 1,880 |
| 2016 | 9,743 | +418.24 per cent | 11,623 |
| 2017 | 14,608 | +49.93 per cent | 26,231 |
| 2018 | 17,687 | +21.08 per cent | 43,918 |
| 2019 (1st quarter) | 5,990 | +35.69 per cent | 49,908 |
| 2019 (end-of-year estimation) | 17,970 | (estimated until the end of 2019 based on 1st quarter) | 67,878 |

Source: ImproveMyCity analytics reports.

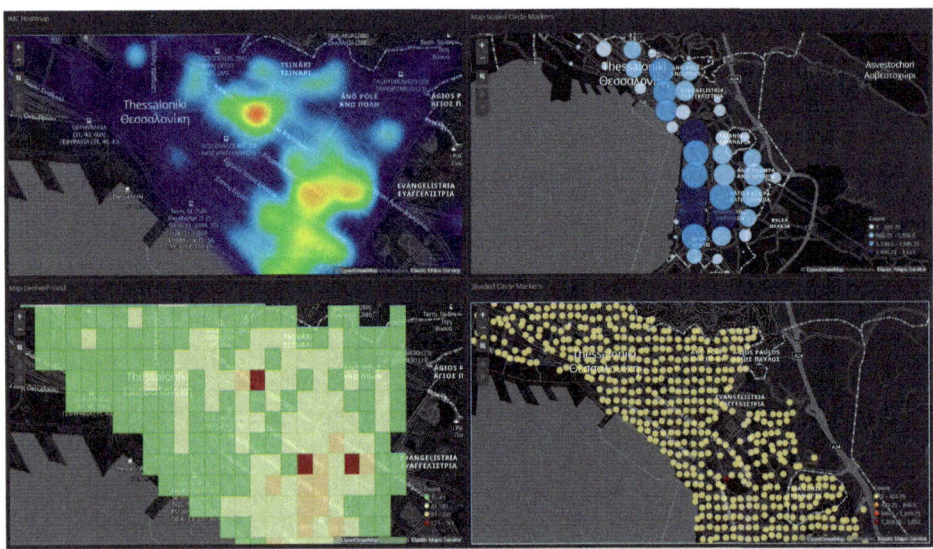

**Fig. 9.1** Interactive map-based visualisations from ImproveMyCity (IMC). Clockwise from top left: heat map, scaled circle markers, shaded markers and geohash grid. Exported by authors using the IMC analytics. Basemap © OpenStreetMap contributors. Visualisations created with Kibana from Elasticsearch BV ('Elastic').

The majority of registered users submitted up to two issues, but there are some outstanding cases worth mentioning. A single citizen reported 328 issues over a period of three years. There are also user groups and non-governmental organisations that are using common accounts to report problems in a coordinated effort, such as the group 'Friends of the Historical City Centre' which had reported 483 issues up to early 2019. Having a number of organised citizen groups using the application for reporting problems is encouraging and denotes strong community-driven user engagement, as broadly supported by research in the context of community informatics. Gurstein (2000), for example, shows how communities harness information and communication technologies to further their social development efforts. Another remarkable outcome is the fact that 93.5 per cent of the citizens whose reports received at least one positive vote by other users of the IMC system have either submitted additional new issues or responded to existing ones with at least one comment (these IMC features are detailed in Section 3). This indicates that interactivity and responsiveness, as suggested by Phillips and Orsini (2002), are critical factors in encouraging and strengthening citizen participation in applications such as IMC.

## 3. The IMC map-centric application

IMC is a software platform open to any individual or consortia to use and contribute to under the Affero General Public License. It was originally introduced in April 2012 (Tsampoulatidis et al. 2013) and since then it has had more than twelve thousand downloads. IMC is offered under the freemium business model, meaning that simple and basic services are offered for free but more advanced services and features are offered at a premium. Official support and customised versions were being offered in about 35 municipalities worldwide as of early 2019.

IMC is a modular platform, and its web-based front-end interface (an indicative instance is shown in Figure 9.2) supports different themes and layout templates so that administrators can customise the interface. Several themes are freely available, including the official one which is based on material design guidelines from Google (2019). There are themes available which are keyboard friendly, use high-contrast colours and are compliant with the WCAG 2.0 level AA accessibility standard (W3C Web Accessibility Initiative 2019). The front-end interface is map-centric, using either a Google Maps or OpenStreetMap background map. Small cities and rural areas usually prefer Google Maps due to better coverage of

**Fig. 9.2** The web-based IMC application front end.
Source: Powered by improve-my-city.com. Basemap © Google Maps.

road names and numbering. Maps are used to show the location of issues and to allow users to report the precise location of a new issue.

Citizens can use IMC not only to collect data but also to view reported issues on the map, to vote for them to show their support or agreement and to discuss reported issues with other citizens and public administration officials. Discussion is held publicly (under moderation) or privately. This is decided by the administrators during the initial set-up of the platform. Citizens receive notifications automatically on each action concerning their reported issues. In addition to reporting issues, citizens can suggest solutions and ideas for improving their neighbourhood and collect positive votes from other citizens to gain attention and support (Tsampoulatidis et al. 2013).

The organisational structure of each municipality is set a priori in the platform so that the reported issues are automatically routed to the relevant local authority department based on the geographic location of the reported issue and the chosen category. Commonly used categories specified by the administrators are: road maintenance, parks and green spaces, cleaning and recycling, public spaces, buildings and structures, and crime and antisocial behaviour.

IMC is available as an extension package for open-source content management systems such as Joomla! and WordPress and in the cloud as Software-as-a-Service (SaaS), based on a mix of Laravel, Node, React,

Kafka and Elasticsearch technologies. The service model of IMC is based on three main pillars: reporting, administering and analysis.

## 3.1 Reporting issues

Although anonymous reporting is possible, the overwhelming majority of IMC installations worldwide prefer citizens to be registered because local authorities prefer to interact nominally with their citizens, and obligatory registration eliminates spamming. Reporting is possible through the IMC web-based application and the mobile app. The Android and iOS apps follow the native design interface guidelines and principles as defined by Google (2019) and Apple (2019). Special characteristics for each operating system, such as the 'Peek Preview' for the iOS and the floating action button for Android, slightly differentiate the two versions. Login via social network accounts and customised authorisation schemes, such as the Italian Public System for Digital Identity, are both supported. The mobile app allows the uploading of multiple photos directly from the mobile device's camera and the geolocation is automatically pulled from its GPS sensor. Since the offline-first design approach (Biørn-Hansen, Majchrzak and Grønli 2018) is followed during the implementation phase, the mobile app can fully operate offline, and it is synchronised with the server automatically when Internet access becomes available.

Push notifications allow local authorities to notify citizens more directly, not only regarding their reports, but also about cultural events, local news and announcements. The first and second screenshots in Figure 9.3 depict the slight interface design differences between iOS (first) and Android (second). The third screenshot shows the clustered markers on the map and the use of custom marker icons which are based on issue categories. The progress and actions towards resolving an issue are displayed as a vertical timeline which is depicted in the fourth screenshot.

Apart from reporting new issues and browsing existing ones or reading notifications, news and announcements, citizens are also able to: (1) edit their own reports if their status is still unmodified; (2) comment publicly or privately; (3) vote positively by giving stars to other issues; (4) filter by area, status, category and ownership; and (5) apply text-based searches.

## 3.2 Administering issues

Management and routing of incoming issues is performed through the back-end administration interface that serves as an integrated management system and which also includes some basic interactive statistics.

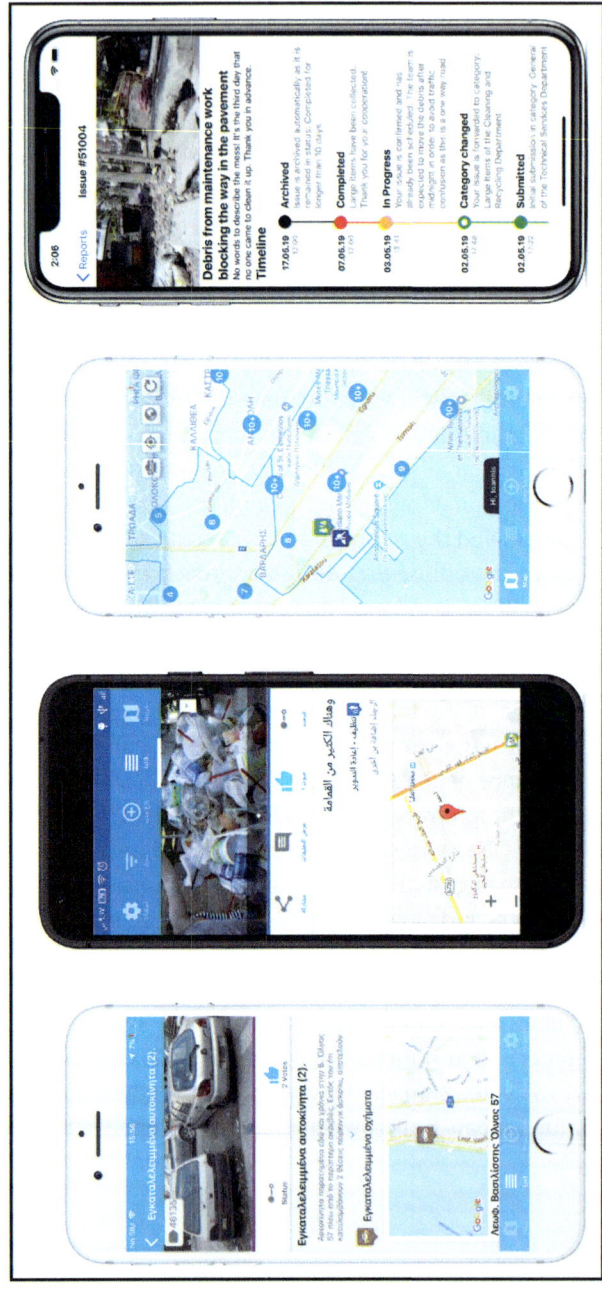

**Fig. 9.3** A set of screenshots from the IMC mobile app depicting the differences between iOS and Android. The map, issue details and timeline features are shown as well.
Source: Powered by improve-my-city.com. Basemap © Google Maps.

IMC, unlike other solutions, does not rely only on emails for communication. As Barbeau (2018) justifies, managing the issue life cycle via email is inefficient, since email is not the most suitable medium for tracking individual issues, including information about the assignees and the actions taken, until an issue is resolved. For this reason, each civil servant who belongs to one or more departments has their own credentials to log in to their personal administration user interface (UI). Based on available permissions, the administration UI is dynamically adapted, hiding unnecessary complexity and therefore making the interface more usable for civil servants. Direct communication with citizens is feasible through a rule-based automated notification system which is triggered by certain actions, for example on changing an issue's category or department, or upon updating the status of an issue. Indicative status levels include the following: submitted – acknowledged – in progress – solved – closed – archived. All actions taken on an issue are logged and displayed on a timeline.

The integrated reporting mechanism allows civil servants to check the overall condition of the city by applying composite filters by category, department, area, date range and other criteria. This integrated reporting mechanism can then be used to inform supervisors by sending them automated emails, notifications or, in some cases, a printed paper form. Scheduled to-do reports on a daily, weekly and monthly basis help civil servants assign jobs to their subcontractors more easily. Technically, the back-end provides the Representational State Transfer (REST)-based application programming interface – this is used by the mobile apps, the basic statistics module, the front-end and third-party systems such as e-protocol systems, geographical information systems, maintenance and asset management systems and others.

## 3.3 Analysis of issues

IMC further supports data analysis through visual analytics which employ map-based visualisations and spatio-temporal filters, graphs, interactive diagrams and tailor-made mechanisms to enable data fusion from public authorities. It also uses external open geospatial data sets in order to provide decision makers with valuable insights and to improve operational activities around the city. Discovering hidden patterns through geographic data aggregation and visualisations and translating these patterns into knowledgeable insights is feasible through the highly interactive and customisable IMC analytics dashboards. These dashboards offer local authorities the necessary tools to identify areas with increased numbers of reported issues and underperforming departments due to heavy workload or seasonal burden on city infrastructures, and generally facilitates the process of turning simple observations into well-advocated decisions. Figure 9.4 depicts an indicative dashboard from the municipality of Thessaloniki installation.

The target group of the IMC analytics are civil servants, as well as citizens who are interested in detecting patterns about the functioning of their city and wish to investigate additional information further that may help them interpret these discovered impressions. Apart from the IMC-produced data, the analytics dashboard can combine and merge external data sets that are typically generated by governmental departments, either regional or federal, and may be related to: (1) infrastructure such as location of schools, hospitals, parks and public facilities; (2) economic data such as average income per area; (3) health-related data such as satisfaction indicators of health-care services; (4) environmental data such as air and water pollution and electricity consumption; and (5) municipal police-related data such as crime level, traffic and parking spaces, and much more. This type of information can be of great interest, especially when viewed and explored in combination with citizens' submitted issues, and it may be used to reveal insights which would have been otherwise impossible or extremely difficult to observe. For instance, by overlaying geographic information related to the location of schools, hospitals and other critical infrastructures, or by comparing traffic-related information along with a density heat map of submitted reports of the relevant categories, critical decisions such as which issues to resolve first, how to plan city resources for next year or where to invest more for infrastructures can be made with better evidence (Vasilopoulos 2017). An indicative example of such data fusion is the combination of the location of reported issues concerning illegal parking with that of registered parking spots. This evidence would support the policymakers of the municipality in redefining the ratio between residents' and visitors' parking spots in specific areas of the city.

## 4. IMC as a geographic citizen science application

In this section, we provide anecdotal evidence from more than six years of use of the IMC application. The emphasis is on the obstacles and side effects of turning citizens into the living sensors of their city and transforming local authorities to be able to accommodate and care for their input smoothly.

**Fig. 9.4** IMC analytics dashboard with an indicative set of various visualisations also combining external data sources which are interactively interconnected. Dashboards can be embedded to existing websites or included in reports. Basemap © OpenStreetMap contributors, Elastic Maps Service.

## 4.1 A view from the perspective of politicians and civil servants

To overcome preliminary concerns associated with the adoption and utilisation of highly open and transparent geographic citizen science applications such as IMC which are used to report non-emergency issues and eventually incorporate them into existing governmental ecosystems, there is a need to recognise and take a series of actions. First, there is a need for firm political intention and commitment by higher-ranked municipal politicians during the earliest stages of the platform's initiation. Second, key personnel and supervisors need to be well trained on the various aspects of the system and should be able to transfer this knowledge, tailored to the ongoing workflow of each municipal department, to their colleagues. Therefore it is essential to adjust the platform rules and settings to operate consistently with the municipality's ongoing business organisational flow and structural hierarchy in order to make the adaptation of the new technology smoother. Third, civil servants need to appreciate that including the new platform into their day-to-day workflow will benefit them directly, since their work will become more efficient and robust. They also need to recognise that citizens are part of the solution, since they are actually acting as the living sensors of the city.

In order to ensure the long-term successful utilisation of applications such as IMC by local authorities, the following conditions are necessary: (1) the creation of a new centralised administrative structure in the form of an exclusive department which is responsible for the overall unified management and support of the application, (2) the creation and standardisation of relevant procedures and the adoption of an internal accountability process and (3) the continuous support of procedures, rules and structures by supervisors in order to ensure compatibility and integration with existing ICT systems.

## 4.2 The problem of duplication

One of the challenges in the utilisation of IMC is dealing with multiple reports on the same issue. Not only in the case of Thessaloniki, but also in the majority of IMC installations in Greece and abroad, citizens are reporting the same issue over and over again to put pressure on local authorities and to prioritise their requests. On top of this, multiple reports on the same issue are submitted with minor differences in the way the title and location of the issue are described, and it is not always clear that the same problem has already been reported. To deal with the problem of duplicated reports, during the submission of a new report, users are prompted to fetch from the server all issues in the same category within the last 10 days from a radius of about 20 metres and to return a list of potential problems reported at the same location. The list should contain the title, part of the description and a photo. Greyed-out markers of the suggested issues appear on the input map to denote potentially related reports. The citizen can then follow the link to an existing issue and give a positive vote to show their support instead of reporting the same issue again. It should be kept in mind, though, that acquiring a significant number of votes may take several days or weeks, leading to ineffective and delayed responses from civil servants. As a solution to this problem, Masdeval and Veloso (2015) propose applying a dynamic text analysis on the title and description of the reports in order to help prioritise issues more quickly. A priority level (e.g. normal, urgent) could also be set by citizens during submission and be evaluated by civil servants during the moderation phase.

## 4.3 Engaging the citizens

Promoting the application via advertising or word of mouth and making citizens aware of its benefits in order to convince them to join the platform is very important. Skarlatidou et al. (2019, 2) highlight that 'motivating users to remain active, ensuring that users can effectively use the applications, and guaranteeing satisfaction of use, should be central in the design and development' of citizen science applications. The retention of volunteers who use and contribute to IMC is essential. In the case of the municipality of Thessaloniki, as previously noted, several volunteer groups have created IMC accounts to coordinate their submissions, with a massive positive impact on the way the application is utilised in the long term. Focusing on similar initiatives should therefore be a top priority. In this vain, introducing new features such as push notifications about local events, participatory actions and official announcements alongside a mechanism to invite citizens to act on specific topics (e.g. weekend reporting on accessibility issues on your neighbourhood) can further enhance citizen participation and keep users engaged. Gamification is another technique which is broadly being used for motivating and retaining volunteers. Nevertheless, it should be noted that gamification in the context of IMC was not favoured by local authorities as a potential engagement strategy. It was therefore removed from the provided IMC functionalities.

## 4.4 Moderation

Although moderation complicates the workflow processes because an extra step is added to the authority's workload, almost all municipalities prefer to enable the moderation mechanism. When a citizen reports an issue, it does not become publicly available until a civil servant validates its content and ensures that the issue complies with the terms of use. As a matter of good practice, the reported issue becomes immediately visible to its owner, who is allowed to make changes (e.g. fix typos or include further information) while the status of the issue remains unchanged.

## 4.5 UI design recommendations

In this section, we provide a list of design recommendations that have emerged from the feedback we have collected from users of the IMC web-based application and mobile app through Google Analytics and in-app analytics and through user feedback from the open-source community, forums and live chat sessions.

When a new issue is submitted using the mobile IMC app, the initial geolocation is pulled from the device's GPS sensor. Users have the option to make minor adjustments to their location or to set a different location by tapping on the map. A marker is pinned to the new position set by the user, which is reverse geocoded to a readable address. Usability tests demonstrated that this method is not optimal. Due to the small real estate of mobile screens, tapping on the desired location is difficult because the thumb hides the desired area. The suggested method is to put a marker on the centre of the map which remains still, and let users pan the whole map until the static marker is on their preferred geolocation.

User feedback further revealed that IMC mobile app users prefer to submit new issues by following a series of steps rather than by filling in a single form. This way, scrolling is avoided, and form validation becomes more user friendly. Moreover, the problem of having to interact with a map which includes unknown road names is not usually an issue in big cities, but it is still a problem for rural areas. For this reason, the UI design should make it obvious that the address field can be also filled in with free text and not only via the map's reverse geocoding mechanism. Lastly, it is suggested that markers on the map are displayed in clusters based on the zoom level, especially when more than 50 locations are displayed at the same time. Another solution is to paginate the results to display fewer markers on the map.

## 5. Citizens become partners

Typically, the lack of a unified system for monitoring and managing incoming non-emergency issues prevents the formation of a comprehensive overview of the city because it limits civil servants to monitoring: the volume and type of incoming requests, the average response time, the time needed for an issue to be fully resolved, the geographic distribution of incoming issues and their relationship with other spatial characteristics (e.g. periodic city events), which can be revealed through the use of more sophisticated spatial analysis techniques. To address this, IMC supports the processing of the collected data and corresponding solutions so that civil servants (and citizens) can draw meaningful conclusions and gain a deeper insight into the way non-emergency issues are being managed at city level. This analysis includes the processing of quantitative characteristics of the submitted reports, such as the number of reports on specific geographic areas and the number of votes and comments, in order to reveal information about the intensity and frequency of reported problems. Citizens' subjective attitudes and perceptions can be further analysed using sentiment analysis, which is available as an IMC plug-in, and which can be used to analyse comments and textual descriptions of the reported issues. This in turn can help identify citizens' reactions as positive, negative or neutral.

## 6. Lessons learned

- Geographic citizen science in the context of non-emergency issues reporting has had a significant impact on the way the city of Thessaloniki is managed in both short-term planning and long-term policymaking.
- Using standards, such as openAPI, geoJSON, WCAG and OAuth, and applying well-accepted design techniques and guidelines such as the 'design for all' approach (O'Ferrall 2019) should be a priority in order to provide a user-friendly interface and seamless integration with third-party systems.
- Ensuring citizens can actually use the application and the retention of volunteers should be central concerns in the design choices and development process of geographic citizen science applications. IMC provides themes that incorporate best practice towards better user

experience such as keyboard-friendly interfaces and offline use of the mobile app.
- IMC relies heavily on principles of openness and transparency, which we found to be fundamental for the smooth operation and adoption for applications of this type. Municipalities which try to limit transparency towards practices they feel comfortable implementing (e.g. by displaying to users only their own submitted issues and not showing issues reported by other citizens) have a negative impact and should be discouraged.
- Less technically able citizens (e.g. older people) and those with disabilities who could potentially face problems using such applications should not be left behind. Besides applying best accessibility practice in the UI, alternative channels of communication such as telephone support must also be available. IMC introduced the 'Call Centre administrator group' which allows telephone operators to add issues that are reported by phone directly into the IMC workflow.
- High transparency can also have negative effects. It is essential that personal data are secured and kept confidential and that a series of actions for this is put in place. These actions include the censoring of photos containing sensitive information and the moderation of issues that directly or indirectly refer to physical persons or legal entities.

## References

Apple. 2019. Apple's human interface guidelines for iOS. Accessed 6 April 2020. https://developer.apple.com/design/human-interface-guidelines/ios/overview/themes/.

Barbeau, Sean J. 2018. 'Closing the loop: Improving transit through crowdsourced information', *Transportation Research Record* 2672: 224–34.

Biørn-Hansen, Andreas, Tim A. Majchrzak and Tor-Morten Grønli. 2018. 'Progressive web apps for the unified development of mobile applications'. In *Web Information Systems and Technologies*, edited by Tim A. Majchrzak, Paolo Traverso, Karl-Heinz Krempels and Valérie Monfort. WEBIST 2017. Lecture Notes in Business Information Processing Book Series, Vol. 322. Cham, Switzerland: Springer.

Buell, Ryan W., Ethan Porter and Michael I. Norton. 2018. 'Surfacing the submerged state: Operational transparency increases trust in and engagement with government'. Harvard Business School Technology and Operations Mgt. Unit Working Paper No. 14-034.

Chatfield, Akemi T., and Christopher Reddick. 2017. 'Customer agility and responsiveness through big data analytics for public value creation: A case study of Houston 311 on-demand services'. *Government Information Quarterly* 35: 336–47.

Google. 2019. Google's material design principles. Accessed 21 September 2020. https://material.io/design/introduction/#principles.

Gurstein, Michael. 2000. *Community Informatics: Enabling communities with information and communication technologies*. Surrey, BC: Technical University of British Columbia.

Haklay, Mordechai (Muki). 2013. 'Citizen science and volunteered geographic information: Overview and typology of participation'. In *Crowdsourcing Geographic Knowledge: Volunteered geographic information (VGI) in theory and practice*, edited by Daniel Z. Sui, Sarah Elwood and Michael F. Goodchild, 105–22. New York: Springer.

Komninos, N. 2018. 'Internet platforms, disruptive innovation and smart growth'. Presented at 20th Scientific Conference of the Hellenic Regional Association. Regions at a Turning Point. Athens, Greece, 4–5 June 2018.

Masdeval, Christian, and Adriano Veloso. 2015. 'Mining citizen emotions to estimate the urgency of urban issues', *Information Systems* 54: 147–55.

O'Ferrall, Elizabeth. 2019. 'Accessibility of products and services following a design for all approach in standards'. In *Proceedings of the 20th Congress of the International Ergonomics Association (IEA 2018)*, edited by Sebastiano Bagnara, Riccardo Tartaglia, Sara Albolino, Thomas Alexander and Yushi Fujita. Advances in Intelligent Systems and Computing, Vol. 824. Cham, Switzerland: Springer.

Pánek, Jiří. 2016. 'From mental maps to GeoParticipation', *The Cartographic Journal* 53: 300–307.

Phillips, Susan D., and Michael Orsini. 2002. 'Making the links: Citizen involvement in policy processes'. Discussion Paper, Canadian Policy Research Networks. Ottawa, Canada.

Skarlatidou, Artemis, Alexandra Hamilton, Michalis Vitos and Muki Haklay. 2019. 'What do volunteers want from citizen science technologies? A systematic literature review and best practice guidelines', *Journal of Science Communication* 18: 1–23.

Tsampoulatidis, Ioannis, Dimitrios Ververidis, Panagiotis Tsarchopoulos, Spiros Nikolopoulos, Ioannis Kompatsiaris and Nicos Komninos. 2013. 'ImproveMyCity – An open source platform for direct citizen–government communication'. Presented at the 21st ACM International Conference on Multimedia – Open Source Software Competition, Barcelona, Catalunya, Spain, 21–25 October 2013.

Vasilopoulos, A. 2017. 'Study on web-based application for citizen's request in the municipality of Thessaloniki'. MSc diss., University of Macedonia.

W3C Web Accessibility Initiative. 2019. Web Content Accessibility Guidelines (WCAG). Accessed 5 March 2020. https://www.w3.org/WAI/standards-guidelines/wcag/.

# Part III
## Geographic citizen science with indigenous communities

# Chapter 10
# Developing a referrals management tool with First Nations in northern Canada: an iterative programming approach

Jon Corbett and Aaron Derrickson

## Highlights

- There is a need for an open-source digital tool to support Canadian First Nation communities to communicate more effectively with industries in regard to proposed resource extraction projects.
- Gather was the result of an iterative, responsive and 'just-in-time' approach to co-design a geographic citizen science data gathering and management tool.
- The Gather software and the latest version of the tool diverged from the programming team's initial ideas as a result of the co-design process.
- Implementation challenges tended not to be technical or design focused. Rather, they related to the individuals and organisations involved in the project, and the sensitivity of the information being handled.

## 1. Introduction

All proposed resource development projects in Canada, whether a new mine or a major new piece of infrastructure, are required to consult with the indigenous parties that will be impacted by the work. This referral process has emerged as the result of a series of precedent-setting court cases which found that the Provincial and Federal Crown[1] have a legal duty to consult and, where necessary, accommodate First Nation[2] communities (Harris 2006) when development activities are being carried out within their traditional territories (Ecotrust Canada 2017). Regardless of scale, the First Nation who has rights on the land where the development will take place is given the opportunity to examine the proposed project for potential adverse environmental, economic, social, heritage and health impacts that may occur during the project life cycle. The process is rigorous and often involves extensive documentation provided by the proponents, who in turn draw on the expertise of specialist consultants. In British Columbia, once a referral is submitted, the First Nation must respond with written comments within a 20-day period. If the review cannot be completed within this time frame, the government notes that the proponent has fulfilled their obligatory duty to consult with the community.

With the continued and growing presence of large-scale resource development (particularly mining and forestry operations) in northern Canada, First Nation communities are becoming overwhelmed by the obligation to manage, review and respond to these impact assessment proposals (Power 2017). First Nation leaders and community lands departments recognise a clear need to research, design, develop, implement and evaluate affordable tools that could streamline the duty to consult between government, proponents and communities, as well as facilitate community decision making related to the referral process. In 2015, several First Nation communities and their representatives approached the Spatial Information for Community Engagement (SpICE) lab at the University of British Columbia seeking the development of a web-based tool that might be used to improve their capacity to understand the extent of all proposed resource developments, as well as manage and respond to the referrals. The project also involved the co-design, development and implementation of a mobile app to enable community members to collect spatial information on their contemporary use of lands and resources. This information in turn can be accessed and viewed using a map interface by community lands department members and community leaders in order to inform lands-related decision-making processes. This chapter describes the development and initial implementation of 'Gather: The referral management tool' that we co-designed to address these needs.

## 2. Supporting First Nations' land-management needs

### 2.1 Context

Speaking in general terms, First Nations throughout Canada have a unique, respectful and stewardship-focused relationship with the lands on which they live (Berkes, Folke and Colding 2000; Turner, Ignace and Ignace 2000). This relationship has developed over millennia. It directly contributes to the social, cultural, economic, subsistence, health and spiritual well-being of First Nation communities throughout the country (Berkes 2017). Traditional knowledge, languages, cultural practices and oral traditions are all connected to the land (Alfred and Corntassel 2005). European colonisation and settlement in Canada have profoundly challenged this relationship. Over the past 125 years, resource industries have harvested and sold natural resources from First Nation lands (Angell and Parkins 2011). These businesses have become major drivers for the Canadian economy and employment.

The Royal Proclamation, signed in 1763 by King George III, has helped shape the legal relationship between the Crown and First Nations. The Proclamation implicitly recognises First Nations as owners of their lands, and in doing so, it provides the basis of the legal recognition of their rights to land (Borrows 1994). In 1973, the Supreme Court of Canada, through the Calder decision, recognised that aboriginal title existed in law, and therefore could be enforced (Foster, Raven and Webber 2011). That decision was followed in 1997 with the Delgamuukw and Gisday'way, as well as Sparrow and Tsilhqot'in, decisions that found that aboriginal title was something substantive and robust and should be considered 'a right to the land itself' (Morse 2017). These court cases are significant because they mean that the provincial and federal governments now have a legal duty to consult and, where necessary, accommodate First Nation groups when development activities are being carried out within their traditional territories.

It is important to note that although the duty to consult can be conflictual in nature (Zietsma et al. 2002; Hayter 2003), most levels of government as well as industry leaders have accepted that consultation with First Nations is a legal, necessary and important aspect of doing business with First Nations (Joseph 2015) on First Nation territory. Furthermore, many businesses conduct their own engagement process with indigenous communities as a part of their project planning before they apply for regulatory approvals (Canadian Chamber of Commerce 2016). This often involves establishing relationships with community decision makers and including them in initial project planning processes and developing impact and benefit agreements (Gogal, Reigert and Jamieson 2005; Caine and Krogman 2010). This is done to help avoid the delay or cancellation of projects that might occur if consultation only takes place during the formal referral process.

The consequence of this legal requirement to consult is that many small First Nations, often operating with limited staff and resources, have been overwhelmed by the number of referrals that they receive daily. It has proven to be a major logistical and administrative challenge to organise, prioritise, analyse and respond to these referrals in a meaningful and effective way (Ecotrust Canada 2017). This is especially the case in smaller, more geographically remote communities where most resource development projects take place in Canada. For example, in 2014, Saulteau First Nations (SFN), a community in northern British Columbia, received more than 3,500 applications referred by federal, provincial and local governments. The current procedural requirements within the regulatory process oblige SFN to assign significant resources to manage, review and respond to each referral. This process also necessitates understanding the spatial extent of the proposed development intervention and how it potentially impacts the traditional and contemporary uses of the land. Presently, the capacity for SFN to acknowledge the infringement of their indigenous and treaty rights from a proposed development in an effective and timely way is both limited and costly.

A number of proprietary software tools have been, and continue to be, developed in response to this challenge. A report written in July 2017 by Ecotrust Canada and the Aboriginal Mapping Network identified eight software applications used in 44 different communities around the province of British Columbia. Most of these tools included a mapping component to the software, but none directly linked their software platforms to community-contributed geographic citizen science information. The report further identified several critical challenges to implementing and using this software. These were related to access to training, usability, licensing costs, updates and software bugs. There are currently no open-source tools to facilitate the management of the referral process at the community level, nor are there any examples of where software has been co-developed from inception with the communities who use them.

In 2017, community members from the Wabun Tribal Council (WTC), SFN, the Firelight Group and the SpICE lab began to co-develop a web-based collaborative tool referred to as 'Gather: The referrals management tool' (hereafter 'Gather'). From its inception, we intended Gather to be an open-source, free to implement, easy to set up, intuitive to use, extendible and integrated contemporary geographic citizen science tool. It was designed to capture data that could provide evidence to the government and industry that community members are still active land stewards and that resource extraction activities would impinge on their current, and not just historical, livelihood activities. At the time of writing this chapter, we have finished an initial draft of Gather and its associated smartphone data-collection applications. Because of this, we can only talk about the development of the tool and share some of the challenges and barriers to the tool's design, development and pilot launch. We do not yet have any specific examples of interaction and uptake in the field.

## 2.2 Partners

The WTC is the regional representative for the First Nations of Brunswick House, Chapleau Ojibway, Flying Post, Matachewan, Mattagami and Beaverhouse. These communities are located in north-eastern Ontario (see Figure 10.1). The WTC's Board of Directors comprises the chiefs of the six communities. The WTC work in the fields of health, education, economic development and resource development. WTC staff are responsible for negotiating mining development agreements in collaboration with community leaders and acting as a point of contact for project proponents and as a liaison in communications between government, industry and the communities.

SFN are located in Moberly Lake, northern British Columbia (see Figure 10.1) and are a Treaty 8 First Nation. Treaty 8 territory covers approximately 840,000 km$^2$ in what is now northern Alberta, north-eastern British Columbia, north-western Saskatchewan and the southernmost portion of the Northwest Territories. The Treaty provides the SFN membership with (among other things) the constitutionally protected right to hunt, fish and trap, and to gain a livelihood from the lands and resources within Treaty 8 territory. As SFN notes in its 2015 Comprehensive Community Plan, 'Practicing our Treaty Rights provides our people with the means for a rich spiritual, social, and economic life. The land and the activities carried out upon the land connect our people to their past and provide them with the resources they need to build a healthy, stable, culturally rich future' (Saulteau First Nations 2015, 7). Although SFN were a key project partner during the initial stages of the project design, staff turnover has meant that they are no longer involved in the ongoing development of Gather.

The Firelight Group are a consulting group that works with indigenous and local communities throughout Canada and internationally. They

**Fig. 10.1** Location of Gather project partners.
Source: author.

work in collaboration with many First Nation communities to provide research, policy, planning, negotiation and advisory services. Their work focuses on culture, health, socio-economics, ecology and governance to support the rights and interests of indigenous communities. Firelight are driven by the principles of participation and capacity building. They funded the initial stages of developing Gather through their Social Return fund.

The SpICE lab, based at the University of British Columbia, Okanagan, partners with Canadian and international communities to co-develop, deploy and evaluate digital participatory mapping tools. The lab's partnerships are framed within the practice of community-based research and represent a collaborative enterprise between researchers and community members. The SpICE lab's research programme explores questions related to how digital mapping technologies and associated processes impact indigenous and vulnerable communities, and whether these technologies can effectively capture – and add value and authority to – local knowledge.

## 2.3 Project methodology

Because of the nature of the partners involved, the project is grounded in the paradigms of indigenous methodologies (Kovach 2010; Smith 2013) and community-based research (Israel et al. 2001; Minkler and Wallerstein 2011). This means that all aspects of this project (design, evaluation, extension and outreach) are conducted in a reciprocal and an empowering manner; the outcomes are of tangible benefit to the partnering First Nations; and community members feel a strong sense of ownership over the co-design process and the final technology. In concrete terms, this has meant that the programming team responded directly to the needs and concerns of our community partners. The community partners became the principal architects and designers for Gather. This co-design approach involved all actors in the project. We did not record any of our meetings, as our intent was not for our process to be considered a research exercise. Rather, we focused on the design of the tool itself and its functionality. For this reason, we are not interested in conveying the personal thoughts or details of the meetings, or in the geographic citizen science data gathered using the tool. Once the tool is operational, we will consider evaluating its usefulness, in which case we will be bound by our university's research ethics board requirements. However, we are not yet at that point of deployment in the community.

Over the summer of 2017 (May–August), we co-designed and co-developed the first iteration of Gather with the WTC and the lands managers from the Beaverhouse First Nation, Brunswick House First Nation, Chapleau Ojibway First Nation, Flying Post First Nation and Matachewan First Nation, as well as the referrals team from SFN. We held a five-day workshop in Timmins and a two-day workshop in Chetwynd with representatives from the lands department. We used the materials produced from these sessions to co-design the first draft of Gather. We continued to meet regularly through videoconferencing. We also met for a third one-day workshop in Winnipeg, which was held prior to the Indigenous Mapping Workshop (November 2017). During these community workshops, participants discussed their specific concerns and needs, including how the software should look and function to address the varying requirements of each community.

Our design and development process was based on a 'just-in-time' iterative programming approach (WikiWikiWeb 2005). In other words, during the workshops, we would focus on developing only the immediate requirements and functionality identified as being important by the participants. This meant that in the mornings, we discussed and planned the application's functionality and workflow. During the afternoon and evenings, the programming and digital design team coded Gather. Each day, the team reported back, were offered additional suggestions and then adapted the draft tool to meet those further recommendations. By the end of the workshops, we had developed a partially functioning version of Gather. As a result of this approach, participants were considerably invested in the design of the tool and continued to work with the programming team on its subsequent (and ongoing) development. Overall, it was a successful and fulfilling – though at times exhausting – experience in community engagement.

## 3. Creating Gather

### 3.1 Initial intent

Prior to commencing the design and development workshops, we envisioned the creation of a web-based platform that would focus on two user groups: members of First Nation lands departments and industry proponents. The tool would allow First Nation lands department members to manage existing community traditional land-use data and also to view the spatial intersection of proposed development projects and traditional and contemporary community land uses. The software and associated databases could sit on University of British Columbia, community-located or cloud-based servers, depending on the server management capacity within the community and the availability of high-bandwidth hardware. Industry proponents would upload all the project proposal documentation, including letters and permits, as well as a SHP (a common spatial file type used by GIS software) or KML (the spatial file type used in Google Maps and other web mapping applications) file delineating the spatial extent of the project. Lands department members would then overlay the industry spatial extent file on top of their traditional use data and thus have a clear visual representation of the impact of proposed projects on the social, cultural, economic, subsistence, health and spiritual well-being of the community and its membership (see Figure 10.2).

We wanted the user interface (UI) to be straightforward and intuitive, and for the tool itself to facilitate a semi-automated workflow built around a structured set of predetermined steps. As specific steps are completed in the lands department's workflow, automated messages are sent

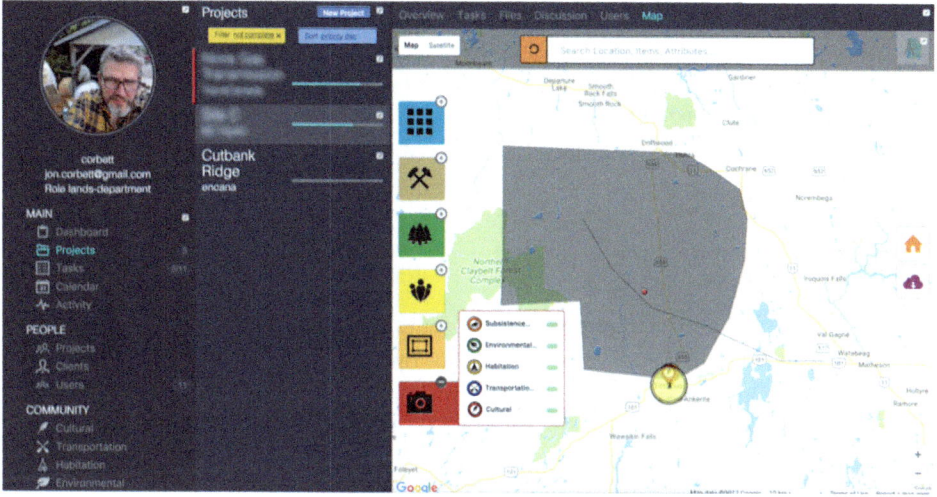

**Fig. 10.2** Hypothetical example of the extent of an industry-proposed project, represented by the grey polygon, overlaying community-contributed geographic information.
Source: Gather app. Credit: Spatial Information for Community Engagement Lab (SpICE). Basemap © Google Maps.

to the industry proponent who can monitor their proposed project as it moves through the community decision-making process. Thus, both the lands department – through a more efficient referral management system – and industry – through having an increased ability to monitor their individual proposals – benefit from the tool.

Our design and development workshops largely supported our initial ideas, with one notable exception. Lands department members clearly recognised the need to include all members of the community in the project in order to encourage their engagement in the referrals process and to contribute their own information related to contemporary community use of the lands and resources. This helped to articulate clearly the need for Gather to be usable for three unique user groups, each with their own distinct set of needs and ways in which they interact with the system. In other words, there was a need for Gather to enable:

- Community members to volunteer and selectively share information pertaining to their contemporary use of the land through an intuitive mobile phone-based geographic citizen science data-collection app;

- Community lands department managers and technicians to review, manage, delegate and respond to referrals, and produce reports that help clarify how proposed projects impact both traditional and contemporary land uses; and
- Industry to standardise their referral submission process and track the review of their referrals within the community's workflow.

### 3.2 UI design

Gather's initial UI design was loosely built around previous online participatory geographic applications developed over the past 12 years in the SpICE lab. The referrals document management functionality was built directly into a map interface and made accessible through a single web page view (see Figure 10.3). The specific interactive functions available through the UI varied according to the three unique user groups (lands department, community member and industry). At its most basic level, this meant that lands department members would see and interact with a set of controls, queries and functions that supported the management and response to referrals; these controls were not visible to the other two user groups. Lands department members could also view commu-

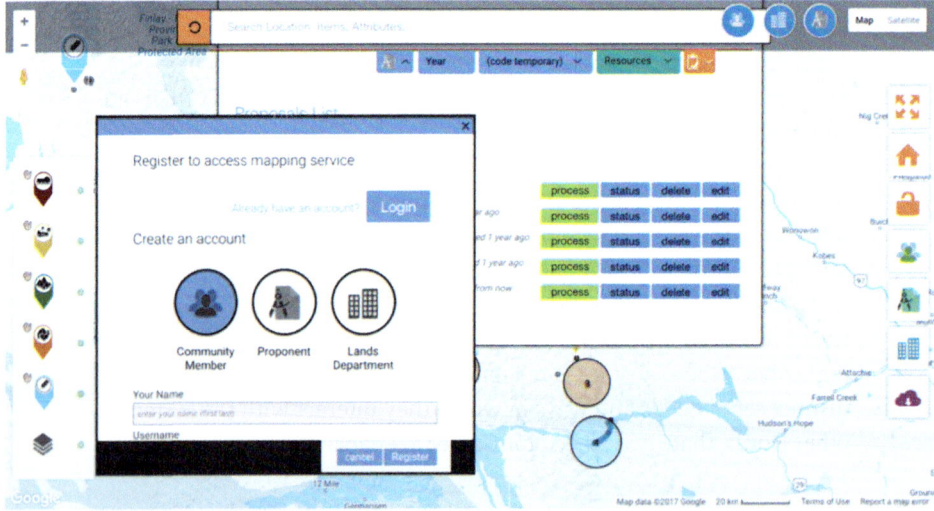

**Fig. 10.3** Referrals management tool – initial map-centric information management interface.
Source: Gather app. Credit: SpICE. Basemap © Google Maps.

nity geographic citizen science contributions. Community members could see and comment on their own and other community member contributions, but they could not interact with the referral data. Industry could only view their own referrals and no other information.

The programming team's prior experience developing online mapping tools meant that our initial draft of Gather was map-centric. Referral data were managed entirely through the map interface, and data-management tools were associated with icons and drop-down menus built around the outside edges of the map (see Figure 10.3). After the community co-design workshops, the UI was redesigned to focus on projects, tasks and deadlines (see Figure 10.4). The map still plays an important role, but its significance is muted, and it is only accessible when viewing information about a specific project. This change was made because of the limitations of the map interface for filtering and viewing large numbers of proposed projects (sometimes in the thousands).

## 3.3 Mobile app

At the outset of the project, the programming team were focused on developing a tool that would be exclusively browser based and used primarily by lands department members and secondarily by industry and government proponents. This browser-based approach meant that the tool would function on any operating system and not require any specialised software, and that it would support the ability to share the same information seamlessly between different user groups, as well as have a low entry cost. After working directly with the lands managers, we modified the system to include the development and integration of a mobile geographic citizen science app to enable community members to volunteer (i.e. capture and share) examples of contemporary land use from the territory using their own smartphones. Data could be collected by community members at any time of the year and act as a repository of current and relevant community land-use practices. This required developing both iOS and Android apps that could be used by resource managers, hunters and community members to record and share their activities.

Lands managers considered this to be of the upmost importance because the data could be used to provide evidence to the government and industry that community members are still active land stewards and that any resource extraction activity would impinge on their current, not just historical, livelihood activities. However, the design, development and use of a geographic citizen science app to support data collection in northern Canada present their own sets of challenges. These include

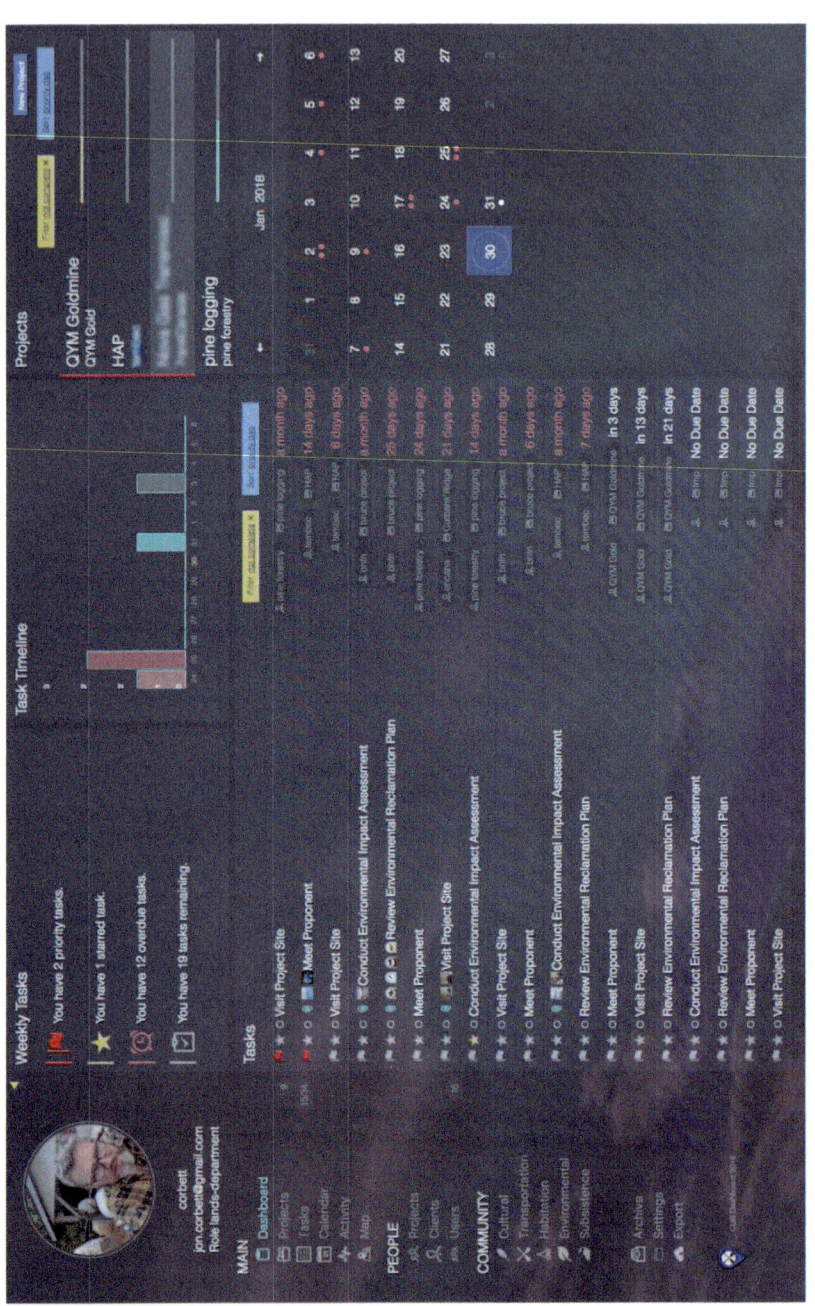

**Fig. 10.4** Redesigned referrals management tool interface.
Source: Gather app. Credit: SpICE.

limited (and largely non-existent) connectivity, the need to program for a broad range of devices (due to both the age of the devices and the presence of multiple operating system versions) and issues related to testing, training and app distribution. We recognise that these are common issues in information and communications technology for development projects around the world (Rashid and Elder 2009; Aker and Mbiti 2010) that have been addressed through the development of tools, such as Sapelli or ODK (Open Data Kit), which are open source, offer mobile data collection and can operate offline. However, we chose to develop our own suite of apps, partially to develop the long-term skills and capacity in the SpICE lab to offer these services to our partners, but also to ensure data consistency and interoperability with our project servers, databases and design components.

## 4. Challenges to the community's interaction with the tool

Throughout the design, development and initial implementation stages of Gather, we have both realised and constantly been reminded of the needs and challenges of working with multiple users handling sensitive indigenous information in digital form in a politically contentious environment. It should be further noted that each community partner has differing experiences with industry proponents, which in turn means that each partner has a specific set of needs; this is reflected in the variations in functionality required for the tool. We discuss these barriers and challenges in this section.

### 4.1 The challenge of getting people to use the software

The project has benefitted from a high level of participation by First Nation land managers and community leaders. These groups were closely involved in the co-design of both the overall project as well as the specific software functionality. They are champions of the software both within the community as well as in other indigenous communities in Canada through presenting the tool at gatherings, such as the Indigenous Mapping Workshop. However, anecdotal evidence shows that it is hard for them to introduce and encourage the use of mobile geographic citizen science apps within their own communities. This is partially due to limited connectivity and many people in remote areas not having access to high-speed mobile Internet. However, it is also difficult to persuade people that their contributions are important and useful for lands department managers and can be used to inform community decision making. This remains a constant challenge. Furthermore, many of the community members that continue to use the land for hunting, fishing and medicine gathering tend to be older and are often less likely to use a mobile phone. There is a need for the champions to communicate better the purpose of data gathering to community members, as well as to express the importance of the community's contributions and how their data will be used.

## 4.2 Tension with commercial operations

One of the challenges we had not anticipated at the beginning of the project was the tensions that emerged between our development team and commercial referral management software providers. One business contacted us several times about Gather's functionality, and expressed concern about the composition of our project partnerships. They were especially uneasy about the role of the Firelight Group in the project, who they felt had a conflict of interest and an unfair business advantage through their involvement with our university-based programming team. This did help us understand that within the university, we have advantages regarding institutional support and access to resources that are not available to many small businesses. The business was also concerned that the software being made available as an open-source product that is freely available might undermine the business model of small companies offering referral management tools to other First Nation communities. This tension is hard for us to reconcile. On one hand, we understand the importance of not undermining small business, but on the other, we recognise that many small rural First Nations are overwhelmed with the number of referrals, and they lack the financial and human resources to respond effectively to this challenge. After discussions between project partners and the university, we held true to our original objectives of making Gather open source and freely available. We also invited the concerned business to examine and use our open-source software to augment their own offerings.

## 4.3 Difficulty of creating generic software applications

Given the varying histories of the relationships between indigenous peoples, industry and the Crown, as well as the nature and/or scope of differing proposed projects, it is unrealistic to expect that the referrals process will be same for First Nations throughout Canada. The WTC, for example, had a close and long-term working relationship with several mining interests in their region. Therefore, they expected Gather to be a useful tool to store and access existing project referral documentation and engage community members through the mobile app. They did not view Gather as a tool to structure their relationship with existing partnering industries. In contrast, SFN deals with a far greater number of industry proponents, many of whom they do not have a relationship with. They hoped that Gather would provide a structured approach to dealing with industry. The SFN Lands Department already had a clear set of regulations and protocols which systematised the ways in which they managed and responded to referrals. SFN wanted Gather to emulate these protocols in order to create greater efficiencies in their dealings with industry proposals and the Crown.

In past projects, the SpICE lab has focused on custom code design for projects that involve a limited number of partners. However, because of the varying circumstances and different needs within different communities involved in this project, there was limited agreement about the specific functions required to manage large numbers of referrals. For some of our project partners, Gather provides a rudimentary set of limited functions; for others, the software provides too many functions, making it overly complicated and difficult to implement. It has become important to be able to articulate and communicate clearly who the software is targeting and how they might use it. We decided it should be aimed at small communities with limited resources; it is not designed for large communities that receive many thousands of referrals a year.

## 4.4 Mistrust between representatives and communities

There is often a naivety among university researchers working with First Nations, particularly in regards to understanding ongoing community–government tensions, such as those found when addressing issues of land claims and land-use studies, as well as internal community tensions related to the identification of community priorities through to the nature of the information that members should be volunteering. This complexity increases as more communities become involved in the design and development of the technology. The programming team did not have the luxury of remaining uninvolved in these tensions. We often had to engage in soft mediation during decision-making sessions, which usually involved offering technical solutions to human concerns. This was particularly the case around how community information is managed and accessed and by whom.

## 4.5 Limitations of our map-centric approach

As noted above, this project required us to rethink our cartographic perspective. The referrals management tools were originally built directly into a map interface and made accessible through a single web page view, but the space constraints of this UI were too limiting. Nor did the tight UI allow for the management of specific tasks related to projects. Our redesigned UI tried to mimic the steps of the SFN's existing workflow and tasks associated with managing each referral using a series of tabs with the addition of a task management tool (see Figure 10.4). Ultimately, this meant that the interface was considerably more textual in nature, and the role of the map was diminished. This forced us to rethink our approach to the significance of the geographic component and the geographic citizen science data. This has also helped us to increase the range of software services that we can offer in other community-engaged research projects.

## 4.6 Challenges of including government and industry

At the beginning of the project, we envisioned that Gather would be as useful for government and industry proponents as for First Nations. The tool currently supports the uploading of industry referrals as both digital text documents and spatial files (SHP and KML). However, we have found it difficult to secure the participation of government and industry in the project. Often, these actors are innovation averse and tied into their own existing data-management systems. However, perhaps the explicit role reversal of Gather – in other words, First Nations taking control of the means and processes by which information is shared – might act as a further challenge to these proponents' involvement. In the next phase, we will more directly consult industry and develop strategic partnerships to ensure that the interface and workflow align with their existing data protocols and processes.

## 5. Conclusions

When we set out to design Gather, we did not intend that this would be a commercial undertaking. Our motivations were driven by the clear need to develop open-source software that can be used in small rural First Nation communities to help them address the overwhelming nature of the referrals process. Many of these communities do not have either the financial or human resources to be able to deploy and use often complex proprietary software. Our community partners also clearly recognised the need for innovative approaches to include their membership base in contributing information about how they currently use their lands and resources. This was a vital step in informing referral decision making, but also served to include the broader community in land-management processes. We therefore feel that the components of Gather, including an intuitive interface, the associated mobile apps and an industry upload section, mean that the tool is relatively easy to use once it is set up. However, we recognise that several critical obstacles to successful and long-term usability remain. The principal issue is that of connectivity. Many areas in northern Canada remain outside of mobile phone coverage. Many communities find it hard to maintain servers and server architecture within the community. Therefore, if communities want to manage the referrals effectively, they have often turned to commercial solutions. However, the cost of these solutions, as well as other challenges related to usability, make it a difficult choice for many small communities to make.

Despite the barriers identified, the project has drawn considerable interest among our project partners, as well as other First Nation communities throughout Canada. We anticipate having an operational version of the software available to communities by mid-2020 using an open-source licence. We continue to be somewhat concerned about the ongoing sustainability of the software because, as with any open-source project, there will be a need to update both the usability as well as the security required by the software in the future. We will therefore likely consider some form of donation model from our users. However, this will not be obligatory. In the meantime, we have secured grant funding for the next three years (until 2022) and will continue to improve functionality and scale up and make Gather available to whomever might want to use the software.

## 6. Lessons learned

- Be prepared to let go of your preconceived notions of what is needed. Just because it worked in the past or in another community does not mean that it will work in new contexts.
- Be sufficiently agile to develop tools that suit the needs, capacity and limitations of the community with whom you work and regions in which you are based.
- Try to remain separated from enduring and embedded politics, recognising the need to seek to develop relationships with all potential actors, not just the easy ones.

- Ensure that all partners have a collective and common understanding of the purpose(s) of the project. Sometimes, it is useful to document this in a project charter so that if there are disruptions in the project, this road map may help reorient and reinvigorate your efforts.

## Acknowledgements

We would like to acknowledge respectfully the ideas, input and enthusiasm of the lands managers from the Beaverhouse First Nation, Brunswick House First Nation, Chapleau Ojibway First Nation, Flying Post First Nation and Matachewan First Nation, as well as the members of Wabun Tribal Council and the Resource Department of the Saulteau First Nations. We would also like to formally recognise our project funders, the Firelight Group, Mitacs, the Real Estate Foundation of British Columbia and the Mitsubishi Corporation Foundation for the Americas.

## Notes

1. In Canada, the Crown is the source of sovereign authority and a part of the legislative, executive and judicial powers that govern the country (Harris 2006). The term is commonly used to refer to the functions of the government.
2. First Nations is a term used to describe the indigenous peoples of Canada. They have been present on the land since time immemorial. There are more than 600 First Nations communities in Canada, speaking more than 100 distinct languages.

## References

Aker, Jenny C., and Isaac M. Mbiti. 2010. 'Mobile phones and economic development in Africa', *Journal of Economic Perspectives* 24: 207–32.
Alfred, Taiaiake, and Jeff Corntassel. 2005. 'Being indigenous: Resurgences against contemporary colonialism', *Government and Opposition* 40: 597–614.
Angell, Angela C., and John R. Parkins. 2011. 'Resource development and aboriginal culture in the Canadian north', *Polar Record* 47: 67–79.
Berkes, Fikret. 2017. *Sacred Ecology*, 4th ed. New York: Routledge.
Berkes, Fikret, Carl Folke and Johan Colding. 2000. *Linking Social and Ecological Systems: Management practices and social mechanisms for building resilience*. Cambridge: Cambridge University Press.
Borrows, John. 1994. 'Constitutional law from a First Nation perspective: Self-government and the Royal Proclamation', *University of British Columbia Law Review* 28: 1–47.
Caine, Ken J., and Naomi Krogman. 2010. 'Powerful or just plain power-full? A power analysis of impact and benefit agreements in Canada's north', *Organization and Environment* 23: 76–98.
Canadian Chamber of Commerce. 2016. *Seizing Six Opportunities for More Clarity in the Duty to Consult and Accommodate Process*. Ottawa: Canadian Chamber of Commerce.
Ecotrust Canada. 2017. *Referrals Software an Analysis of Options*. Portland, OR: Ecotrust Canada and Aboriginal Mapping Network.

Foster, Hamar, Heather Raven and Jeremy Webber. 2011. *Let Right Be Done: Aboriginal title, the Calder case, and the future of indigenous rights*. Vancouver, Canada: UBC Press.

Gogal, Sandra, Richard Reigert and JoAnn Jamieson. 2005. 'Aboriginal impact and benefit agreements: Practical considerations', *Alberta Law Review* 43: 129.

Harris, Carolyn. 2006. Crown. Accessed 21 September 2020. https://www.thecanadian encyclopedia.ca/en/article/crown.

Hayter, Roger. 2003. '"The War in the Woods": Post-Fordist restructuring, globalization, and the contested remapping of British Columbia's forest economy', *Annals of the Association of American Geographers* 93: 706–29.

Israel, Barbara A., Amy J. Schulz, Edith A. Parker and Adam B. Becker. 2001. 'Community-based participatory research: Policy recommendations for promoting a partnership approach in health research', *Education for Health (Abingdon, England)* 14: 182–97.

Joseph, Bob. 2015. 12 Common mistakes in First Nation consultation. Accessed 21 February 2017. http://www.ictinc.ca/blog/12-common-mistakes-in-first-nation-consultation.

Kovach, Margaret. 2010. *Indigenous Methodologies: Characteristics, conversations, and contexts*. Toronto, Canada: University of Toronto Press.

Minkler, Meredith, and Nina Wallerstein. 2011. *Community-Based Participatory Research for Health: From process to outcomes*. San Francisco, CA: John Wiley.

Morse, Bradford W. 2017. 'Tsilhqot'in Nation v. British Columbia: Is it a game changer in Canadian aboriginal title law and Crown–Indigenous Relations?', *Lakehead Law Journal* 2: 65–88.

Power, Alex. 2017. 'Duty to consult' a cruel joke if First Nations can't handle the load. Accessed 12 February 2020. https://thetyee.ca/Opinion/2017/01/16/Duty-Consult-Cruel-Joke/.

Rashid, Ahmed T., and Laurent Elder. 2009. 'Mobile phones and development: An analysis of IDRC-supported projects', *The Electronic Journal of Information Systems in Developing Countries* 36: 1–16.

Saulteau First Nations. 2015. *Our Comprehensive Community Plan*. Moberly Lake, Canada: Saulteau First Nations.

Smith, Linda Tuhiwai. 2013. *Decolonizing Methodologies: Research and indigenous peoples*. London: Zed Books.

Turner, Nancy J., Marianne Boelscher Ignace and Ronald Ignace. 2000. 'Traditional ecological knowledge and wisdom of aboriginal peoples in British Columbia', *Ecological Applications* 10: 1275–87.

WikiWikiWeb. 2005. Just in time programming. Accessed 2 March 2019. http://c2.com/xp/JustInTimeProgramming.html.

Zietsma, Charlene, Monika Winn, Oana Branzei and Ilan Vertinsky. 2002. 'The War of the Woods: Facilitators and impediments of organizational learning processes', *British Journal of Management* 13: S61–74.

# Chapter 11
# Lessons from recording Traditional Ecological Knowledge in the Congo Basin

Michalis Vitos

## Highlights

- Providing indigenous communities with information and communications technology tools and methods for collecting and sharing their Traditional Ecological Knowledge is increasingly recognised as an avenue for improvements in environmental governance.
- Usability engineering methods for data-collection interfaces for non-literate forest communities can facilitate geographic citizen science and information sharing with various stakeholders.
- Working in 'extreme' environments necessitates adopting a thoroughly flexible approach to the design, development, introduction and evaluation of technology and the modes of interaction it offers.

## 1. Introduction

Sustainable management of natural resources is one of the most crucial challenges of our times. Particularly when considering the management of rainforest environments, local and indigenous communities often possess unique knowledge about the natural resources on which their livelihoods depend. This Traditional Ecological Knowledge (TEK) is increasingly recognised as critical for sustaining these resources (Huntington 2011; Danielsen et al. 2014). Recent technological developments and growing acceptance of different forms of knowledge mean that geographic citizen science is seen as a promising solution to achieve long-term management of key environments with greater respect for, and an active role accorded to, local communities (Bonney et al. 2014).

TEK needs to be understood and recorded. It is, however, often difficult to capture it in a digital format, especially given the environment in which many of the communities that hold it live in and their lack of technical knowledge. Yet, if we achieve capturing local TEK in a digital form, there is greater potential to reach a wider audience and subsequently to use it to inform local actions and even policymaking. For instance, stakeholders, such as logging companies that operate in the rainforest, require accurate mapping of local resources that are important for local communities financially and culturally, so that these resources are excluded from future cutting sessions. Capturing this information in a digital form and communicating it with relevant stakeholders can therefore be crucial in the way logging companies (and other natural resources extractors) operate and the ways local resources are managed.

An effective approach to capturing TEK is through the use of geographic citizen science tools, which usually offer a geographic component (or a mapping user interface (UI)) to support data collection and visualisation of spatial and non-spatial TEK. While these tools and participatory mapping in general have a long tradition of utilisation in Western societies (e.g. for planning purposes), individuals and communities in developing countries have distinct skills in terms of literacy, familiarisation with technology and access to information and communications technology (ICT) which create barriers and challenges in the ways these tools are designed and utilised. This is particularly pertinent when working with forest communities, as they generally consist of mostly non-literate individuals who lack prior exposure to ICT and mapping technologies. This context introduces a range of sociocultural, practical, methodological and interaction challenges which are discussed in this chapter (see also Irani et al. 2010; Vitos et al. 2013; Stevens et al. 2014).

It is generally accepted that understanding the needs of and working with end users is essential in order to ensure the successful use of geographic data-collection tools, which can be achieved through participatory and culturally informed methods of community and user engagement, interaction design and (iterative) evaluation. However, working with users in remote places of developing countries presents several logistic, organisational, legal, financial and security-related challenges (Vitos et al. 2013). To address these challenges, usually the collaboration of locally situated intermediaries is required, which often severely limits the time that can be spent with the actual communities during the design and development process.

This chapter provides unique insights from the work that the Extreme Citizen Science group at University College London carried out with a number of communities; demonstrates how a geographic data-collection tool was designed, evaluated and used in these contexts; and emphasises the use of methodologies from the fields of human–computer interaction (HCI) and ICT for development (ICT4D). Specifically, this chapter shows how action research (Reason and Bradbury 2008) methodology is applied to introduce, evaluate and adapt ICT systems with communities in the Congo Basin that have little or no formal education and a lack of prior exposure to technology.

In Section 2, the context of the case study is sketched out, followed by descriptions of the technology (Section 3) and the UIs that we designed to enable these communities to capture their TEK. In Section 4, a series of usability experiments are discussed, which are used to demonstrate how forest communities interact with Sapelli, our geographic data-collection tool, while examining the practical and interaction challenges encountered. Finally, Sections 5 and 6 summarise our conclusions and highlight the lessons learned.

## 2. Traditional Ecological Knowledge in the Congo

Our three case studies took place in the Congo Basin and focused on enabling local communities to participate in socio-environmental monitoring schemes which aim to capture local TEK. The Congolese forest in Central Africa is estimated to host more than 29 million rural people, including up to half a million indigenous people, whose livelihood is closely related to the forest, which provides food, fuel, fibre and a wide range of other ecosystem services (Lewis and Nelson 2006).

Although the Congo Basin is internationally recognised as a unique biodiversity hot spot which has a direct impact on climate change, its forestry and resource extraction sectors are rapidly growing (Lewis 2012). The promotion of private-sector investment to meet adjustment targets and the Millennium Development Goals, in combination with the high demand of tropical timber, led to a massive increase of the forestry sector (Lewis 2012). As logging roads open up increasingly remote regions to commercial activities, more and more of the forest's resources are drawn into (inter)national trade networks, and forest people see their resource base diminishing.

However, the legal context has improved since 2010 when the Republic of the Congo (RoC) signed a Voluntary Partnership Agreement with the European Union, which includes a series of principles regarding the active involvement of local communities in the management of forest concessions. Moreover, sustainable, responsible logging companies such as CIB have sought to acquire Forest Stewardship Council (FSC) accreditation. FSC is an international non-profit non-governmental organisation (NGO) that promotes sustainable management of forestry resources, and requires logging companies to respect the rights and resources of indigenous and local forest communities. Within this context, the Extreme Citizen Science research group (ExCiteS) collaborated with different NGOs in the area to provide them with digital solutions that would enable communities to participate in mapping their resources. By directly involving community members, NGOs and logging companies, the aim is to improve their local understanding regarding valuable resources.

For the development and evaluation of the proposed mobile tools, we followed a user-centred design (UCD) approach. Typically, the UCD process consists of multiple experimental iterations which require the engagement of end users. However, access to end users in developing countries is not always possible due to various barriers such as: the distance, costs and logistics of organising a field trip and contacting participants; stakeholder expectations; cultural differences; and time constraints. In an attempt to overcome these challenges, we adopted a flexible approach to assist the design and evaluation of the proposed technical solution. For the actual user involvement, we organised two major lengthy field visits, each consisting of multiple design and development sub-iterations and working with different communities, as well as on-the-spot creation and evaluation of new software features and interaction prototypes.

The approach for engaging with the communities and introducing our tools was adapted from previous projects conducted in Congo (Lewis 2012) and refined in response to local conditions, as is common practice. Upon arriving in a community, we always began with a thorough process of free, prior and informed consent (FPIC; Lewis 2012), introducing ourselves first to the local chiefs or authorities and then to a wider assembly of the local population. Once we received consent and established a community protocol, we inclusively engaged community members in a participatory exercise. To ensure that the gathered sample was representative, we sought to involve both adult males and females of various ages and different ethnicities.

A bottom-up action research approach means that communities decide the project scope and design the project details. After the initial FPIC process (see Chapter 4), where the community decides whether they would like to participate and, if so, what the project aims will be, the next phase emerges, which is an exercise in iterative participatory design. Once the purpose of collaboration was agreed by the indigenous community (e.g. in this case study, the collection of data to monitor logging activities), potential users were then engaged in developing the interface design. The first step is to define the types of information to be collected. Working with a prototype, the key measurements, environmental parameters or local observations to be made are discussed. Participants comment on their ability and willingness to provide the information, as gathering observations of illegal activities may pose a risk to personal safety for those collecting the data, and whether they consider it relevant to do so. We encouraged people to suggest other types of information that they considered important. In this conceptual phase, we also endeavoured to include representatives from local stakeholders (i.e. a local NGO monitoring logging concessions) to be involved in the process of capturing the data which would be collected later. This phase spanned over many days and involved different communities before we concluded with a commonly understood set of concepts, categorisation and representations (e.g. terminology or iconography) which are discussed in detail in Sections 3 and 4.

## 3. The Sapelli platform

Sapelli is a mobile data-collection and -sharing platform designed with a particular focus on non-literate and illiterate users with little or no prior ICT experience. The platform plays a central role in ExCiteS' mission which is to develop theories, tools and methodologies to enable any community, anywhere, to engage in geographic citizen science. In this section, we describe how three digital tools were designed and tested: (1) Sapelli's pictorial interface, (2) an audio feedback feature and (3) a tangible interface for geographic data-collection purposes.

### 3.1 Pictorial interfaces: development of the Sapelli collector interface

The Sapelli collector application was designed based on Jerome Lewis's prior work (Lewis 2012). In 2012, the ExCiteS team started gathering a list of technical and non-technical requirements for the development of a smartphone-based geographic data-collection platform to be used across collaborations with forest communities which at that time were mostly based in the Congo Basin, Central Africa. Requirements included: improving usability for non-literate people through the use of a pictorial-based interface (i.e. use of images instead of text); providing a flexible way to define and modify surveys; and providing built-in functionality to enable offline and autonomous data synchronisation via SMS and the Internet. During the requirement-gathering process, several existing tools to support the collection of TEK were investigated, including CyberTracker (CyberTracker 2015), EpiCollect (Aanensen et al. 2009) and Open Data Kit or ODK (Anokwa et al. 2009). None of these platforms fully met the previously described requirements, and therefore ExCiteS set out to develop a new geographic data-collection tool, which was named Sapelli after the endangered sapelli tree (*Entandrophragma cylindricum*) which grows in the Congo Basin rainforest (Stevens et al. 2014).

Sapelli was developed to run on Android phones and tablets. It was designed to be generic and open source and to facilitate geographic data collection across language or literacy barriers through highly configurable pictogram-driven UIs. The application executes surveys described in a bespoke XML-based language which was designed to be highly readable and simple enough for anyone with basic computer skills but no prior programming experience to learn in a few hours.

A typical Sapelli survey takes the form of a pictorial decision tree. The decision tree represents an ontology of things that will be collected, or issues on which we want to collect information, with a predefined set of answers organised in a hierarchical structure. The leaves represent the most specific answers or classifications, while the in-between nodes represent categories or groups. Users navigate the decision space by repeatedly 'tapping' images to select child nodes until they reach a leaf node (Figure 11.1b). Sapelli supports multiple decision trees in sequence, and thus it can collect answers for multiple questions. For example, a community might want to collect information about plants and animals (and thus will have the two top-level categories), followed by images of the plants and animals that are most important locally, such as an antelope or a banana tree. Once a category is selected, a set of questions about the data point needs to be answered. Was it seen or heard? Can you take a picture? Finally, the location is recorded.

## 3.2 Development of an audio feedback feature

An important barrier that was identified while working with users to use the first iteration of Sapelli referred to pictograms (pictoral icons) and how easy it was for the user to understand, recognise and recall the pictogram

(a)

**Fig. 11.1** Sapelli platform. (a) Participant using the application. © (2014) IEEE. Reprinted with permission from: M. Stevens, M. Vitos, J. Altenbuchner, G. Conquest, J. Lewis and M. Haklay, 'Taking participatory citizen science to extremes'. Credit: Michalis Vitos using Sapelli app, UCL Extreme Citizen Science research group (ExCiteS).
(b) Decision tree designed in collaboration with Forests Monitor, CAGDF and local communities. Credit: Sapelli platform UCL Extreme Citizen Science research group (ExCiteS).

provided. To improve recognition and recall, we developed and evaluated an audio feedback feature. Research has shown that providing information across different human senses can have an impact on a participant's performance (Brewster, Wright and Edwards 1993). Thus, the audio feature enables descriptions of various items on the display to be played back by the application in order to aid the user's understanding of the interface. A similar approach was employed by Parikh and Lazowska (2006) and Parikh et al. (2006), where the researchers used audio clips to assist rural users in India to perform microfinance transactions.

Sapelli was extended to support audio descriptions, using pre-recorded audio files to accompany each of the UI elements. As explained in the previous section, in a Sapelli-based survey, the pictorial decision tree, which is used to structure the available options, represents a specific 'question' (e.g. map the agricultural resources of the community or indicate points of conflict between the community and the logging com-

(b)

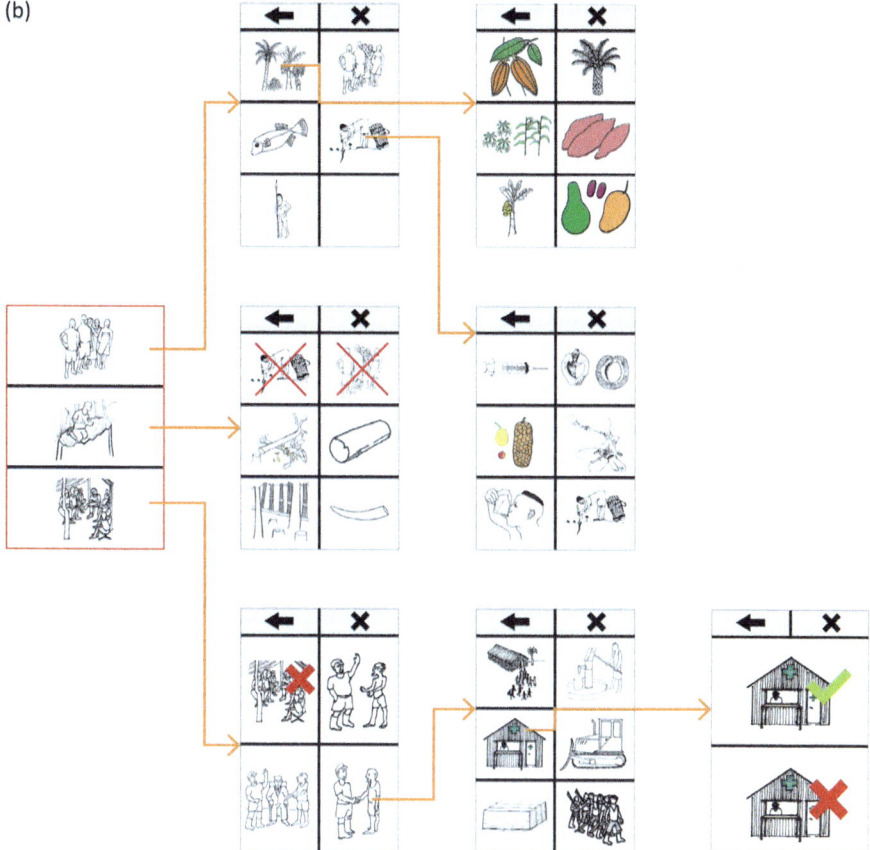

**Fig. 11.1** (continued)

pany). When the user navigates the interface, every time a new screen is accessed, an audio file narrates the overall question; each pictogram on each screen is also sequentially explained by audio playbacks, while an animation signals the pictogram the user selected to be described with the audio feature.

## 3.3 Tangible interfaces: development of the Tap&Map tool

Even though pictorial interfaces and audio feedback reduce the accessibility barriers introduced by textual interfaces, they still do not provide a universal solution to accommodate the needs of all potential users. This is because many users, especially those who lack formal education or experience in using digital technologies, face difficulties in using Sapelli.

The main barriers that the ExCiteS team identified after close observation and follow-up interviews with users in the field during the audio feedback experiments were: (1) fear of using the technology, (2) difficulties in navigating the decision-tree-based interface and (3) inappropriate or incorrect categorisation used in the decision tree's hierarchical structure.

For these reasons, the author of this chapter developed and evaluated an alternative tool to allow users to collect geographic data which aimed to eliminate categorisation and navigational structures and to reduce the need for extensive interaction with the device. The idea of developing a tangible interface to meet these aims was further based on the recent growing interest in forms of interaction that combine physical objects and graphical interfaces (Jensen 2012). Providing a link with the real world and building on users' knowledge concerning how to interact with tangible objects can improve participants' confidence (Rekimoto, Ullmer and Oba 2001).

The tangible interface, Sapelli Tap&Map, consists of two elements: (1) a series of cards, each with a pictogram representing a point of interest to be mapped; and (2) a smartphone app. Each card is equipped with a near field communication (NFC) tag and therefore acts as a tangible UI. The app reacts when one of the 'control' cards is touched on the device. For example, when a participant wants to record a point of interest, such as a banana tree, they (1) select the appropriate card from a stack of cards (Figure 11.2b) and (2) touch the card to the phone while standing as close as possible to the site to be mapped (Figure 11.2c). The device

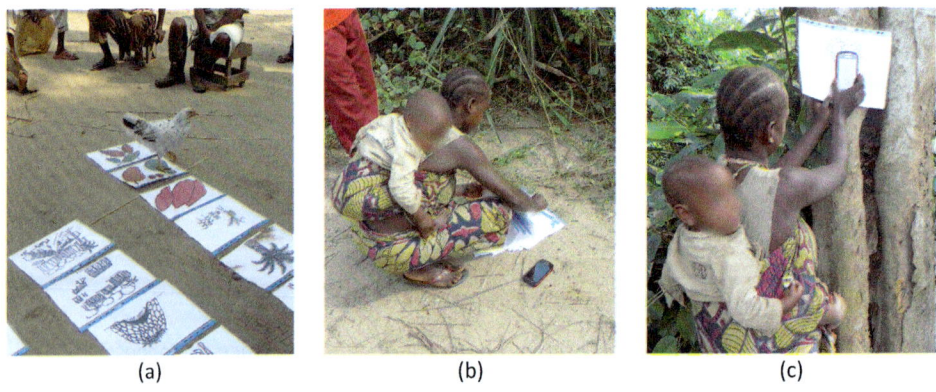

**Fig. 11.2** Prototype version of Tap&Map. (a) Printed prototype cards. (b) Picking the appropriate card. (c) Mapping a medicinal tree. Credit: Sapelli platform UCL Extreme Citizen Science research group (ExCiteS).

then reads the user's location from the Global Positioning System (GPS) sensor and stores it, along with other necessary metadata. We hypothesised that in comparison to on-screen decision trees provided by the Sapelli data-collector interface, it could enable a simpler and more intuitive way for non-literate participants to collect geographical data and map their local resources.

## 4. Collecting TEK in the field: evaluation and challenges

In this section, anecdotal evidence demonstrates how people used our tools in the field and the barriers that influenced these interactions. During the first field trip in the Congo Basin in 2013, the goal was to investigate the technical feasibility of Sapelli and the appropriateness of pictorial decision trees in terms of usability, effectiveness and user experience. During our fieldwork, we identified a series of interaction challenges and cultural differences in terms of evaluating software and conducting structured usability experiments.

As a consequence, and building on the knowledge that we gained from the first field visit, during our second field trip in the Congo Basin in 2015, we drafted strategies to conduct more successful usability field experiments. It was during the second field visit that we further designed and implemented the audio feedback feature which is used in Sapelli. Based on the usability results of our experiments with the pictorial decision trees and audio feedback, during the same field trip, it was further decided to explore physical, tangible interfaces in geographic data collection, which resulted in increased participant performance and satisfaction, as described in more detail in the following sections.

### 4.1 Evaluating the pictorial interface of Sapelli collector

Since pictograms and decision trees are a central component of Sapelli data collector, we worked with Baka hunter-gatherers in the field in training exercises to introduce them to the pictograms which were initially printed on large flash cards (Figure 11.2a). To ensure the pictograms were clearly understood, first we asked the community which gathered to participate in the experiment to guess what each image represented and whether they thought it was relevant. This had the added advantage of making the experiment fun and accessible to all community members. We took notes of any potential interaction barriers (e.g. unclear or misunderstood pictograms) and incorporated suggestions for alterations or additions.

The decision tree was subsequently updated incorporating all feedback and the community's suggestions before visiting the next community in the same area with which we further evaluated the interface.

In this second experiment, we introduced the smartphones to the communities (in Gbagbali, Kabo, Matoto and Sembola), and demonstrated how to navigate the interface. We asked the users who participated to familiarise themselves with the pictograms and the structure of the decision tree. It was vital to contextualise these activities to make sure people understood the context and the purpose of the data-collection activity in which Sapelli is used and why and how it may be relevant to them. Therefore, a group of people was trained, and we asked small teams of men and women to take the smartphones (i.e. Samsung Galaxy Xcover running Android) into the surrounding area and map some of their nearby natural resources. Throughout the process, we continued gathering feedback and suggestions for possible improvements.

Interaction with the devices, specifically with the touch-screen smartphones, proved challenging and frustrating for some people. Participants were unsure how long they needed to press on a pictogram, and assuming that their tap was not registered, they tapped twice or more on the same interface spot, which subsequently resulted in accidentally navigating deeper into the decision tree, and in a sense they 'got lost'. We sought to redesign and address this problem immediately by introducing a short waiting animation to show that a tap had been registered and that the new screen would appear shortly, which was successful.

Since the pictograms were always co-designed with the community which will be collecting the geographic data, most were easily recognised. Nevertheless, the interpretation of some pictograms was a major challenge. While the majority of the pictograms were intended to be interpreted literally – for example, a pictogram of a banana tree was used to represent a banana tree – the implementation of a hierarchical structure in organising a decision tree meant that there were also pictograms which were used to represent abstract categories or groups of more than one item. It was mainly these categorical pictograms and their use which proved to be much more challenging. For instance, a pictogram in the first step of interacting with the interface would represent a specific type of activity, for example hunting or fishing. Tapping on the relevant pictogram would take the user to the next screen, offering a choice of hunting or fishing species that would be available to select from based on the user's first selection. It was mainly the pictograms in the first step which were taken literally and were therefore confusing for participants.

Many participants had great difficulty understanding the overall hierarchical structure and how to navigate through it using forward or backward steps, represented by a left arrow for going backwards and choosing the next pictogram in the decision tree for moving forward. These results are in line with previous research undertaken by Medhi et al. (2013) in Bangalore, who found that the level of formal education was positively correlated with cognitive skills such as conceptual abstraction and categorisation, and thus with the ability to apply these skills when navigating hierarchical interfaces, even when they are text free.

Towards the end of our fieldwork, we also carried out a more structured usability experiment which included exposing users to a set of pre-defined scenario tasks which were used to evaluate the interface's efficiency, accuracy and recall. Participants were presented with hypothetical scenarios and asked to use the application to act, depending on the scenario. We quickly realised that conducting experiments outside of a controlled environment poses critical challenges. The communities we were working with are highly cooperative and communal. Therefore, evaluations that include only a single individual using the application were perceived as strange and awkward. Interrupting or preventing people from assisting each other during the usability experiment was considered offensive. For the majority of the participants, the scenarios were too abstract and the evaluation too intimidating to cooperate in a similar way to the Western practice of usability evaluations. As a result, many of the structured usability testing participants performed poorly when using standard measures.

## 4.2 Evaluating Sapelli's audio feedback feature

The audio feedback feature, which included audio in the local Mbendjele language, was evaluated in four different communities (Gbagbali, Kabo, Matoto and Sembola) in northern RoC, with a total of 48 adult participants (24 males and 24 females), who were selected on a voluntary basis. Aged between 18 and 69 years ($M = 34.5$, $SD = 12.5$), the majority of the participants had received no formal education (50 per cent), 38 per cent had primary school education, 6 per cent had secondary education, while another 6 per cent stated they had received some education but could not specify which level they had reached.

The goal of the audio feedback experiments was to evaluate participants' accuracy and recall by providing them with a set of representative scenario tasks to complete using both versions, with and without audio feedback (Vitos et al. 2017; Vitos 2018). Our experience from the previous usability experiments (Section 4.1) demonstrated that tasks which are based on hypothetical scenarios (e.g. 'Suppose you are walking in the forest and you find a medicinal tree, how would you use the smartphone to capture that tree?') did not work well. We therefore decided to ask our participants to perform five practical tasks, which included collecting data for five nearby resources under different top-level categories of the decision tree. All the selected points were valuable resources for the community which they wished to protect from damage from any future logging activities (e.g. medicinal trees, the local cemetery, cacao trees, etc.).

We followed the same FPIC and community protocol process as before to introduce the project, the project scope and our technology. Once people were comfortable with the application, 'tapping' the pictograms and moving between screens, we asked them to participate in pairs. The methodological choice to group participants and ask them to work together was mostly based on our previous experience from conducting usability tests in the field where it was revealed that it was relevant to the local cultural context to ask people to work together. Research assistants in the field facilitated the translation, note taking and video recording of the participants' interactions with Sapelli. In each site, participants were asked to describe the point of interest in front of them (e.g. a medicinal tree) in order to ensure that they understood its significance, and then use the Sapelli collector to record its type and location.

Following completion of the experiment, structured and semi-structured interviews with participants took place. During the interviews, the researchers tried to facilitate a discussion on the usability of and user satisfaction with the Sapelli collector, and to identify the barriers that caused poor performance on certain tasks.

The participants completed 240 tasks without and 240 tasks with audio feedback enabled. When using the version without audio assistance, they performed 177 successful observations (73.75 per cent), while when using the version with audio assistance, they performed 185 successful observations (77.08 per cent). The audio prompt thus seemed to be effective in slightly improving participant accuracy. However, performing a paired $t$-test revealed that the mean increase in accuracy ($M = 0.16$, $SD = 1.15$) was not statistically significant ($t(47) = 1, p = 0.32$).

Interestingly, the results indicate that the success rates in the use of decision trees are correlated with the literacy level of participants and their prior exposure to technology. In Gbagbali and Kabo, two remote communities with lower literacy levels, the success rate in the use of Sapelli's decision trees, without audio, was 63.3 per cent. When asked if they had ever used a mobile phone before (feature phone or smartphone), only 8 per cent of participants replied positively. In Matoto and Sembola, communities with higher education levels, which are located closer to the logging company's camp and which have easier access to technology

(33 per cent of participants claimed that they had used a phone before), the success rate was 84.1 per cent.

However, there was a significant difference between the two interfaces in terms of user experience and user satisfaction. During the interviews, 33/48 (69 per cent) participants stated that they preferred the version with the audio feedback. The main reason was that they found the audio feature in their local language entertaining and reassuring. For many participants, the version with audio prompts had a pedagogic element, as it provided them with knowledge about the pictograms and the project. For others, the audio feedback was a good way of verifying what they already knew, and of giving them reassurance that they were selecting the appropriate pictograms. One interesting case involved an older woman who stated that she liked the audio version because her bad eyesight did not allow her to distinguish the pictograms clearly. Thirteen (27 per cent) participants stated that they liked both versions and could not decide on one. Finally, two participants preferred the version without audio, since they already knew the answers, and they considered the audio prompt to be distracting.

## 4.3 Evaluating the tangible interface of Tap&Map

The main barriers identified after close observation and follow-up interviews during the audio feedback experiments were again a fear of using the technology, difficulties in navigation and inappropriate categorisation. The problems that participants had with the structural organisation of the decision tree seemed to be twofold. First, participants had difficulty understanding the abstract hierarchical structure and the pictograms used for navigational purposes. For instance, the function of the pictograms for navigating back to the previous interface (left arrow) and for cancelling an observation (cross; Figure 11.1b) were not clear to all participants, and they were rarely used. Although these pictograms were grouped together at the top of the screen and had a different look and feel to the 'normal' Sapelli pictograms, it was clear that participants did not understand their role in navigation and misinterpreted them as ordinary pictograms that should be used for mapping resources.

Second, the categories themselves and the pictograms designed to represent them were difficult to interpret. As explained earlier, designing pictograms for the categories was a major challenge, as some of the pictograms were intended to be interpreted literally, while others were meant to be more abstract and represent categories. During the field experiments, it became clear that category examples were often interpreted literally.

Following a rapid prototyping approach, we decided to implement and evaluate the Tap&Map prototype in two communities (Matoto and Sembola). Thirty-two adult participants (15 males and 17 females) took part in the study, and were aged between 18 and 61 years ($M = 28.8$, $SD = 11.4$). With regards to education, 44 per cent of the participants had no formal education, 47 per cent had a primary school education and 9 per cent had a secondary education.

The printed cards which were used were the same as those used for training purposes during the audio feedback experiments (Figure 11.2a). The procedure was very similar to that followed for the audio feedback experiments. We used the same five nearby resources, and participants were asked to describe specific points of interests they encountered. Then, they were given a stack of shuffled cards and a Samsung Xcover 2 smartphone with the Tap&Map functionality loaded. The participants' task was to map the resources by finding and selecting the appropriate card, placing the card as close as possible to the resource, touching their phone on the card, waiting for the GPS screen and then the success screen, which was a photo of a person giving them the thumbs up.

Over a period of two days, 32 participants completed 160 tasks using Tap&Map, with an impressive success rate of 97.50 per cent. During the interviews, the participants were very enthusiastic about the Tap&Map prototype, and unanimously agreed that this version was faster, easier and more comfortable for them to use compared to Sapelli. All agreed that they had no difficulties in selecting the appropriate pictograms and performing the tapping exercises. All four failed attempts with Tap&Map occurred when participants tried to map their village. In the scope of the project, participants could map their village and declare whether it was a Baka village or a Bantu village. In all four instances, participants chose the wrong village pictogram, instead, for example, choosing a random pictogram, which suggested that they understood the process but could not distinguish between the pictograms. This was an indication that the pictograms symbolising Baka and Bantu villages were problematic and had to be redesigned.

## 5. Discussion

### 5.1 Interfaces for data collection

One interesting finding of our research is that pictograms, especially when these are designed as part of a participatory iterative design process with the involvement of the indigenous community, can be an appropriate way to visualise various measurements, environmental parameters or local observations for which data will be collected. This aligns with relevant literature in other interaction contexts of non-literate and illiterate people with technological tools, which also suggest the use of pictograms to represent different actions (Medhi, Gautama and Toyama 2009). It was nevertheless a surprising finding that pictograms which were used to represent abstract concepts such as categories were not received well by non-literate participants, although these were also co-designed with the end users. This finding suggests that pictograms should be carefully selected or designed in such contexts to represent specific objects.

Another important finding concerns the use of pictorial decision trees which were found to be less suitable for non-literate participants. It was found that pictograms were problematic when they were used to represent categories or navigation. The overall hierarchical structure, in combination with the abstract or metaphorical nature of certain pictograms, was a significant interaction barrier for non-literate participants. It is therefore our recommendation that decision trees should be avoided in cases where low literacy or non-literacy prevails. Several studies have highlighted that a lack or absence of education has an impact on the development of cognitive abilities such as conceptual abstraction and categorisation (Medhi et al. 2013; Vitos 2018). Nevertheless, it should be noted that our findings on the appropriateness of pictorial decision trees contradicts the results of relevant literature in the field of ICT4D, where in many initiatives decision trees were employed and used by non-literate or semi-literate communities.

As an alternative solution, tangible or audio interfaces can be used to improve participants' performance and satisfaction. For example, the use of Tap&Map led to higher levels of confidence and performance and enhanced the overall user experience. It further helped to overcome the categorisation and abstraction issues that pictorial decision trees were introducing. Interestingly, physical interfaces performed equally well among semi-literate and non-literate participants. However, tangible interfaces come with many logistical issues of designing and managing physical objects, while they also have a negative impact on efficiency (time needed for a participant to complete a task).

Audio interfaces significantly increase participants' satisfaction, and subsequently they can increase their engagement with the project. Audio feedback can positively impact the training and mapping sessions, since it provides participants with a playful and comforting system. However, audio interfaces do not have an impact on participants' performance. They can even negatively influence efficiency and can be annoying for highly trained participants, or they may be 'risky' to use in certain contexts where the use of audio may pose a threat to the user's well-being (e.g. when mapping wildlife crime).

Data-collection schemes should be adapted regularly, depending on users' needs and requirements. All the suggested solutions have a notable limitation: the options offered for data collection are strictly predefined by the options identified during the design phase of the project or survey. In other words, participants are restricted to the pictograms offered in the pictorial decision tree survey, or by the NFC cards they are given for use with a physical interface, and they rarely have the option to map 'other' geographic features. This limitation means that a monitoring project should be regularly followed up and continually adapted to match additional needs and requirements. If participants are unable to capture all the data they need, they might lose interest in the project, or this might cause friction between the community and different stakeholders.

## 5.2 Contextual understanding and evaluation methods

Conducting usability experiments in remote locations, such as the Congolese forest, has specific implications which need to be taken into account. An important methodological challenge in this respect was caused by our lack of understanding of local culture and conditions, which resulted in the design of unsuccessful individual usability testing sessions. As a session progressed, bystanders – and even translators – would often help participants when they struggled to understand or perform the tasks in question, while preventing people from doing so was considered to be offensive. In our last field trip, we sought to remedy this issue by establishing a strict protocol with our research assistants, and we conducted the experiments during a walk in the nearby forest to separate participants from the rest of the community. However, instead of working in individual sessions, we grouped participants in pairs, which matched their local cultural context and improved the quality of their feedback.

Another key challenge was the design of appropriate tasks for the evaluation experiments. During our first field trip, we introduced scenario tasks to participants in the form of short hypothetical stories. However, such scenario-based tasks did not translate well from a local cultural perspective. Participants were often unsure what actions were required of them and needed to be talked through the steps of each task explicitly. On our second field trip, we tried to target this issue by engaging participants in real-life practical tasks, such as mapping nearby resources.

Finally, our usability experiments showed that techniques such as thinking aloud should be avoided, while interviews should be used with caution, since participants tend not to provide negative feedback regarding the prototypes, as they consider this to be impolite (courtesy bias; Vitos 2018). Alternatively, we advocate for field observations and field evaluations as the most appropriate methods for the given context, cross-referenced with other evaluation methods, such as text logging within the application.

## 6. Lessons learned

- There is a correlation between the performance rates of decision trees and the literacy levels of participants, and we therefore suggest using pictorial decision trees with semi-literate users.
- Pictograms should be co-designed with participants, and they can be used to represent concrete measurements such as environmental parameters or local observations for data to be collected.
- Categorical or abstract pictograms might pose challenges to low-literate participants and should be avoided.
- Audio feedback increases participants' satisfaction because it is reassuring, pedagogic and entertaining, but it should be used with caution, as it might also be irritating and disturbing for experienced participants, or risky in certain situations.
- Tangible interfaces seem to be the most appropriate interaction mode for non-literate participants, who have had minimal exposure to technology, in terms of performance – although their utilisation comes with logistical and practical implications.
- Ethnographic approaches, such as observational evaluations as well as semi-structured interviews, can prove appropriate to gain a better understanding of major usability issues.

## Acknowledgements

The research that underlines this project was supported by the European Union's ERC Advanced Grant project European Citizen Science: Analysis and Visualisation (under Grant Agreement No. 694767).

## References

Aanensen, David M., Derek M. Huntley, Edward J. Feil, Fada'a Al-Own and Brian G. Spratt. 2009. 'EpiCollect: Linking smartphones to web applications for epidemiology, ecology and community data collection', *PLoS One* 4: e6968.

Anokwa, Yaw, Carl Hartung, Waylon Brunette, Gaetano Borriello and Adam Lerer. 2009. 'Open source data collection in the developing world', *Computer* 42: 97–9.

Bonney, Rick, Jennifer Shirk, Tina Phillips, Andrea Wiggins, Heidi Ballard, Abraham Miller-Rushing and Julia Parrish. 2014. 'Next steps for citizen science', *Science* 343: 1436–7.

Brewster, Stephen A., Peter C. Wright and Alistair D. N. Edwards. 1993. 'An evaluation of earcons for use in auditory human–computer interfaces'. In *CHI '93 – Proceedings of the SIGCHI Conference on Human Factors in Computing Systems*, 222–7. New York: ACM.

Cybertracker. 2015. Accessed 22 March 2015. http://cybertracker.org/.

Danielsen, Finn, Per M. Jensen, Neil D. Burgess, Indiana Coronado, Sune Holt, Michael K. Poulsen, Ricardo M. Rueda et al. 2014. 'Testing focus groups as a tool for connecting indigenous and local knowledge on abundance of natural resources with science-based land management systems', *Conservation Letters* 7: 380–9.

Huntington, Henry P. 2011. 'Arctic science: The local perspective', *Nature* 478: 182–3.

Irani, Lilly, Janet Vertesi, Paul Dourish, Kavita Philip and Rebecca E. Grinter. 2010. 'Postcolonial computing'. In *CHI '10 – Proceedings of the 28th International Conference on Human Factors in Computing Systems*, 1311–20. New York: ACM.

Jensen, Kasper Løvborg. 2012. 'Sensible smartphones for Southern Africa', *Interactions* 19: 66.

Lewis, Jerome. 2012. 'Technological leap-frogging in the Congo Basin. Pygmies and Geographic Positioning Systems in Central Africa: What has happened and where is it going?', *African Study Monographs* 43: 15–44.

Lewis, Jerome, and John Nelson. 2006. Logging in the Congo Basin: What hope for indigenous peoples' resources and their environments? International Work Group for Indigenous Affairs (IWGIA) 4/06. Accessed 21 September 2020. https://www.iwgia.org/images/publications//IA_4-06_Congo.pdf.

Medhi, Indrani, S. N. Nagasena Gautama and Kentaro Toyama. 2009. 'A comparison of mobile money-transfer UIs for non-literate and semi-literate users'. In *CHI '09 – Proceedings of the SIGCHI Conference on Human Factors in Computing Systems*, 1741–50. New York: ACM.

Medhi, Indrani, Meera Lakshmanan, Kentaro Toyama and Edward Cutrell. 2013. 'Some evidence for the impact of limited education on hierarchical user interface navigation'. In *CHI '13 – Proceedings of the SIGCHI Conference on Human Factors in Computing Systems*, 2813–22. New York: ACM.

Parikh, Tapan, Paul Javid, K. Sasikumar, Kaushik Ghosh and Kentaro Toyama. 2006. 'Mobile phones and paper documents: Evaluating a new approach for capturing microfinance data in rural India'. In *CHI '06 – Proceedings of the SIGCHI Conference on Human Factors in Computing Systems*, 551–60. New York: ACM.

Parikh, Tapan, and Edward Lazowska. 2006. 'Designing an architecture for delivering mobile information services to the rural developing world'. In *Proceedings of the 15th International Conference on World Wide Web, WWW '06*, 791–800. New York: ACM.

Reason, Peter, and Hilary Bradbury. 2008. *The SAGE Handbook of Action Research: Participative inquiry and practice*. London: Sage.

Rekimoto, Jun, Brygg Ullmer and Haruo Oba. 2001. 'DataTiles: A modular platform for mixed physical and graphical interactions'. In *CHI '01 – Proceedings of the SIGCHI Conference on Human Factors in Computing Systems*, 269–76. New York: ACM.

Stevens, Matthias, Michalis Vitos, Julia Altenbuchner, Gillian Conquest, Jerome Lewis and Muki Haklay. 2014. 'Taking participatory citizen science to extremes', *IEEE Pervasive Computing* 13: 20–9.

Vitos, Michalis. 2018. 'Making local knowledge matter: Exploring the appropriateness of pictorial decision trees as interaction style for non-literate communities to capture their traditional ecological knowledge'. PhD diss., University College London.

Vitos, Michalis, Julia Altenbuchner, Matthias Stevens, Gillian Conquest, Jerome Lewis and Muki Haklay. 2013. 'Making local knowledge matter: Supporting non-literate people to monitor poaching in Congo'. In *Proceedings of the 3rd ACM Symposium on Computing for Development, ACM DEV '13*. New York: ACM.

Vitos, Michalis, Julia Altenbuchner, Matthias Stevens, Gillian Conquest, Jerome Lewis and Muki Haklay. 2017. 'Supporting collaboration with non-literate forest communities in the Congo-Basin'. In *Proceedings of the 2017 ACM Conference on Computer Supported Cooperative Work and Social Computing – CSCW '17*, 1576–90. New York: ACM.

# Chapter 12
# Co-designing extreme citizen science projects in Cameroon: biodiversity conservation led by local values and indigenous knowledge

Simon Hoyte

## Highlights

- Indigenous and local communities possess a wealth of knowledge that is largely neglected in conventional science but is essential to tackling worsening environmental issues sustainably.
- Extreme citizen science can provide the necessary approaches to assist otherwise excluded communities in collecting data that are based on local values and concerns but which can also be meaningful to a range of decision makers.
- A process of co-design, including that of the user interface, project management and data utilisation, ascertains that locally built citizen science projects can successfully empower some of the most marginalised communities to become involved in scientific data collection.
- Participant interaction should shape the entire process of co-design, rather than being subsequently observed.

## 1. Introduction

For centuries, a separation between nature and culture has informed the thinking on how to conserve the planet's biodiversity. In Africa, local perceptions and knowledge of environments are still too often considered irrelevant in conservation, despite these communities inhabiting key conservation areas and interacting daily with the fauna and flora so highly prized by the international community. Led by ideas of socio-ecological systems and damaged by hard-hitting allegations of human-rights abuses, organisations and governments are under increasing pressure to adopt community-centred approaches to protect biodiversity (Tauli-Corpuz, Alcorn and Molnar 2018). Accordingly, community-based natural resource management (CBNRM) projects are increasing around the world, with the application of environmental monitoring by local and indigenous peoples becoming more accepted in mainstream science and policy (Ferrari, de Jong and Belohrad 2015; Danielsen et al. 2018). Geographic citizen science, as a form of community-based monitoring, has emerged as a key methodology to both empower community members to collect locally relevant geolocated data themselves and provide the quantity of local data required to address heightening environmental change (Ballard, Phillips and Robinson 2018). Indigenous knowledge of place-based societies is recognised as particularly valuable due to its temporality and extensive detail of landscapes and ecological systems that may lack sufficient coverage within conventional scientific research (Gadgil, Berkes and Folke 1993; Alexiades et al. 2013). Over recent years, such knowledge has been effectively brought to the forefront through participative mapping techniques, proving to be an effective tool not only for engaging indigenous people in data collection around CBNRM, but also for co-designing projects built on indigenous knowledge, value systems and priorities, rather than those of external researchers.

The Extreme Citizen Science research group (ExCiteS) at University College London has designed extreme citizen science projects alongside diverse groups of indigenous communities in both the Amazon and Congo Basin rainforests. Since August 2016, the icon-based geographical data-collection platform Sapelli (for more information about Sapelli, see Chapter 11) has been implemented using smartphones alongside eight rainforest communities in the south and east regions of Cameroon, in partnership with the Zoological Society of London (ZSL), to address issues of wildlife crime, indigenous marginalisation and environmental injustice (Hoyte 2017). Cameroon, akin to all states in Central Africa, largely rejects the involvement of local and indigenous communities in environmental monitoring in favour of so-called conservation from above, making the current mapping particularly important as a means to demonstrate the necessity of bottom-up approaches for sustainable and just forest management (Pyhälä, Osuna Orozco and Counsell 2016; Adams 2017).

The context and requirements of this case study – namely, working with non-literate participants and engaging in community-led design from the beginning – were sufficiently challenging for traditional citizen science methodologies that an extreme citizen science method was adopted, aiming to incorporate participants with any education level from problem definition through to action (Stevens et al. 2014). Utilising methodologies of participative mapping, the work embraces participative action research to both conduct anthropological, environmental, geographical and human–computer interaction research and effect change on the ground. This chapter gives an insight into co-designing citizen science projects alongside non-literate participants in a challenging environment, the difficulties that may arise and mechanisms to overcome them. Key lessons are provided for researchers and practitioners seeking to initiate similar work, and remain relevant for those carrying out community monitoring projects more broadly.

## 2. The forest people

### 2.1 Cultural and social context

Cameroon harbours extraordinary biological diversity: western lowland gorillas, forest elephants, chimpanzees, mandrills, okapi and leopards reside in this portion of the Congo Basin rainforest. It is one of the last places on earth where such a diversity of megafauna exists in the wild, but it is being rapidly depleted by the illegal wildlife trade (IWT) and extractive industries (N'Goran, Nzooh and Le-Duc 2017). Equally rich is Cameroon's cultural diversity, with more than 280 languages and a multitude of forest peoples who have developed unique ways to thrive alongside the forest's plants and animals in complex relationships. The carving up of the forest in the south and east into logging and mining concessions, agriculture plantations, safari hunting zones and protected areas, allocated largely for short-term economic gain, has spelled disaster for indigenous forest communities, leading to forced relocation and economic and social discrimination. Despite some progressive legislation, wildlife and forest conservation in Cameroon is predominantly 'bioimperial', excluding communities and ignoring indigenous knowledge (Pyhälä, Osuna Orozco and Counsell 2016). International non-governmental organisations (NGOs) dominate conservation work, favouring the application of Eurocentric models of conservation and 'wilderness' preservation through strictly protected areas (Pemunta 2018). These models are heavily critiqued among anthropologists, political ecologists and human rights advocates due to their incompatibility with and potential damage to local African contexts which differ enormously from European and

American human–environment interactions. However, they continue to take precedence in the methodologies of international conservation NGOs working in Africa (Lewis 2016).

Indigenous Baka hunter-gatherers have occupied the forests of the Congo Basin for a substantial portion of human history, but in Cameroon, they have been forcibly evicted from their ancestral lands to roadsides, providing space for people-free conservation areas and extractive industries (Nguiffo 2003). In northern Congo, Baka people around the proposed Messok Dja protected area claim they are currently facing similar threats of becoming 'conservation refugees' as a result of forced eviction from their ancestral forest (UNDP Social and Environmental Compliance Unit 2018). Reports of abuse at the hands of NGO-funded protected area enforcement units ('eco-guards') are widespread across different Central African hunter-gatherer, or BaYaka, peoples, including the Baka of Cameroon and Mbendjele of northern Congo, who represent 'soft-targets for violent visitations' (Lewis 2016, 379; Survival International 2017). This militarisation of conservation, also termed 'green militarisation', has been widely criticised for leading to 'green violence' and rendering conservation work unfeasible due to human rights abuses (Büscher and Ramutsindela 2016).

Ethnoecological research has given insights to the extensive knowledge of plants and animals held by the Baka. More than 650 plant species were identified by one community, and in another, medicinal plant knowledge alone encompassed 624 species (Hattori 2006; Fa, pers. commun., 2019). The Baka language contains, for example, more than 19 words for gorilla and 28 for elephant, depending on exactly the state of the animal, their behaviour, their relationship to others and many other factors. Sadly, as Fikret Berkes bluntly notes, it largely remains 'difficult for people from "advanced" cultures to accept the idea that people from "primitive" cultures might know something scientifically significant' (Berkes 2012, 14). There is a strong case that the Baka represent conservation's most important allies. Yet, exclusion and discrimination are turning these allies into enemies of conservation via the IWT. Now that access to forest resources is restricted, communities are all too vulnerable to wildlife traffickers who coerce indigenous hunters into poaching. As a Baka man put it, 'We cannot be happy with them [the wildlife traffickers], as people are coming from outside and exploiting the community'.

A mounting body of research is showing that initiatives which include communities in conservation are more likely to succeed, whereas those working against local values and knowledge are more likely to fail (e.g. Olsen et al. 2001; Porter-Bolland et al. 2012; Homewood 2017).

Biocultural conservation is based on ideas about interconnectivity rather than incongruity between biological and cultural diversity, and may represent the best chance to integrate local values and knowledge into conservation initiatives. For the focus region of the current case study, there is recognition among anthropologists and some local practitioners that the region's rich biological diversity is inseparable from the practices, knowledge, values and beliefs of the forest communities, where one cannot thrive without the other (Lewis 2003).

## 2.2 Collaborative intelligence at its most extreme

Extreme citizen science, a technique intent on reducing power hierarchies, may act as an ideal vehicle for materialising ideas of biocultural conservation. Collaboration with communities as citizen scientists can not only produce far more data, but also maintain the data quality of that taken by conventional scientists whilst empowering users – a method which is also described by the Citizen Science Global Partnership as 'collaborative intelligence' (Danielsen, Burgess and Balmford 2005; Haklay 2013; Bowser and Brocklehurst 2018). Adopting an extreme citizen science approach, as in the current case study, shifts the focus to embracing diverse knowledge systems and the values of participants as required for effective biocultural approaches to conservation. The objective of extreme citizen science is to empower participants to be able to:

- Frame environmental problems in their own terms;
- Be supported to elaborate scientifically valid data collection protocols to provide evidence of the problems identified;
- Present the results in formats that all key participants, including ecosystem managers, can read;
- Facilitate informed decision making by all concerned parties based on addressing the problems and trends identified;
- Continue to monitor in real-time the efficacy of actions or interventions taken to address the problems identified (Lewis 2012, 39–40).

In the context of south-eastern Cameroon, many Baka communities feel a great sense of injustice towards external wildlife traffickers pillaging forest resources, and consistently express a desire to be involved in tackling such activity. A lack of empowerment and appropriate tools are most often cited as the greatest barriers to achieving involvement in conservation, offering a genuine opportunity for an extreme citizen science methodology initiated by local people.

The case study detailed here explores a geographic citizen science project collaborating primarily with non-literate Baka hunter-gatherers to collect geographic data using smartphones on illegal wildlife crime and animal monitoring. It also includes a number of Bantu-speaking semi-literate farming communities, recognising that multiple local stakeholders hold important knowledge about the forest's ecology and illegal activities. All in all, approximately 50 participants have been directly involved, the data covering a total geographical area of roughly 953 km$^2$. In the dense rainforests of southern Cameroon, the collection of IWT and animal monitoring data by indigenous and local communities may be the only viable mechanism to obtain the IWT and ecological information necessary for effective forest management in the future. In relation to IWT data, deeply entrenched corruption among forestry officers, police officials, ministers and other government agents seriously threatens their motivation to join anti-trafficking efforts. And for ecological data, embracing local ecological knowledge is more efficient both financially and practically than conventional sampling techniques such as line transects, and opens up access to immeasurably more spatial and temporal detail (Danielsen, Burgess and Balmford 2005).

The opportunity to collaborate with non-literate and semi-literate rainforest communities brings not only exciting possibilities, but also a multitude of technological and practical challenges. Such challenges, covering trust building, participant consent, community leadership, interface design and utilisation, data sovereignty and data verification and feedback, will be discussed here, exploring some of the innovative solutions developed in conjunction with participants.

## 3. Constructing the foundations: preparation for the Sapelli application

### 3.1 Building trust

Extreme citizen science relies on communities defining the initial problem themselves. The problems being faced, however, particularly among marginalised communities, are likely to be wrapped up in local politics and difficult to discuss. In the case of illegal wildlife crime, which is of huge concern to the communities involved in the current research, voicing concern could put local people in trouble with corrupt officials. On the other hand, honestly explaining the complex ways in which community members interact with illegal wildlife crime could result in disciplinary action by conservation agencies or harassment by eco-guards. Trust that the researchers are who they say they are, and that they will honour the agreements made over the course of project design, requires real effort by research staff and is often lacking in NGO-run community projects.

Building trust has been achieved by spending time in communities, not working but staying overnight in villages, cooking and eating together; learning the vocabulary of local languages; dancing and singing (both very important activities to Cameroonians); and showing an interest in personal stories, histories and opinions. Such activities may seem trivial but are seldom done by those working with communities. In this regard, actions certainly speak louder than words, as one elder told me: 'I know you work elsewhere too, but you decided to stay over here which I'm very happy about!' Unfulfilled promises by NGO staff have created a legacy of distrust in Cameroon. As a result, promises were not given to communities except those which could be delivered, for which communities showed great appreciation. Through building trust, issues in relation to the forest and its conservation were discussed extensively, leading to a set of specific priorities from each community which informed the design and implementation of the projects.

## 3.2 Community consent, management and leadership

Deceiving local and indigenous communities into agreeing with projects is commonplace for big industry, governments and some international organisations. To avoid this, a process of free, prior and informed consent (FPIC) has been developed and enshrined into many international agreements, including the United Nations Declaration of the Rights of Indigenous Peoples and the legally binding United Nations Convention on Biological Diversity. Unfortunately, FPIC is not always adhered to properly, and communities, particularly those which are marginalised and illiterate, continue to be exploited.

The current project employed a process of FPIC outlined by Lewis and Nkuintchua (2012) ensuring that communities:

- Thoroughly understand the objectives and aims of the project;
- Recognise the potential benefits and potential risks of partaking;
- Understand their ownership of the data, and agree with how their data will be used and shared; and
- Know their right to change or withdraw from the project at any point, and the possibility of having their data erased.

The FPIC mechanism is not finished once an agreement is signed. Rather, it is an ongoing process which responds to the project's progress and any challenges that arise.

A key element of extreme citizen science is for communities to be able to manage and lead projects themselves. Where marginalisation and disempowerment are prevalent, attempts to encourage self-management may be more difficult and viewed with suspicion. Baka peoples in Cameroon are among the most marginalised in the world and, as a result of deeply entrenched power dynamics, are often not prepared for self-management and project leadership. Accordingly, in addition to the FPIC process, communities are empowered to take ownership of the project through the creation of a community protocol. This protocol is in the form of a series of discussions revolving around what type of data will be collected and by whom, how the community will organise data collection and maintenance of the technology if it is to be left in the village, how and with whom the data should be shared, and in what form participants will be remunerated. Through the protocol, each community selects a team of up to eight to become participants; one smartphone is shared among the team. Akin to FPIC, community protocols are fluid, with participants regularly changing details of who keeps and charges the device and which members of the community are involved. Data sovereignty is an important aspect of extreme citizen science. However, to what extent this is genuinely achieved is arguable. The data and the responsibility of utilising them are, after all, on the laptops of the researchers.

The community protocol established in the current study was a space where ideas and suggestions by both the participants and researchers could be shared, with the aim of different skill sets and knowledge complementing each other. These protocols are taken very seriously by both the researchers and the communities: 'It is not good when people do not use the phone well. There is a protocol that we have agreed on and when people use other colours [referring to the ID system] and things like this it is not good'. All eight communities had previously expressed concern over illegal wildlife crime, and so collecting IWT-related data was a focus that all communities agreed with. A participant in one community proposed collecting monitoring data on animals, suggesting that it would be useful to know the spatial distribution of animals. The rest of the community agreed to add this function, as did all other communities when discussing the idea – a good example of project co-design.

The methodology of participative mapping was recommended by researchers to all the communities as a potentially powerful format through which the knowledge and concerns of community members could be conveyed to decision makers in a format understood by all. The idea of cartography appealed to participants because complaints about wildlife crime and other activities occurring in the forest were often dismissed by forest managers, citing the lack of evidence and specificity that maps could provide.

Respecting a community-led approach, where the projects and data belong to community members, led to conversations about the utilisation of the data. Wanting to have as much impact as possible but weary about corrupt figures and undesirable outcomes, all the communities decided that the data should be analysed by the researchers and could be shared with relevant staff from the Zoological Society of London and TRAFFIC, but only forwarded on to trusted officials of the Cameroonian Ministry of Forestry and Wildlife. The ambitions for sharing data centred on reducing wildlife crime, supporting community-led conservation approaches and demonstrating the ability of communities to act as collaborators.

## 4. Interface design and interaction challenges

As a result of a co-designed, community-led process, participant interaction with the technology heavily shaped the design in an iterative process. As little as possible was predetermined to maximise flexibility for community input. Therefore, the design and interaction with the technology will be discussed here together.

### 4.1 Device interaction

In terms of experience with Western mobile technology, communities in the remote rainforest of south-eastern Cameroon are likely to be some of the most removed. None of the participants had previously used touchscreen phones. Indeed, two of the villages had never used mobile phones at all. A period of training was undertaken whereby community teams were introduced to the phone – how to hold it, how to use the touch screen, the location of the camera, speaker, microphone, and so forth. During this process, it became clear that specific device models were required, namely those which were rugged (waterproof, shockproof and dustproof), featured physical buttons rather than on-screen ones and could be used discreetly (no bright colours). Handsets with a touch hypersensitivity option proved more popular, especially among elders whose fingers are highly callused as a result of life in the forest. For those with the roughest fingers, even hypersensitive screens were non-responsive; licking fingers or using knuckles and noses proved to be effective (and very humorous) solutions. Rounds of testing led to the adoption of the Samsung XCover 2, 3 and 4 as the most appropriate models.

To address community concerns about access to the data by outsiders, the application SureLock (42Gears Mobility Systems, Manchester, UK) was used to lock the devices on the Sapelli project. This was an important addition not only for security, but also to ease interaction with the devices. On a locked device, any physical button pressed, other than the power and volume buttons, has no effect.

## 4.2 Icon co-design

Sapelli, the icon-based data-collection tool designed by ExCiteS, has been developed to enable anybody, regardless of education level or technological ability, to initiate a citizen science project (Stevens et al. 2014). Scientific data collection almost always involves textual and/or numerical interaction, excluding those without the necessary textual literacy skills. As the Baka participants in the current project are largely non-literate, an icon-based design is necessary. Iconography, however, is locally variable, and in order to make sense to local people, a process of icon co-design was carried out in each community, whereby if participants were confused by existing icons, they were encouraged to adapt or design icons themselves. Most often this included changing existing icons to be more locally relevant or replacing foreign icons such as ticks (as a confirmation step) with those understood locally. Clearly, the detail in the design matters: participants mocked a pre-designed icon representing leopard skins because the tail was still attached, and others insisted on adapting the poacher icon to feature trousers rather than shorts, exemplifying the importance of local input in the co-design process. The means by which participants could create new icons was left deliberately vague in order to embrace alternative approaches. In one village, a Baka man asked if he could design his icon with a stick in the sand, rather than using a pen on paper, explaining that he felt more comfortable this way (Figure 12.1). Viewing community-designed icons on the phone interface was met with great excitement and joy. This not only reduces the likelihood of icons being pressed in error, but also has a clear empowerment aspect, even before any data have been collected.

Icons were arranged in the form of a pictorial decision tree, whereby an initial icon choice leads to the presentation of further choices and so on in a tree-like structure, although we were aware that this is not neces-

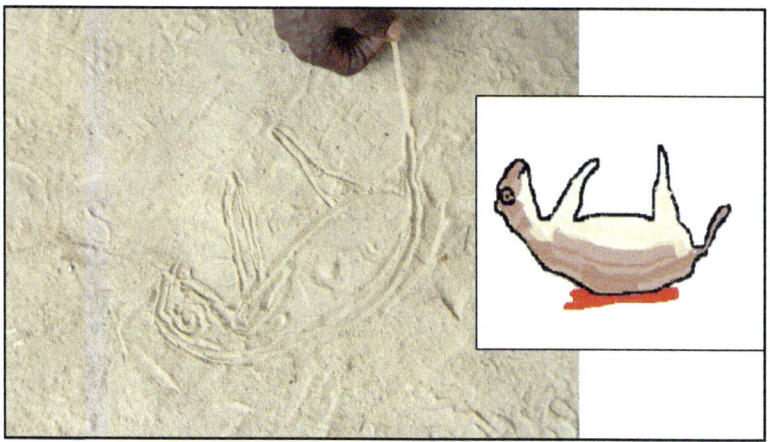

**Fig. 12.1** Community members lead icon design in whichever form they see fit (left). Credit: Photograph taken by Simon Hoyte 2018. The digitised icon (right). Credit: 2018 Bemba II village; Sapelli platform UCL Extreme Citizen Science research group (ExCiteS).

sarily the most effective interface type and depends primarily on the specific context (Vitos 2016). As a result of the decision tree layout on medium-sized smartphones, categorical icons were needed, as well as those with a literal use. This presented one of the biggest difficulties with the user interface: how to represent multiple things with a single icon. Of two trialled solutions – choosing one intermediary-level icon as a top-level icon to represent them all or, where possible, fitting reduced versions of all intermediary-level icons into one top-level category icon – fewer errors are made with the latter, but this remains an option only if there are a small number of intermediary icons. Experiments with top-level icon design interfaces were constantly scrutinised by the participants themselves. When one participant was asked, 'Which of the category icons would you press to report a killed gorilla?', a long period of struggle ensued, ending with the incorrect icon being chosen. However, the question rephrased as 'Could you report a killed gorilla?' prompted a quick response to use the correct 'killed elephant' top-level category icon to subsequently select the 'killed gorilla' intermediary-level icon. It therefore seems that at least in some cases, the format of the top-level icon is not important, as participants are memorising the necessary path to find the desired icon, rather than recognising the category icons.

The ability to accumulate evidence within the geographic data points that are collected was considered a priority by communities. As such, media functions were discussed, and subsequently photo and audio functions were designed and integrated into the Sapelli application. The empowering aspect of having the option to take photos and audio recordings themselves, with most participants having never taken a photo or heard their voice recorded before, was clear, even before any real data had been collected. Interaction with these functions in the trial phase quickly revealed the importance of indicators to show that an action was ongoing. Whilst interacting with the photo function, participants often continued pressing the 'take photo' icon repetitively, unaware that the process was already ongoing, which led to deleting photos by accident. The introduction of a loading icon (in the form of a spinning wheel) reduced the incidence of this drastically. Using the audio recording function, participants would consistently start speaking into the device before recording had begun. Some of those who waited to speak would wait until they saw the audio visualisation bars, but these do not light up until audio is detected creating a catch-22-like situation. Adapting the interface so that the audio icon turns red when recording is ongoing aided both groups of participants (Figure 12.2).

**Fig. 12.2** A red indicator to show when audio recording is ongoing helped to reduce confusion and mistakes by participants. Credit: Photograph taken by Simon Hoyte 2017 (right); Sapelli platform UCL Extreme Citizen Science research group (ExCiteS; left).

As a result of this iterative design process and the associated improvements, non-literate participants could successfully take accurate reports whilst also recognising limitations with regard to transmission: 'It is very easy to use, except if data did not fall'.

Security was raised by community members as a fundamental part of the project. As one elder stated, 'We can be involved in such an activity, but anonymity is of intense importance'. Accordingly, ideas of security were discussed together in relation to user interface. An anonymous ID system was suggested by researchers, to which participants agreed and posited a range of things, from favourite trees to colours, to represent each person. In the end, all communities decided a colour-coded ID scheme would be the most effective.

## 4.3 Design outcomes of Sapelli and mapping interface

As detailed over the previous sections, the user interface of Sapelli, the utilising of maps and the management of each project was decided over an iterative process of discussion and interaction. Since August 2016, Sapelli has been initiated in eight communities, each with their own unique project. To display the outcome of co-design, one project is shown as an example from start to finish in Figure 12.3.

Community protocol discussions led to agreements that collected data points would be most effective if uploaded to a mapping interface. Community Maps developed by Mapping for Change, a start-up formed from ExCiteS, was utilised for the online geographic information system (GIS; Figure 12.4), with data sent automatically at predefined intervals using a mobile data connection. To account for poor connection, data were also sent via SMS to a receiver phone located in Yaoundé which then uploaded the data to Community Maps. This function is enabled by Sapelli utilising binary SMS which breaks up and sends the textual data into continuous small chunks (no maximum file size), automatically importing into Sapelli on the receiver phone. To date, more than 620 data points and 560 media files have been sent successfully to Community Maps using either mobile data or SMS.

## 4.4 Verification and feedback

For participants to verify the data they are collecting themselves is an aspect which must be taken seriously if the process is to be community led. Figure 12.3 demonstrates how participants can verify the media they are taking, as well as confirming that the overall report is good to

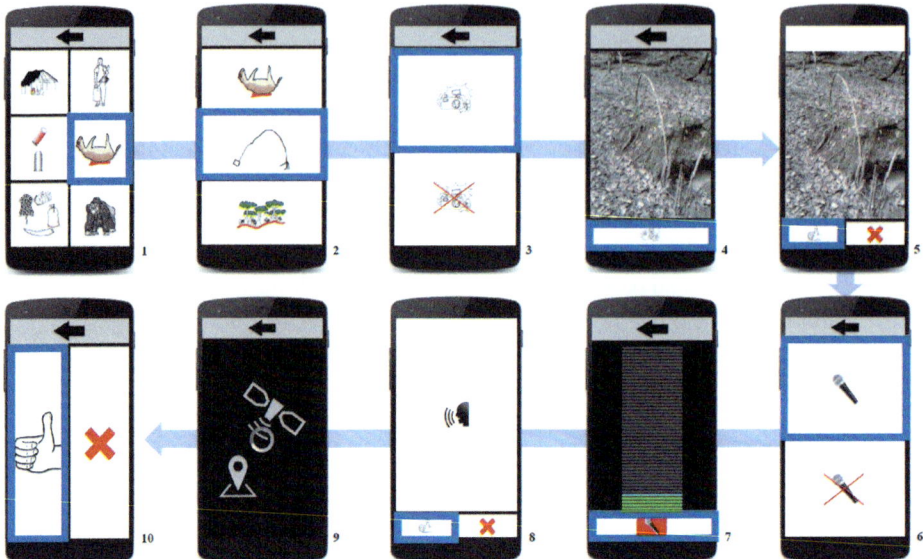

**Fig. 12.3** Example of the participant interaction required to take a data point. The participant begins by selecting a top-level category (1; 'killed animal'), followed by an intermediary-level icon (2; 'trap'). The choice is then presented to take a photo or not (3; the latter represented with a red cross). The participant can take the photo using the 'take photo' icon at the bottom of the screen (4), and verify it on the succeeding screen (5), followed by the option to take an audio recording (6). Audio is captured by pressing the 'take audio' icon (7), and can be subsequently verified (8). Sapelli then takes a Global Positioning System point automatically (9), after which the participant confirms the report (10). Credit: 2018 Bemba II village; Sapelli icons – Sapelli platform UCL Extreme Citizen Science research group (ExCiteS).

submit – an aspect they found useful. However, the verification of reports with the researchers after they had been received proved more challenging. The media attached to reports served as a form of verification, but those without media could only be verified through discussion with participants or other local people – a task not so simple in remote forest. Verification has, however, become easier over time, as specialists have naturally emerged in each community through using Sapelli, leading to reduced erroneous reports and specific participants who can be consulted to confirm validity.

Interaction with the data was limited, as reports were stored on the device's hard drive, protected by a code lock, whilst the online GIS required

**Fig. 12.4** An example of the Community Maps online geographic information system used for the project. The geolocated data points are displayed on the satellite map by their icons (right). On selecting a point (yellow circle), the relevant data are presented (left). Credit: Bemba II village. Community Maps platform © 2019 Mapping for Change. Basemap © Mapbox © OSM contributors.

a password and administrator permission to access. Far from being exclusionary, these procedures were requested by participants in order to keep sensitive data secure. Due to the risks associated with IWT data, presenting data back to participants created difficulties. However, this was carried out by providing participants with overall statistics of their project and, for non-sensitive data, printed-out copies of photos taken through the Sapelli interface. On receiving these forms of feedback, participants expressed joy and felt more motivated to continue. Feedback of data using the mapping interface is yet to be trialled. However, experimentation and the current research that it is being carried out by ExCiteS at UCL will eventually contribute to the creation of interactive maps which can be manipulated and understood by non-literate users.

## 5. Conclusion

In many ways, initiating a citizen science project with non-literate forest communities in arguably one of the most remote inhabited regions of the world is asking for problems. Yet, it is these communities which possess the necessary knowledge to manage the rainforest of south-eastern

Cameroon sustainably. Utilising an extreme citizen science methodology, where project objectives, design, management and application fully support community-led processes, has provided the space for overcoming many challenges which may have otherwise been deemed insurmountable. The communities that collaborated in the project, particularly Baka, are readily dismissed as lazy and often labelled as criminals. Prior to the project's initiation, many practitioners and NGO staff warned that communities would not be able to collect data and that devices would be stolen. After two and a half years of the project running, all devices are in good condition and functioning. This is a testament to the power of communities, no matter how remote or inexperienced with mobile technology, to be able to collect important data if local ideas, values and knowledge are respected and they are encouraged to lead project design. As demonstrated here, co-designed projects should not involve a compromise of conventional scientific standards or quality in order to be inclusive, but rather they can offer the opportunity for local knowledge to be equally valued.

The small but increasing interest in citizen science in government policy and NGO work in Africa is encouraging. However, collaboration alongside Africa's indigenous peoples remains very rare. A vast reserve of local scientific knowledge of some of the continent's most important regions for biodiversity is being excluded. Difficulties in accessing indigenous knowledge systems using citizen science can be eased through utilising an extreme citizen science approach based on local priorities. Such priorities must not be seen as irrelevant, but rather incorporated into biodiversity conservation in a form of biocultural conservation, whereby the interconnectivity between communities and biodiversity is celebrated. This type of interaction runs contrary to the majority of environmental projects carried out in Africa. However, at a stage where both biological and cultural diversity are facing severe threats after decades of top-down interventions, adopting an approach based on listening to one another presents a realistic path to sustainability.

## 6. Lessons learned

- Spending time building trust is a prerequisite for honest discussions, particularly around sensitive topics, genuine collaborative work where power hierarchies are reduced and community-led projects that entail extensive independent data collection.

- Community-led co-design can lead to reduced participant error, addressing of local issues, inclusion of local knowledge and alternative knowledge systems, greater empowerment and heightened motivation to continue involvement.
- Icon-based rather than text-based design can enable effective data collection by non-literate and illiterate communities which may otherwise be overlooked by conventional science.
- If only the bare necessities of the project are pre-decided before consulting communities, sufficient space is provided for participants to shape the practicalities and management of the project in their own terms.
- Participative mapping provides a platform through which local knowledge can be shared with other actors in a way that is meaningful for all.
- Community protocols are an effective mechanism to discuss, share and agree on how the project and data will be managed. However, this does not necessarily lead to data sovereignty, and more work is required to enable participants to analyse and share their data themselves.

## Acknowledgements

The research that underlines this project was supported by the European Union's ERC Advanced Grant project European Citizen Science: Analysis and Visualisation (under Grant Agreement No. 694767) and the Darwin Initiative Project 23-001.

## References

Adams, William M. 2017. 'Conservation from above: Globalising care for nature'. In *The Anthropology of Sustainability*, edited by Mark Brightman and Jerome Lewis, 111–26. London: Palgrave Macmillan.

Alexiades, Miguel N., Charles M. Peters, Sarah A. Laird, Citlalli L. Binnqüist and Patricia N. Castillo. 2013. 'The missing skill set in community management of tropical forests', *Conservation Biology* 27: 635–7.

Ballard, Heidi L., Tina B. Phillips and Lucy Robinson. 2018. 'Conservation outcomes of citizen science'. In *Citizen Science: Innovation in open science, society and policy*, edited by Susanne Hecker, Muki Haklay, Anne Bowser, Zen Makuch, Johannes Vogel and Aletta Bonn, 254–68. London: UCL Press.

Berkes, Fikret. 2012. *Sacred Ecology*. Oxon, UK: Routledge.

Bowser, Anne, and Martin Brocklehurst. 2018. Introducing the citizen science global partnership. Accessed 5 March 2020. https://www.citizenscience.org/wp-content/uploads/2018/02/EoE-Webinar-Citizen-Science-Global-Partnership.pdf.

Büscher, Bram, and Maano Ramutsindela. 2016. 'Green violence: Rhino poaching and the war to save Southern Africa's peace parks', *African Affairs* 115: 1–22.

Danielsen, Finn, Neil D. Burgess and Andrew Balmford. 2005. 'Monitoring matters: Examining the potential of locally-based approaches', *Biodiversity and Conservation* 14: 2507–42.

Danielsen, Finn, Neil D. Burgess, Indiana Coronado, Martin Enghoff, Sune Holt, Per M. Jensen, Michael K. Poulsen and Ricardo M. Rueda. 2018. 'The value of indigenous and local knowledge as citizen science'. In *Citizen Science: Innovation in open science, society and policy*, edited by Susanne Hecker, Muki Haklay, Anne Bowser, Zen Makuch, Johannes Vogel and Aletta Bonn, 254–68. London: UCL Press.

Ferrari, Maurizio F., Caroline de Jong and Viola S. Belohrad. 2015. 'Community-based monitoring and information systems (CBMIS) in the context of the Convention on Biological Diversity (CBD)', *Biodiversity* 16: 57–67.

Gadgil, Madhav, Fikret Berkes and Carl Folke. 1993. 'Indigenous knowledge for biodiversity conservation', *Ambio* 22: 151–6.

Haklay, Mordechai (Muki). 2013. 'Citizen science and volunteered geographic information: Overview and typology of participation'. In *Crowdsourcing Geographic Knowledge: Volunteered geographic information (VGI) in theory and practice*, edited by Daniel Z. Sui, Sarah Elwood and Michael F. Goodchild, 105–22. New York: Springer.

Hattori, Shiho. 2006. 'Utilization of *Marantaceae* plants by the Baka hunter-gatherers in Southeastern Cameroon', *African Study Monographs* 33: 29–48.

Homewood, Katherine M. 2017. '"They call it Shangri-La": Sustainable conservation, or African enclosures?'. In *The Anthropology of Sustainability*, edited by Mark Brightman and Jerome Lewis, 91–110. London: Palgrave Macmillan.

Hoyte, Simon. 2017. 'Indigenous elephant hunters and extreme citizen science', *Anthropolitan* 14: 44–5.

Lewis, Jerome. 2003. 'From abundance to scarcity. Contrasting conceptions of the forest in Northern Congo-Brazzaville, and issues for conservation'. Paper presented at the Canadian Anthropology Society Meeting, Dalhousie University, Halifax, Nova Scotia, 8–11 May 2003.

Lewis, Jerome. 2012. 'Technological leap-frogging in the Congo Basin. Pygmies and Geographic Positioning Systems in Central Africa: What has happened and where is it going?', *African Study Monographs* 43: 15–44.

Lewis, Jerome. 2016. 'Our life has turned upside down! And nobody cares', *Hunter Gatherer Research* 2: 375–84.

Lewis, Jerome, and Téodyl Nkuintchua. 2012. 'Accessible technologies and FPIC: Independent monitoring with forest communities in Cameroon', *Participatory Learning and Action* 65: 151–65.

N'Goran, Paul K., Laurent D. Z. Nzooh and Stephane Y. Le-Duc. 2017. *The Status of Forest Elephant and Great Apes in Central Africa Priority Sites*. Yaoundé, Cameroon: World Wide Fund for Nature (WWF).

Nguiffo, Samuel. 2003. 'One forest and two dreams: The constraints imposed on the Baka in Miatta by the Dja Wildlife Reserve'. In *Indigenous Peoples and Protected Areas in Africa*, edited by John Nelson and Lindsay Hossack, 195–214. Moreton-in-Marsh, UK: Forest Peoples Programme.

Olsen, Kristin B., Henry Ekwoge, Rose M. Ongie, James Acworth, Ebwekoh M. O'Kah and Charles Tako. 2001. *A Community Wildlife Management Model from Mount Cameroon*. London: Rural Development Forestry Network.

Pemunta, Ngambouk V. 2018. 'Fortress conservation, wildlife legislation and the Baka pygmies of Southeast Cameroon', *GeoJournal* 84: 1035–55.

Porter-Bolland, Luciana, Edward A. Ellis, Manuel R. Guariguata, Isabel Ruiz-Mallén, Simoneta Negrete-Yankelevich and Victoria Reyes-García. 2012. 'Community managed forests and forest protected areas: An assessment of their conservation effectiveness across the tropics', *Forest Ecology and Management* 268: 6–17.

Pyhälä, Aili, Ana Osuna Orozco and Simon Counsell. 2016. *Protected Areas in the Congo Basin: Failing both people and biodiversity?* London: Rainforest Foundation UK.

Stevens, Matthias, Michalis Vitos, Julia Altenbuchner, Gillian Conquest, Jerome Lewis and Muki Haklay. 2014. 'Taking participatory citizen science to extremes', *IEEE Pervasive Computing* 13: 20–9.

Survival International. 2017. How will we survive? The destruction of Congo Basin tribes in the name of conservation. Accessed 7 February 2018. https://assets.survivalinternational.org/documents/1683/how-will-we-survive.pdf.

Tauli-Corpuz, Victoria, Janis Alcorn and Augusta Molnar. 2018. *Cornered by Protected Areas: Replacing "fortress" conservation with rights-based approaches helps bring justice for indigenous peoples and local communities, reduces conflict, and enables cost-effective conservation and climate action*. Washington, DC: Rights and Resources Initiative.

United Nations Development Programme (UNDP) Social and Environmental Compliance Unit. 2018. Registration card – SECU0009. Accessed 3 August 2020. https://info.undp.org/sites/registry/secu/SECUPages/CaseDetail.aspx?ItemID=27.

Vitos, Michalis. 2016. 'Making local knowledge matter: Design and evaluation of ICT tools for forest monitoring in the Congo-Basin'. PhD diss., University College London.

Chapter 13
# Community monitoring of illegal logging and forest resources using smartphones and the Prey Lang application in Cambodia

Ida Theilade, Søren Brofeldt, Nerea Turreira-García and Dimitris Argyriou

## Highlights

- Indigenous and local people are increasingly recognised as playing an important role in the global environmental science policy arena. Participatory environmental monitoring is promoted as a cost-effective approach to collect and report data on environmental trends and support decision making while providing social co-benefits to local people.
- The use of geographic citizen science applications for data collection has opened up new opportunities for communities wishing to engage in environmental monitoring. While geographic citizen science applications assist in data collection and analysis, the use of technology may present a barrier to broad community involvement.
- Using a geographic citizen science tool to collect data on forest crimes and forest resources in Cambodia showed that community members could collect large amounts of geographic data regardless of their gender or age. The documentation facilitated advocacy and awareness-raising on social media and helped petition the government of Cambodia to protect Prey Lang forest officially.

# 1. Introduction

Tropical forests are home to a large proportion of the global terrestrial biodiversity, act as important carbon sinks helping to regulate the world's climate and are essential to millions of livelihoods across the tropics. However, tropical forests remain under pressure from alternative land uses and unsustainable resource use (Hansen et al. 2013). Since the United Nations climate change conference (Conference of Parties or COP 14) in Poznan in December 2008, Reduced Emissions from Deforestation and Forest Degradation (REDD+) has been at the heart of the UN Framework Convention on Climate Change (UNFCCC) strategy for mitigating global climate change through the protection and expansion of forests as carbon reservoirs and sinks (UNFCCC 2010, Decision 1/CP.16). REDD+ was adopted by UNFCCC at COP 21 in Paris in 2015 (UNFCCC 2015, Decision 1/CP.17). Simply put, REDD+ provides funding and processes that pay for forest protection in developing countries by acknowledging their carbon capture capacities.

While various types of community-based monitoring of forests have proven effective at informing management decision at local levels (Danielsen et al. 2010), the inclusion of community-based monitoring in the UNFCCC REDD+ strategy would require data collected by local stakeholders to inform global forest management policies (Boissière et al. 2017). A key question is how to standardise data collection so that it can feed into monitoring at national and international levels.

The Intergovernmental Panel on Climate Change (IPCC) guidelines require the use of activity data (changes in extent of areas affected) and emission factors (changes in carbon stock within areas) to estimate emissions at a national level (Herold et al. 2011). Remote sensing is increasingly used but requires calibration by on-the-ground monitoring of emission activities (IPCC 2006, 4) such as forest inventories. While several types of emission activities from forests are described by the IPCC, such as forest fires and carbon stock changes in dead organic matter and mineral soils, illegal logging and illegal conversion of forest will be the emission activities addressed in this chapter. Local people's involvement in this monitoring of illegal activities has faced a number of obstacles, including: concerns of impartiality, as local people often rely on harvesting of forest products for their livelihoods (Kanninen et al. 2007); questions about local people's authority to engage in law enforcement (Klooster 2000; Kaimowitz 2003); and opposition from local regulatory institutions and forestry officials (Glastra 1999; Tacconi, Mahanty and Suich 2013;

Milne 2015). Most importantly, individual community members potentially run a personal risk (Global Witness 2016) when monitoring the extraction of contested resources or using monitoring as a tool for advocating local forest rights (Tacconi, Mahanty and Suich 2013). As a result, no coherent body of literature exists on locally based monitoring of emission activities, but a number of studies have included documentation of illegal extraction activities in community-based monitoring of forest resources (Clarke, Reed and Shrestha 1993; McCall, Chutz and Skutsch 2016).

Here, we present a case study of how a local community network use smartphones and a geographic citizen science application to monitor and report illegal logging and illegal conversion of forest in Cambodia.

## 2. Collective action and self-governance in the protection of Cambodia's forests

This case study is based on the initiative of the Prey Lang Community Network (PLCN) – a network of villagers, mainly farmers with little or no formal education, who patrol their ancestral forests in the Central Plains of Cambodia. Formed in the early 2000s, as a response to rampant illegal logging, the PLCN was part of a wider network of community groups across the country coined the natural resources protection groups. The PLCN conducted regular forest patrols to protect Prey Lang from illegal logging and poaching. It worked successfully until 2012 when the founder and leader of the natural resources protection groups, Chut Wutty, was murdered while on a trip to document illegal logging. Hereafter, most villages worked in isolation, reports on illegal activities were often lost and the patrols had limited impact.

In 2014, an innovative partnership was formed between an international non-governmental organisation (NGO), Danmission, the University of Copenhagen, a local IT company, the PLCN and two local NGOs, namely Peacebridges Organisation (PBO) and the Community Peace-Building Network (CPN). The project was named 'It's Our Forest Too' after the Prey Lang, which means 'our forest' in the local language. The aim of the project was to engage vulnerable communities in peaceful dialogue for forest protection in Cambodia. One of the unique features of the project was to develop a specially designed geographic citizen science application for smartphones for communities to collect documentation about resources and illegal logging in Prey Lang forest. In February 2015, a PhD student taught more than 100 villagers from the PLCN how to use the Prey Lang application.

The forest patrols remain a central feature of the project. The patrols consist of 15–20 people on motorbikes, covering various sections of the Prey Lang forest. Illegally cut timber is seized and often burned on the site. Logging equipment is confiscated and turned over to the authorities upon completion of the patrol. Offenders are questioned, and reports, including fingerprints of the offenders, are filed and sent to authorities. As part of the 'It's Our Forest Too' project, all patrol members have been trained in non-violence, peaceful methods and conflict resolution following the theory and methods of Galtung (1996). Usually, the PLCN invite apprehended offenders to share a meal. Only after befriending the loggers, many of whom are poor villagers paid off to harvest timber, will the PLCN question the offenders and try to establish links to the kingpins behind the logging. Finally, PLCN members ask the offenders to sign a statement not to return to Prey Lang.

Currently, the PLCN is not formally recognised by the Cambodian government and has no official rule enforcement or sanctioning power. In May 2016, after the PLCN had petitioned the government to protect Prey Lang for 10 years, 432,000 ha of Prey Lang was declared a Wildlife Sanctuary (Figure 13.1). At the same time, the government started to draft a new national environmental legislation, the environmental code, which allows for greater participation of local and indigenous peoples in the management of the country's protected areas.

## 2.1 Forest-dependent communities in Prey Lang

Cambodia has one of the world's highest national deforestation rates (Hansen et al. 2013), mainly driven by large-scale acquisitions of land for agro-industrial purposes, primarily in the form of economic land concessions and mining concessions (Jiao, Smith-Hall and Theilade 2015; Work and Thuon 2017). These have led to large-scale agricultural conversion of forest land and extensive illegal logging operations outside the borders of the officially granted concession areas, which are in conflict with the land law, forestry law and the law on protected areas. Prey Lang forest holds great ecological (Theilade et al. 2011), economic (Jiao, Smith-Hall and Theilade 2015; Hüls-Dyrmose et al. 2017) and cultural (Turreira-García et al. 2017) value. Roughly 250,000 people live within the vicinity of Prey Lang, most of whom rely directly on the forest for their livelihoods. Hence, resin extraction from dipterocarp trees is the main source of cash income for many (Hüls-Dyrmose et al. 2017). Prey Lang is also a source of medicines, food, building materials and firewood. Access to natural resources is customary and without official property rights.

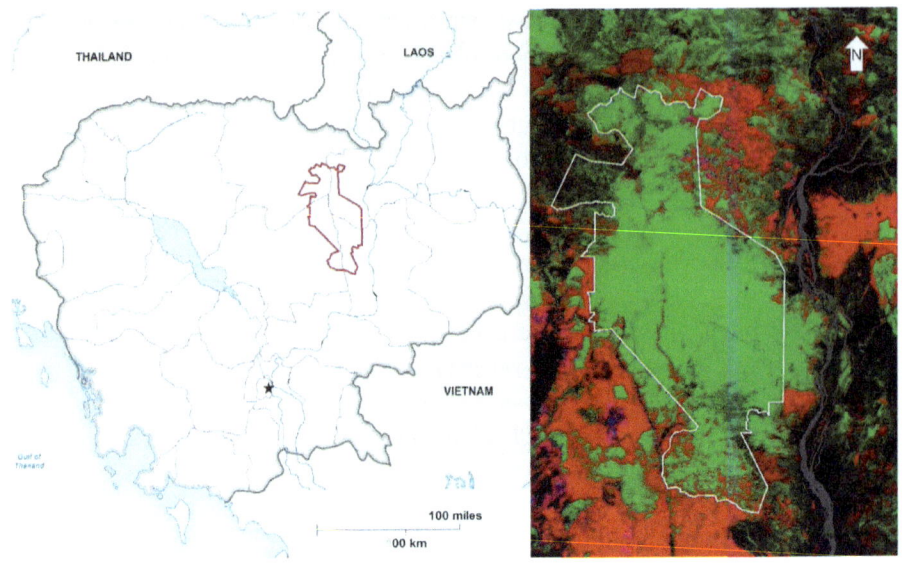

**Fig. 13.1** Map showing Cambodia and Prey Lang Wildlife Sanctuary in red (left), and satellite image showing forest loss 2000–2016 (red), forest cover 2016 (green), forest gain 2000–2016 (blue), both gain and loss 2000–2016 (purple) and other land uses (black). An asterisk shows the location of the capital city, Phnom Penh.
Source: Brofeldt et al. 2018.

The predominant ethnic groups are Kuy (indigenous) and Khmer. In Prey Lang, both ethnic groups practice animism and are culturally and spiritually linked to their forests (Turreira-García et al. 2018).

## 3. Participatory design and development of the Prey Lang application

In August 2014, a five-day initiation workshop was held with partner organisations including 34 participants, selected by the PLCN, from all four Cambodian provinces in which the PLCN operates. The overall aim of the smartphone-based monitoring programme was discussed in a series of focus group discussions. Participatory mapping on printed land use maps was employed to identify forest areas perceived as important and frequently used by communities. Participants were then asked to list the resources and activities they wanted to monitor and to rank these using cardboard cards. Finally, resources and activities were grouped into categories to guide the design of the geographic citizen science application. It was agreed that the PLCN would have ownership of all produced geographic data, and that no data could be shared without their permission. A smartphone app based on the Sapelli platform (Stevens et al. 2014) was chosen, as it was developed specifically for use by local people with limited experience in interacting with technology. The backend of the application was modified by a local IT company based on the PLCN's input during the workshop and its feedback on a first prototype decision tree, which was presented on the last day of the workshop (Brofeldt et al. 2018).

The Prey Lang application compiled three types of information: (1) reference data: as soon as a new data point was created, metadata, including time and date, Global Positioning System coordinates and phone ID, were automatically attached to the observation; (2) primary documentation: upon establishment of a new data point, the patroller documented the observation with a photo and an optional audio recording using the smartphone's built-in camera and recorder; and (3) thematic tag: the patroller tagged the observation using a decision tree with three main categories: resources, illegal activities and reporting to authorities (Brofeldt et al. 2018). Each main category had a limited number of preset subcategories such as trees, animals, forest products and sacred places for the resources category; and logging, conversion of forest for plantations and mines, illegal hunting and illegal fishing for the category on illegal activities.

After the registration of new geographic observations in the field, the data points were automatically uploaded to an online database via the cell phone network. Next, the data were cleaned manually by a database manager in order to remove incomplete, irrelevant and duplicate entries. When the reference data, primary documentation (photo and optional audio recording) and the thematic tag were uploaded correctly, the entry was validated. Additionally, data points that were clearly tagged incorrectly were corrected if possible (e.g. an animal tagged as a plant), and the entry was post-validated. Entries could be excluded due to human and technical errors or lack of relevance.

The first version of the Prey Lang application (Version 1) initially had the possibility of recording: activities such as different kinds of illegal logging, illegal mining, illegal hunting and so on; resources that were considered valuable to the end users such as resin trees, other luxury trees, non-timber forest products and animals; and interactions with government officials at the local and national level and company officials. It was designed to collect quantitative information on a limited number of activities, resources and interactions to minimise complexity and to encourage systematic use during patrolling. Categories in the decision tree were all illustrated by drawings to allow participants with low literacy to navigate the application more easily (Figure 13.2).

Version 1 of the Prey Lang application was tested by a group of 40 PLCN members using 20 Samsung Galaxy devices during a seven-day field trial in December 2014. The participants agreed that Version 1 was too limited in scope, and there was a strong wish to be able to document more categories of forest resources and illegal activities and to be able to give qualitative descriptions of interactions with illegal loggers and authorities. This input was used to inform the design of an updated version (Version 2) of the Prey Lang application, developed in January 2015. The following functionality was added: one-push audio recording option (for documentation using qualitative description); multiple photo recording option (for documentation using more photos for one event); drop-down menus (for decision tree navigation); and free writing (for decision tree navigation).

Data collection using Version 2, running on 36 Samsung Galaxy devices, began in February 2015. Version 2 of the Prey Lang application was in service for 11 months, during which the PLCN continually discussed needs and operational challenges between themselves during patrols and at the quarterly PLCN committee meetings. The feedback was delivered to the project holders by a project officer who took part in patrols and the committee meetings. Requests included the addition of new secondary thematic tags (categories) to allow documentation of activities that did not fit the primary thematic tags, such as whether illegal timber was recorded as planks or stumps. The users also requested that some of the tiers of thematic tags were extended in order to include more information. Finally, they felt that the categories of plants and animals were too broad, and stated that they wished to record species names. The feedback was used to develop Version 3 of the Prey Lang application, which became operational in December 2015. This version featured a significantly extended decision tree as well as some bug fixes. A number of thematic tags were extended to include more tiers. For example, the thematic tag 'NTFPs' (non-timber forest products) was extended to include types of NTFPs (edible wild plants, construction and medicine). Likewise, the thematic tag 'transport' was extended to include information on how illegal timber was transported. Optional scroll-down menus with vernacular names of plant and animal resources were added. This was the first interface where literacy was required to operate the application.

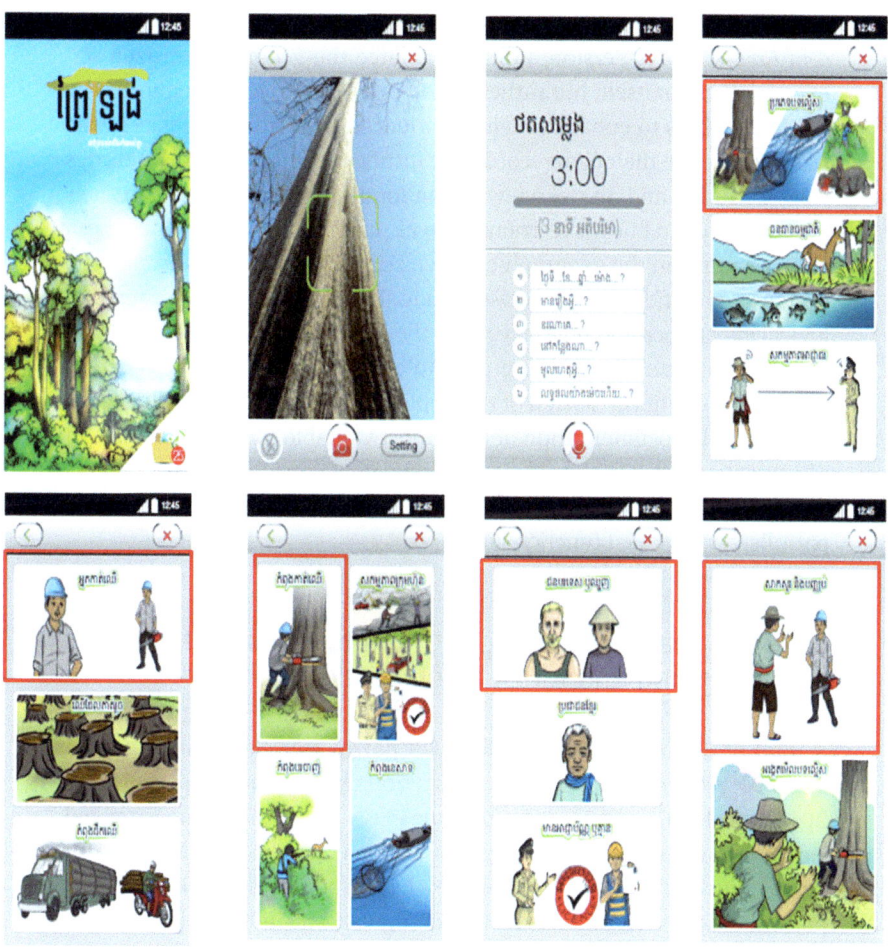

**Fig. 13.2** Interface of the Prey Lang app, showing the opening page, interface for taking photos, interface for recording an audio and the visual representations of the three main categories documented by the app (upper row). The red squares indicate selection of a category (here illegal activities). The lower row of interfaces shows a sequence of the next four points in the decision tree: illegal logging, logging of single tree/stump, offender was a foreigner and patrol member interacted with offender. Credit: Prey Lang data-collection app by Prey Lang Community Network (PLCN).

Version 3 operated for seven months until the release of Version 4 in August 2016. Version 4 did not change the geographic citizen science application itself, but rather added extra functionality requested by PLCN members to enhance their experience. It provided end users with the ability to see their own records, the justification if some of their entries were excluded and an overview of the total number of records in their province. Due to budget restraints, it was not possible to introduce these improvements in the geographic citizen science mobile app. Instead, the options were offered in a browser mode. Unfortunately, lack of technological experience and illiteracy made the new function difficult for PLCN members to use. The guiding principle for all Prey Lang application updates was to fit the design of the application more closely to the needs of the PLCN (documentation of their most valuable forest resources and illegal activities). On the release of every new version, one- to two-day refresher training was conducted in each province to familiarise PLCN patrollers with added functionality.

## 4. Use of the Prey Lang application to document illegal logging and forest resources

A total of 30 male and 6 female PLCN patrol members, ranging from 18 to 61 years of age, used the application in the two-year period from 2015 to 2017. They were selected by the PLCN based on volunteerism and experience with either patrolling or using smartphones. The PLCN members collected the data with the Prey Lang application during existing patrolling activities. These included regular and ad hoc local forest patrols in response to rumours of illegal activities, multiple times every month, and large-scale patrols, covering the Prey Lang core area, three to four times per year.

The primary objective of patrols was to discourage illegal logging by confiscating logging equipment and turning it over to the authorities, along with reports of recorded incidents (Figure 13.3).

Members of the PLCN made 10,842 entries of data on forest resources and illegal logging over the 24-month period. A total of 4,560 entries (42 per cent) were successfully validated by the external data managers, whereas 4,979 entries (46 per cent) were excluded due to technical errors and 1,303 entries (12 per cent) were excluded due to human error. Lack of mobile coverage leading to missing geographic coordinates or photo documentation was by far the biggest obstacle.

**Fig. 13.3** PLCN patrol member documenting transport of illegal timber in Prey Lang Wildlife Sanctuary. Credit: Photograph taken by Ida Theilade, December 2018.

Data recorded with the Prey Lang application during patrols were compiled in biannual monitoring reports by a data manager and students from the University of Copenhagen. The data were published in both English and Khmer, and presented to the general public by PLCN members at press conferences held in Phnom Penh supporting the PLCN's overall advocacy strategy for the protection of Prey Lang.

In January 2017, 24 Prey Lang application users attended a two-day evaluation of the application, and provided input for the development of Version 5 (released in August 2017). The evaluation was conducted using individual questionnaires featuring open-ended questions on experiences that users had in working with the Prey Lang application and participating in the monitoring programme. This was followed by a mediated plenum discussion of the questionnaire results and formulation of recommendations for the Version 5 design (Brofeldt et al. 2018).

Most patrollers (80 per cent) felt that they understood the Prey Lang application and that they were able to use it correctly. Patrollers mentioned challenges associated with learning how to use new functionality added in Versions 3 and 4 in a plenum discussion. About half of the patrollers said that it took them a few failed attempts to navigate the updated decision tree and to use the drop-down menus with local names of trees, animals and resources added in Version 3. They concluded that the new application versions probably led to them uploading some erroneous entries in the first months following release, but all except two users felt that they learned how to use the updated version after that.

Users were asked to free-list challenges encountered and priorities for future Prey Lang application development. Half of the patrollers specifically mentioned problems with uploading data as a key concern. The problems with uploading photos and coordinates were partly fixed in Version 4 by lowering file sizes. Yet, patrollers all mentioned the scarcity of areas with stable mobile phone connection in the provinces as a major cause of this, with one stating, 'It is not possible to get a signal in my village and I have to travel to Thala Barivat in Stung Treng Province to get a signal strong enough to upload my data. Therefore, my phone memory is often full'. Lack of mobile coverage leading to missing geographic coordinates or photo documentation remains the single largest problem for the efficiency of the Prey Lang application to date.

## 4.1 Patrollers' ability to use the smartphone app

In a response to the users' wishes, the complexity of the Prey Lang application, measured as the number of unique end points in the decision tree, became greater with each new version. However, the proportion of submitted entries that were successfully validated increased over time as well, probably due to the introduction of the thematic tags that guided users making an entry. The number of entries that were excluded because of technical errors decreased with the smaller size of photo files in Version 3. However, technical errors remained high at around 40 per cent in Versions 3 and 4, mainly due to problems with mobile coverage in remote areas of Prey Lang that led to missing coordinates and photos.

One third of the patrollers expressed a wish to have access to more functions, including the ability to take videos (in addition to photos and audio recordings), availability of maps to see areas of previous patrols and satellite imagery to see areas of recent forest loss. A few patrollers also requested more species to be added to the drop-down menus to allow for more precision in monitoring of plants and animals. However, the drop-down menus require the user to type first letters of a species and then select the right species from the list provided in writing. This feature was challenging for less literate users and users with poor eyesight. Hence, two patrollers mentioned that 'the app (Version 4) had become too complex and the introduction of drop-down menus to add specific names for trees, animals and resources was the function that was most difficult to use'.

## 4.2 Age and sex

Older people tended to submit more entries than younger people, and men generally submitted more entries than women. We found no significant differences in the proportion of validated entries produced between sex and age groups. A few patrollers in the 36–51 years age group had particularly high validation percentages, but the age group as a whole did not perform significantly better than the 19–24 years reference age group.

## 4.3 Cost

The cost of developing and maintaining the Prey Lang application, including database management and reporting for the first two years of monitoring, was about US$136,500. This is equal to US$0.26/ha monitored/year and about US$30 per validated entry. Approximately one third of the total cost was spent by the PLCN and local partners on patrols and meetings (c. US$51,000), one third by the University of Copenhagen on training, data management and reporting (c. US$45,000) and one-third by the IT company that developed the software (c. US$40,000). Many of the costs of operating the monitoring programme were borne by the PLCN members who volunteered their time on patrols, at meetings and for coordination of activities. These costs are not included in the cost calculations.

## 4.4 Obstacles

During the commune election period in 2017, the government instituted new regulations to control civil society. The result of the commune elections was a major blow to the ruling party. Three months later, the government initiated a crackdown on the opposition, civil society and independent media.

The shrinking civil space has affected the PLCN in a number of ways. The PLCN is now required to seek permits from the Ministry of Environment (MoE) before undertaking patrols, and the new regulations stipulate that rangers from the MoE must be part of the patrols. The permit system has made the organisation of patrols more difficult, and there is a general feeling among PLCN members that the MoE tips off loggers ahead of planned patrols.

The PLCN has responded by seeking permits and collaborating with MoE rangers in the hope that a dialogue and common patrols would further encourage authorities to enforce the forest law. At the same time, the PLCN has continued 'surprise' patrols without contacting the MoE ahead of the patrols or only contacting rangers with short notice. Generally, more illegal logging operations were found when authorities were not forewarned. In some parts of Prey Lang, the PLCN has good cooperation with the forest rangers, while in other parts, cooperation is limited, as rangers are directly or indirectly involved in illegal logging.

Hence, the full potential of ICT in forest monitoring is primarily restricted due to the government failing to provide an enabling environment. Second, mobile coverage was a problem in some of the more remote areas of Prey Lang, which led to many incomplete entries. Other errors were due to bugs in the software of the Prey Lang application, mainly relating to problems of uploading photos and missing coordinates. Finally, a relatively small amount of entries had to be excluded due to human error.

## 5. Conclusion

We found that local communities were able to produce large amounts of geographic data on forest crimes and important forest resources using a smartphone app. They did this at a cost that was only slightly higher than costs in previous community monitoring programmes that did not use any technology. Over the course of the two-year period, the complexity of the smartphone app increased considerably, but this did not negatively impact the quality of data produced. Instead, the data quality increased, as the patrol members gained more experience in using the Prey Lang application. Moreover, it emerged that women and elders were at least as capable of using the application as young men.

We believe that the success of the Prey Lang application is to a large extent due to profound local involvement in the design of the geographic citizen science application, as well as in the planning and execution of the monitoring activities. Other studies have also shown that it is fundamentally important to incorporate local knowledge alongside objectives and priorities stemming from experience with the existing forest patrolling into the design of the monitoring system (Berkes, Colding and Folke 2000; Ens 2012).

The use of a geographic citizen science application increased the reliance on external support for development and maintenance of the application and provision of smartphones, which over time may compromise the sustainability of the community-based monitoring programme. However, if the PLCN become formally recognised by the government of Cambodia as co-managers of the Prey Lang Wildlife Sanctuary, the Prey Lang application would be an important asset helping communities and the MoE to collect and analyse large amounts of forest data that can help inform management decisions.

In 2018, new functionality was added to the Prey Lang application, in collaboration with the international NGO Article 19, which allowed patrol members to report threats, harassment and criminalisation of the users. This is part of an international effort to protect environmental defenders, often indigenous peoples, who are increasingly targeted by governments and agro-businesses when defending their land. A female patroller explained, 'Before I used to receive threats from authorities and loggers after every patrol. The Prey Lang application has improved our safety as offenders know that we may document their threats and report them'.

PLCN members have also expressed a wish to be able to download satellite imagery showing forest loss and added functionality that would allow them to navigate to coordinates of recent tree loss. The ability to download near real-time maps may be populated with GLAD Alerts provided by the Global Forest Watch or near real-time maps generated by the Joint Research Centre of the European Commission and automatically pushed to the patrollers' smartphones. Such new functionality may improve the efficiency of the patrols to intercept illegal logging and mining operations in order to protect Prey Lang forest from further degradation.

## 6. Lessons learned

- Local forest networks have the capacity to collect information on illegal logging and forest resources using geographic citizen science applications.
- It is essential to incorporate local knowledge, objectives and priorities into the design of the geographic citizen science applications used to monitor the environment.

- Simple visual interface is key. Thematic tags should be used to structure data collection and make navigation easy throughout the geographic citizen science application.
- Clear ownership of the data and the geographic citizen science application was critical to the communities that collected the data and in turn increased the sense of responsibility and the quality of the data produced.
- The complexity of the geographic citizen science application did not affect the ability of community patrollers to use the tool.
- The use of a geographic citizen science application in monitoring did not preclude the participation of women and elders. Further, sex and age did not affect users' capabilities in collecting quality data.
- The costs of development and implementation of a geographic citizen science application for monitoring of forest crimes was significantly less than monitoring by professional forest rangers.
- Data collection using a geographic citizen science application facilitated use of results in advocacy, on social media and to petition relevant authorities in the government.
- Community-led monitoring programmes using smartphones may be highly valuable for environmental protection across the tropics and for global conservation and climate-change mitigation efforts.

## References

Berkes, Fikret, Johan Colding and Carl Folke. 2000. 'Rediscovery of traditional ecological knowledge and management', *Ecological Applications* 10: 1251–62.

Boissière, Manuel, Martin Herold, Stibniati Atmadja and Douglas Sheil. 2017. 'The feasibility of local participation in measuring, reporting and verification (PMRV) for REDD+', *PLoS One* 12: e0176897.

Brofeldt, Søren, Dimitrios Argyriou, Nerea Turreira-Garcia, Henrik Meilby, Finn Danielsen and Ida Theilade 2018. 'Community-based monitoring of tropical forest crimes and forest resources using information and communication technology – experiences from Prey Lang, Cambodia', *Citizen Science: Theory and Practice* 3: 1–14.

Clarke, Harry R., William J. Reed and Ram M. Shrestha. 1993. 'Optimal enforcement of property rights on developing country forests subject to illegal logging', *Resource and Energy Economics* 15: 271–93.

Danielsen, Finn, Neil D. Burgess, Per M. Jensen and Karin Pirhofer-Walzl. 2010. 'Environmental monitoring: the scale and speed of implementation varies according to the degree of peoples' involvement', *Journal of Applied Ecology* 47: 1166–8.

Ens, Emilie J. 2012. 'Monitoring outcomes of environmental service provision in low socio-economic indigenous Australia using innovative CyberTracker technology', *Conservation and Society* 10: 42.

Galtung, Johan. 1996. *Peace by Peaceful Means: Peace and conflict, development and civilization*. Oslo, Norway: International Peace Research Institute Oslo.

Glastra, Rob. 1999. *Cut and Run: Illegal logging and timber trade in the tropics*. Ottawa, Canada: International Development Research Centre.

Global Witness. 2016. *On Dangerous Ground – 2015's Deadly Environment: The killing and criminalization of land and environmental defenders worldwide*. London: Global Witness.

Hansen, Matthew C., Peter V. Potapov, Rebecca Moore, Matt Hancher, Svetlana A. Turubanova, Alexandra Tyukavina, Dave Thau et al. 2013. 'High-resolution global maps of 21st-century forest cover change', *Science* 342: 850–3.

Herold, Martin, Rosa María Román-Cuesta, Danilo Mollicone, Yasumasa Hirata, Patrick Van Laake, Gregory P. Asner and Ken MacDicken. 2011. 'Options for monitoring and estimating historical carbon emissions from forest degradation in the context of REDD+', *Carbon Balance and Management* 6: 1–7.

Hüls-Dyrmose, Anne-Mette, Nerea Turreira-García, Ida Theilade and Henrik Meilby. 2017. 'Economic importance of oleoresin (*Dipterocarpus alatus*) to forest-adjacent households in Cambodia', *Natural History Bulletin of the Siam Society* 62: 67–84.

Intergovernmental Panel on Climate Change (IPCC). 2006. *Guidelines for National Greenhouse Gas Inventories – Volume 4: Agriculture, forestry and other land use*. Geneva, Switzerland: United Nations Intergovernmental Panel on Climate Change.

Jiao, Xi, Carsten Smith-Hall and Ida Theilade. 2015. 'Rural household incomes and land grabbing in Cambodia', *Land Use Policy* 48: 317–28.

Kaimowitz, David. 2003. 'Forest law enforcement and rural livelihoods', *International Forestry Review* 5: 199–210.

Kanninen, Markku, Daniel Murdiyarso, Frances Seymour, Arild Angelsen, Sven Wunder and Laura German. 2007. *Do Trees Grow on Money? The implications of deforestation research for policies to promote REDD*. Bogor, Indonesia: Center for International Forestry Research.

Klooster, Daniel. 2000. 'Institutional choice, community, and struggle: A case study of forest co-management in Mexico', *World Development* 28: 1–20.

McCall, Michael K., Noah Chutz and Margaret Skutsch. 2016. 'Moving from measuring, reporting, verification (MRV) of forest carbon to community mapping, measuring, monitoring (MMM): Perspectives from Mexico', *PLoS One* 11: e0146038.

Milne, Sarah. 2015. 'Cambodia's unofficial regime of extraction: Illicit logging in the shadow of transnational governance and investment', *Critical Asian Studies* 47: 200–28.

Stevens, Matthias, Michalis Vitos, Julia Altenbuchner, Gillian Conquest, Jerome Lewis and Muki Haklay. 2014. 'Taking participatory citizen science to extremes', *IEEE Pervasive Computing* 13: 20–9.

Tacconi, Luca, Sango Mahanty and Helen Suich. 2013. 'The livelihood impacts of payments for environmental services and implications for REDD+', *Society and Natural Resources* 26: 733–44.

Theilade, Ida, Lars Schmidt, Phourin Chhang and Andrew J. McDonald. 2011. 'Evergreen swamp forest in Cambodia: Floristic composition, ecological characteristics, and conservation status', *Nordic Journal of Botany* 29: 71–80.

Turreira-García, Nerea, Dimitrios Argyriou, Phourin Chhang, Prachaya Srisanga and Ida Theilade. 2017. 'Ethnobotanical knowledge of the Kuy and Khmer people in Prey Lang, Cambodia', *Cambodian Journal of Natural History* 2017: 76–101.

Turreira-García, Nerea, Henrik Meilby, Søren Brofeldt, Dimitrios Argyriou and Ida Theilade. 2018. 'Who wants to save the forest? Characterizing community-led monitoring in Prey Lang, Cambodia', *Environmental Management* 61: 1019–30.

United Nations Framework Convention on Climate Change (UNFCCC). 2010. Decision 1/CP.16. *The Cancun Agreements: Outcome of the work of the Ad Hoc Working Group on Long-term Cooperative Action under the Convention*. UNFCCC/CP/2010/7/Add.1. Report of the Conference of the Parties on its 16th Session, Cancun, Mexico, 29 November–10 December 2010.

United Nations Framework Convention on Climate Change (UNFCCC). 2015. Decision 1/CP.17. *Adoption of a Protocol, Another Legal Instrument, or an Agreed Outcome with Legal Force Under the Convention Applicable to all Parties*. NFCCC/CP/2015/L.9/Rev.1. Durban Platform for Enhanced Action Conference of the Parties on its 21st Session, Paris, France, 30 November–11 December 2015.

Work, Courtney, and Ratha Thuon. 2017. 'Inside and outside the maps: Accommodating forest destruction in Cambodia', *Canadian Journal of Development Studies* 38: 360–77.

# Chapter 14
# Representing a fish for fishers: geographic citizen science in the Pantanal wetland, Brazil

Rafael Morais Chiaravalloti

## Highlights

- The creation of strictly protected areas in the western border of the Pantanal wetland, Brazil, has led to the physical and economic displacement of local people.
- A geographic citizen science programme was implemented to support local people to represent their customary practices, and to encourage practitioners to incorporate local people's needs better in the conservation agenda.
- Time spent with local people to gain rapport is a fundamental step in the implementation of a successful geographic citizen science programme.

## 1. Introduction

Conservation biology explores ways and means to protect the environment better. It is a scientific discipline that emerged in the early 1960s, and it focuses on 'actions that are intended to establish, improve or maintain good relations with nature' (Sandbrook 2015, 565). Since then, however, the understanding of 'good relations with nature' has changed. It has become increasingly obvious that conservation interventions should include local people's needs in order to meet their goals (Mace 2014).

In many cases, however, conservation practices still face significant challenges in addressing local people's needs. In freshwater fisheries, for example, the majority of management practices proposed and implemented by conservation biologists do not consider fishing strategies that are essential to livelihoods (Chiaravalloti 2017). Common conservation interventions in freshwater systems are based on the idea that people are fixed in time and space (Chiaravalloti 2017), but inland fishers commonly adopt specialised dynamics of use with high mobility; that is, they use a variety of methods of production/extraction at different times and places, including periods of intensive use in response to seasonal abundance, and are flexible to shifts in livelihoods (Abbott and Campbell 2009). Forcing people to live in different ways from those of their customary strategies can lead to physical and economic displacement (Abbott and Campbell 2009).

This chapter focuses on a salient but not extensively studied case of inland fisheries, taking a conservation biology approach to the Pantanal wetland in Brazil. This wetland region is approximately 160,000 km$^2$ in size, and covers parts of three countries in South America (Brazil, Bolivia and Paraguay). It is not only unique in terms of biodiversity, but also it offers several examples of the previously mentioned mismanagement practices of inland fisheries. This chapter aims to demonstrate how a geographic citizen science application and a participatory mapping approach were successfully used to support local people to represent and communicate their customary strategies of natural resource use and management in a scientifically valid form (Chiaravalloti 2017). The main goal is to explore the extent to which these tools can help practitioners to truly incorporate local people's perspectives in the conservation agenda and to reshape local conservation approaches.

## 2. The Pantanal and its protected areas and peoples

For the past 50 years, local fishers in the Pantanal region have been under constant pressure to stop fishing, with decision makers, environmentalists and local businessmen accusing them of overfishing in the region (Alho and Reis 2017). As a consequence, several strictly protected areas have been established in the region, restricting the use of natural resources. The first one was set aside in 1971, the Biological Reserve of Caracará, covering an area of 800 km$^2$. In 1981, the Federal Government replaced the Biological Reserve with the Federal National Park of the Pantanal (Parque Nacional do Pantanal), expanding the protected area to 1,300 km$^2$. In the early 1990s, with support from the non-governmental organisation (NGO) The Nature Conservancy, three additional large farms were converted into privately owned protected area (called a Private Reserve or Reserva Particular do Patrimonio Nacional; Tocantins 2011). Later on, in 2005 and 2006, two Private Reserves were aggregated, leading to the establishment of the environmental group Protection and Conservation Network for the Amolar Region (Rede de Proteção e Conservação da Serra do Amolar). This is a partnership among all protected area managers, which includes the federal agency for protected areas, NGOs and local Forest Policy agents. The partnership's aims are to monitor resource use along 310 km of linear river distance and adjacent channels and to ensure strict conservation measures for 2,620 km² of protected areas in the western border of the Pantanal (Bertassoni et al. 2012).

It is important to point out that local fishers claim that the protected areas in the western border of the Pantanal physically and economically displaced them from their original settlements. According to local people, the first displacement took place in the 1980s, soon after the establishment of the National Park, with recorded incidents of torture and violence perpetrated against them. The second displacement occurred in the 1990s, when the three Private Reserves were created. There are still remnants of their former houses in the area. In fact, the spatial organisation of Settlement 1,[1] which is the closest to the protected areas, is a direct consequence of these displacements. After the second displacement, three different extended families were clustered in a region which covers 0.2 km² and is surrounded by rivers, locally called 'the island' (Chiaravalloti, Homewood and Erikson 2017). This spatial pattern of occupation (i.e. where more than one extended family live together) does not exist in any other community in the western border of the Pantanal (Figure 14.1).

When the protected areas were established, several restrictive laws were imposed upon local fishers. During the 1980s and 1990s, for example, legislation forbid the use of fishing nets in the Pantanal (Catella et al. 2014). At the same time, a new fishing tourism business emerged in the region and rapidly came to dominate the local economy – generating an estimated US$150 million per year (Girard and Vargas 2008). By 1999, in the southern Pantanal (the only region where they record annual data), a total of 59,000 tourists per year came to fish in the region (Catella et al. 2014).

In the face of these new restrictions, local fishermen were driven to seek alternative livelihoods either locally or in nearby cities, with many starting to work in fishing tourism as guides to fishing spots (i.e. *piloteiros*) or as bait suppliers (Catella et al. 2014). The small lungfish Tuvira (*Gymnotus spp.*; 2–42 cm) became the most important bait, and the Pan-

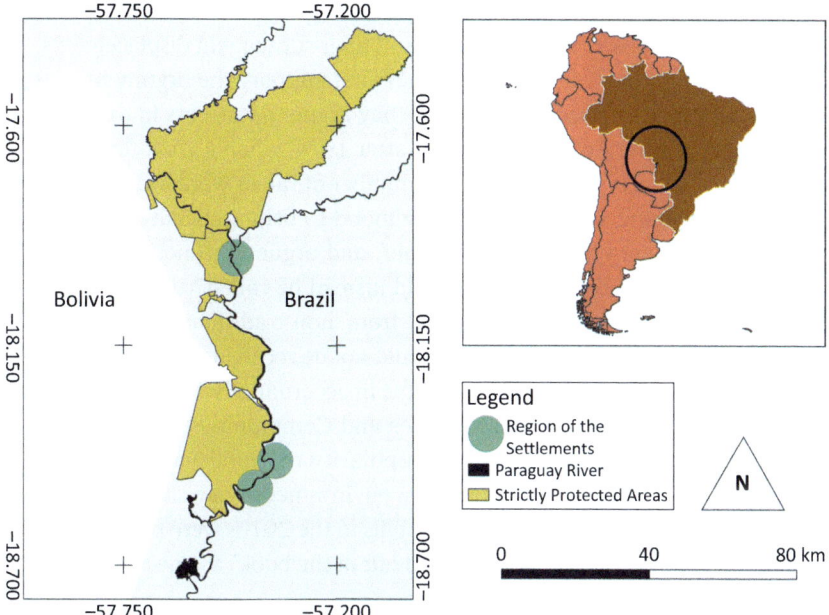

**Fig. 14.1** Current location and extent of protected areas and human settlements. Inset: location of the Pantanal in South America (top right). Credit: Map created by Rafael Chiaravalloti 2019.

tanal crab (*Dilocarcinus pagei*; 5–10 cm) became the second most important, representing 50.1 per cent and 34.2 per cent of the total bait catch, respectively (Moraes and Espinoza 2001).

Suddenly, tourist numbers started to decline. By 2006, tourists in the southern Pantanal dropped to roughly 15,000 people per year (Catella et al. 2014). Local companies claimed that there were no tourists because fish stocks were depleted, reviving accusations of local, small-scale commercial fishermen overfishing. They therefore supported tougher enforcement on fishing quotas, especially with respect to some large fish species, and restrictions on certain types of fishing gear and practices, especially those used by local people (Catella et al. 2014). As a consequence of this, commercial fishermen in the Pantanal today may only practice hook-and-line fishing, and those who continue to gather bait may only do so in southern Pantanal (Catella et al. 2014).

In 2013, several managers of protected areas, scientists and policymakers published a book titled *Biodiversity and Human Occupation in the Pantanal Mato-grossense: Conflicts and Opportunities* (Franco et al. 2013).

The book endeavours to provide a scientific argument for the severe restrictions imposed on local people and their subsequent displacements. Based on one week of fieldwork, they first support the argument that local people were overfishing. Then, they argue that fishers in the western border of the Pantanal appeared after 1974, when a great flood covered part of the region, leading to a number of ranch workers moving to the riverside and switching their livelihood to fishing. Finally, they discuss the concept of 'traditional people', and argue that most of the so-called traditional communities should instead be called 'rural poor' due to their lack of distinctive difference from 'non-traditional people'. The authors argue that the rural poor should not be receiving any public benefits that are given to traditional people in accordance with the Brazilian National Policy of Traditional Peoples and Communities (Chiaravalloti 2019). The book gained so much popularity that policymakers began using its conclusions to propose new environmental legislation for the Pantanal. In fact, many of the fishing bills in the 2010s were proposed 'as a way to address the problem pointed out in the book', as one of the local policymakers commented.

In order to prevent further physical and economic displacements, a local NGO which focuses on human rights and conservation, Ecologia e Ação (ECOA), began to publicise conflicts in the area and support local people in better organising themselves. In Settlement 1, the NGO helped local people to establish a formal association and managed to force the local municipality to build a new school and install public telephone infrastructure. As part of the scientific board of the NGO, I started to get familiar with the conflict and was invited to support them with scientific information and evidence about the socio-ecological dynamics of the area.

For those who were working in the Pantanal, it was clear that Franco et al.'s (2013) conclusions were based on their own views about the region, with little supporting empirical evidence (Chiaravalloti 2016). Their claim that fishers were outsiders was fabricated. First, non-indigenous families have been established in the western border of the Pantanal for more than 150 years, deriving from intermarriage between ex-slaves, Paraguayans and local indigenous families (Da Silva and Silva 1995). Moreover, living across the flooded areas, they carry out different professions but have a primary focus on fishing, with some recorded as selling salted fish in Corumbá city during the early nineteenth century (Silva 1986). In other words, fishing has been part of the local non-Indian people's livelihoods for well over 150 years.

Despite these long-standing patterns of residence, at the time, there was still little to no understanding of the sustainability of local people's current fishing practices. Furthermore, there was little sustained analysis of whether the activities of local people do indeed jeopardise the local ecology. This later became one of the primary research questions for my doctoral research, from which this book chapter draws.

## 3. Mapping the sustainability of fishing

In order to explore natural resource use in the western border of the Pantanal, we used mapping tools that could support collecting geographic data of local people's activities throughout the year. It is important to understand that we did not want to evaluate the ecological sustainability in Pantanal, which would require a different approach based on a long-term ecological study (Cooke et al. 2016). Rather, drawing upon Ostrom's (1990) argument on the sustainability of common property regimes, we hypothesised that local people's embedded rules should ensure the sustainable use of resources. Therefore, the main goal was to uncover local people's resource use strategies, customary use and common property regimes.

At the time, it was clear that people had to go fishing far from their settlements. We did not know how far they were going, if they were going in groups and whether their fishing grounds had physical overlaps with the protected areas' boundaries. Answering these questions, which all have a strong geographic element, could help us evaluate the presence of customary practices, such as the existence of clear boundaries between families or communities.

It was decided to focus on Settlement 1, where currently there are 23 families living in the area and the total population is approximately 100 people. Most of the adults know how to read but cannot write. Most people younger than 20 years of age have attended school, but still have low confidence in writing. Fishing is the main livelihood for more than 90 per cent of households in the settlement (Chiaravalloti 2019).

The tool that was chosen to support the mapping of local people activities was Sapelli. Sapelli is based on extreme citizen science practices and philosophy, and enables local users (regardless of literacy levels) to collect georeferenced data such as boundaries of territories and sites of importance for specific resource use in scientifically robust and locally relevant presentational forms using handheld Global Positioning System (GPS) units and pictogram-driven software. In principle, the tool enables the recording of qualitative data such as human well-being, customary governance and/or natural resource use in a scientifically robust way (Lewis 2012). The mobile app that we used in this study was based on a decision tree logic. People are guided in making a sequence of choices until they reach the final data item in the decision tree for which data are being collected.

In order to build the first Sapelli prototype for the Pantanal case, a preliminary pilot study took place with fishers from Settlement 1 (in July–August 2014). The main goal was to understand all types of resource use that local people carry out in the region, and this information in turn would inform the development of the relevant Sapelli decision tree. The technical development of the decision tree was supported by Gill Conquest, former PhD researcher at the UCL Extreme Citizen Science research group.

The first decision tree included information about all types of activities where people make use of natural resources in the Pantanal; that is, fishing, gathering bait, hunting, harvesting wild rice and collecting wood, straw, fruits, honey and drinking water. People were also asked to record the presence of *dequada*, which is a period of the year when thousands of fish die due to the flooding of vegetation and increase in water temperature. Figure 14.2 shows a zoomed in version of the decision tree from the preliminary pilot, and Figure 14.3 shows a screenshot of the application used during this time. In the last branch of the decision tree, local people were asked to voice record both the name of the place and the number of people who were with them during that specific activity. All decisions (fishing, gathering bait, etc.) were communicated using relevant pictograms, as shown in Figure 14.2.

After reaching the final step of the decision tree, the application automatically searches for a GPS signal. The data collected included the location, day and time, and the selected activity the user was carrying out.

Exploring research questions that require the acquisition of qualitative data, such as customary practices, needs the establishment of a level of trust between the researcher and the interviewees so that accurate responses can be collected. First, participant observation was used as a way to understand local experiences better and to build rapport. This saw me being involved and helping with all daily activities that local people commonly engage in, for example gathering bait, fishing, logging and attending different kinds of meetings and celebrations. Living with them therefore allowed me to build a bond with local fishers, and gradually show and demonstrate the significance of using Sapelli to collect the project's data. Second, semi-structured interviews (SSIs) were also used to gather additional data and to build a better understanding of local peo-

**Fig. 14.2** Part of the first decision tree built for the project. It zooms into specific decisions made regarding resources from 'water'. Another part of the decision tree (not shown) dealt with resources related to 'land', such as wood and honey.
Source: author.

ple's resource management and use from both a historical perspective and a more contemporary perspective.

Finally, paper maps of the region were printed using the new Brazilian collection of RapidEye satellite images which have a five-metre

**Fig. 14.3** Screenshot of the first version of the mobile app. The example shows the branch focused on recording the presence of *dequada* and its extent (large or small). Credit: Sapelli platform UCL Extreme Citizen Science research group (ExCiteS).

resolution on a 1:20,000 scale, and which cover the whole western border of the Pantanal region. They were printed on laminated paper, on which people could draw and edit their drawings. These paper maps were used in a similar way to the Sapelli data-collection application for adding geospatial information that concerns their daily activities. All SSIs were carried out using the paper maps, as shown in Figure 14.4. Through the use of these maps, we were also able to increase the number of people who engaged in the study.

It should be noted that before the project commenced, several ethical consents were sought. First, University College London Anthropology Department Ethics Committee approved a risk assessment and ethics methods procedure, and authorised the research to proceed with fieldwork. Then, following the Brazilian 'Rules of projects for research that involves human beings' (Resolution number 466 from 2012), the project was translated into Portuguese and submitted via a separate ethics application to the National Research Ethics Committee. Then, local NGOs and research institutes located near the study site (such as ECOA, Embrapa Pantanal, UFMS, Acaia and Instituto Homem Pantaneiro) were contacted, and the project was explained to improve their awareness of the purpose of the research and of the form the data collection was going to take, as well as possible outcomes of the project. The same approach was carried

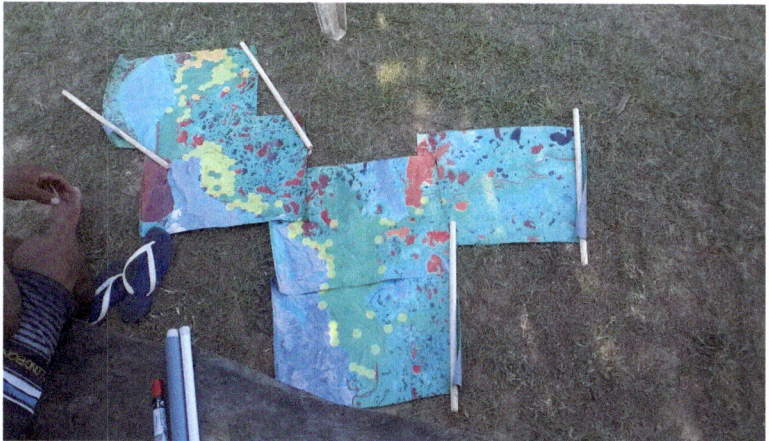

**Fig. 14.4** Photo of a fisher showing locations and types of natural resource use in the region. Credit: Photograph taken by Rafael Chiaravalloti.

out with community leaders. This process was followed by individual informed consent from all participants who were interviewed.

## 4. Interacting with the Sapelli tool for gathering natural resource use data

The Sapelli interface developed for local people in the Pantanal to collect data on their daily activities was installed on Samsung Galaxy XCover smartphones. They are sold as being waterproof and resistant to harsh environmental physical conditions. Given the tough conditions of the Pantanal wetland, they were considered suitable for this case study.

### Step 1: Identifying people to pilot test the tool

In the first meeting with local families living in Settlement 1, after explaining the project and obtaining their individual informed consent for data collection, it was decided that each day, a different family would use the mobile phones to record their daily activities. At the time, we had four mobile devices to use in the Pantanal. The first two weeks of data collection were set as a pilot test to explore whether people had any difficulties using the application. I offered to go with them in their daily activities so that I could provide technical support when needed.

This was the first challenge that I encountered: not all families were comfortable in taking me with them during their activities or in using the mobile phones without any support. Although it was made clear that all data would be kept confidential, people were afraid that I was going to use the information on locations of fishing grounds to inform outsiders and rangers when they were using restricted areas, which could result in fishers receiving a fine of up to R$1,000 (c. £200). During the individual informed consent, it was made clear that people would be able to review the data collected, and agree whether it should be used for publication. All community members agreed to share the data after they were collected. In my opinion, that was due to the majority of the fishing grounds being outside the limits of the National Park, as well as because they understood the importance of showing their territory, using maps, to policymakers and local managers in order gain rights of tenure and use of resources.

The second challenge was related to mobile device use. People were worried about using the mobile devices for fear of breaking or losing them, even though it was made clear that they would not be held responsible for any damage. It was also not possible to charge the phones, since there was no electricity in the settlement, and most families did not have power generators. From the total of 23 household families who were interviewed and invited to use Sapelli, only three agreed to use it.

### Step 2: Adapting the decision tree

In order to test the use of Sapelli fully, I established the following protocol. Each day, I would follow a different family who had agreed to use the mobile phones to collect data. Each family was therefore supported for one out of every three days that they used the device. However, some families did not leave their houses for many days, and when they did, on several occasions they did not take the phones with them. During the two weeks I spent accessing the usability of Sapelli, I nevertheless collected enough data to understand the interface design changes needed in order to improve its usability.

The first interaction barrier encountered was related to the complexity of the initial decision tree. For each new record, people had to navigate across seven or eight interface selections in order to record a single item in a specific geographic location and to collect the necessary information about this particular natural resource use instance. It is not

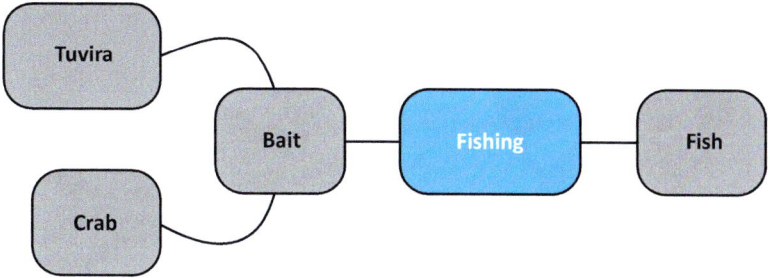

**Fig. 14.5** Final decision tree used for the Pantanal version of the software. Source: author.

easy using a mobile phone on a small boat (or canoe) full of fish or bait and fishing gear. Moreover, the data-collection time actually meant that they had less time to spend doing the activity (e.g. fishing). Finding the right balance of using the application without severely distracting them from pursuing their livelihood was of utmost importance. After initial pilot tests, it was collectively decided that the number of choices or steps they had to make across the decision tree should be reduced.

The first change which was applied to the decision tree referred to the way people carried out the activities. For instance, after choosing 'collecting bait', people would be prompted to identify how they were collecting bait (either 'inside the river' or 'from the boat'). It was decided that not collecting these data would not influence the effectiveness of the main research goal concerned with natural resource use, and therefore it was decided to remove this step from the decision tree.

People were still not comfortable using the updated Sapelli version. They thought it was still too complex. So, it was decided that the main focus of the data-collection activity should be changed. Instead of looking at resource use in general, we decided to concentrate on collecting data about fishing and gathering bait only. Although this was a big change in terms of the underlying research and the purposes for which it was being conducted, we decided that collecting data about fishing activities easily and accurately was a higher priority. As a consequence, a new version of Sapelli was developed which had a much more simplified decision tree, and which, as Figure 14.5 shows, included no more than three steps before reaching the end of the tree. Information about the other activities that we had to remove from the initial Sapelli versions (e.g. harvesting rice, collecting honey or clean water etc.) was then collected via interviews and participatory observation.

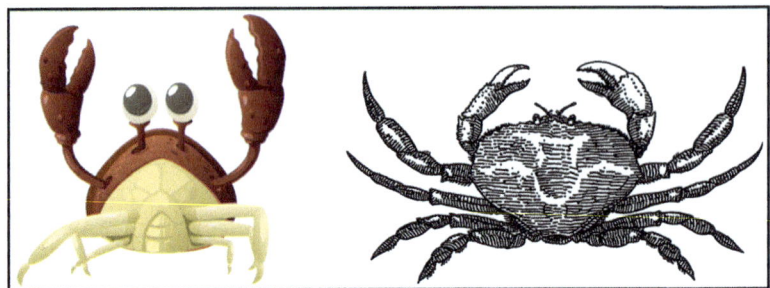

**Fig. 14.6** Figure of a crab initially used to represent the 'gathering crab' in the software (left). Final figure used to represent 'gathering crab' (right). The image is a scientific illustration of *Dilocarcinus pagei* – exactly the same species that local people collect as bait.
Source: Pixabay.com

### Step 3: Adapting the pictograms

After reaching community consensus on the data-collection process in the new Sapelli version, another major interaction barrier emerged. The first version of Sapelli was built using pictograms. That decision was made due to emerging research from other contexts that suggested that people with low literacy skills find it easier to understand pictorial design (Lewis 2012). Interestingly, our experience contradicted those previous findings.

Local people in the Pantanal did not particularly agree with the use of pictograms. First, they thought that the cartoon symbols did not accurately represent the fish or bait they were seeing. Although I pointed out that the pictograms were a way to represent a general concept of fish or bait, they argued that it would be better to have a visualisation which resembled the actual species for which data were being collected. This was one way they felt they could verify the accuracy of the data they were collecting. The pictograms were therefore replaced with scientific illustrations of fish and bait, as shown in Figure 14.6.

### Step 4: Replacing the mobile devices

During the pilot test period, two out of four mobile phones broke, leaving only two mobile devices to use during the data-collection process, which is illustrated by Figure 14.7.

**Fig. 14.7** Sequence of photos showing the same fisherman recording his natural resource use in the Western Border of the Pantanal. In the first two frames, he is fishing, and in the second sequence, he is gathering bait. Credit: Photographs taken by Rafael Chiaravalloti.

## Step 5: Data collection

After defining the content of the decision tree and the final pictograms with local people, we invited two families who had participated in the initial pilot tests to use the remaining two mobile phones to collect data on their fishing or gathering bait activities. Both families collected data over a period of five months, from July to November 2015.

## Step 6: Results of the data collection

Combining the results of the participatory mapping using the satellite imagery, participant observation and the use of Sapelli, we managed to uncover a deep understanding of natural resources customary use in the Pantanal region. The georeferenced data collected throughout the year showed that people do not spend more than a week on the same fishing or gathering bait ground. They move to a new site when the fishing return diminishes (Chiaravalloti 2017). Because the flood pulse keeps moving from north to south, people have to move their fishing sites regularly. Throughout the year, this process creates a rotational fishing system. Moreover, it was observed that several families go together either to fish or to gather bait. This increases the chance of finding a good fishing ground.

This fishing system is very similar to mobile systems used by other communities around the developing world – practices hailed as displaying sustainable management for non-timber forest products (Assies 1997), grazing (Kothari, Camill and Brown 2013), fishing (Berkes 2006), agriculture (Sunderlin et al. 2005) and bushmeat hunting (Kümpel et al. 2009), in line with the biological principles of metapopulation dynamics (Hanski 1998). In principle, and often in practice, mobile exploitation helps avoid exhaustion of natural resources and allows different species populations to recolonise the areas that have been used (Wilson et al. 1994).

Another important aspect that was revealed through the use of Sapelli is the presence of community territory. The data showed that the reciprocity within people from Settlement 1 towards other community members does not extend to people from outside their community. People from Settlement 1 were able to indicate on the maps what they call 'their' area, and demarcate the boundaries which define Settlement 1 resources (e.g. see Figure 14.8). Therefore, each settlement has its own territory, with clear notions as to the numbers of people allowed access, who controls use of specific spots and with whom each person shares information about such spots. Based on the presence of this territory and

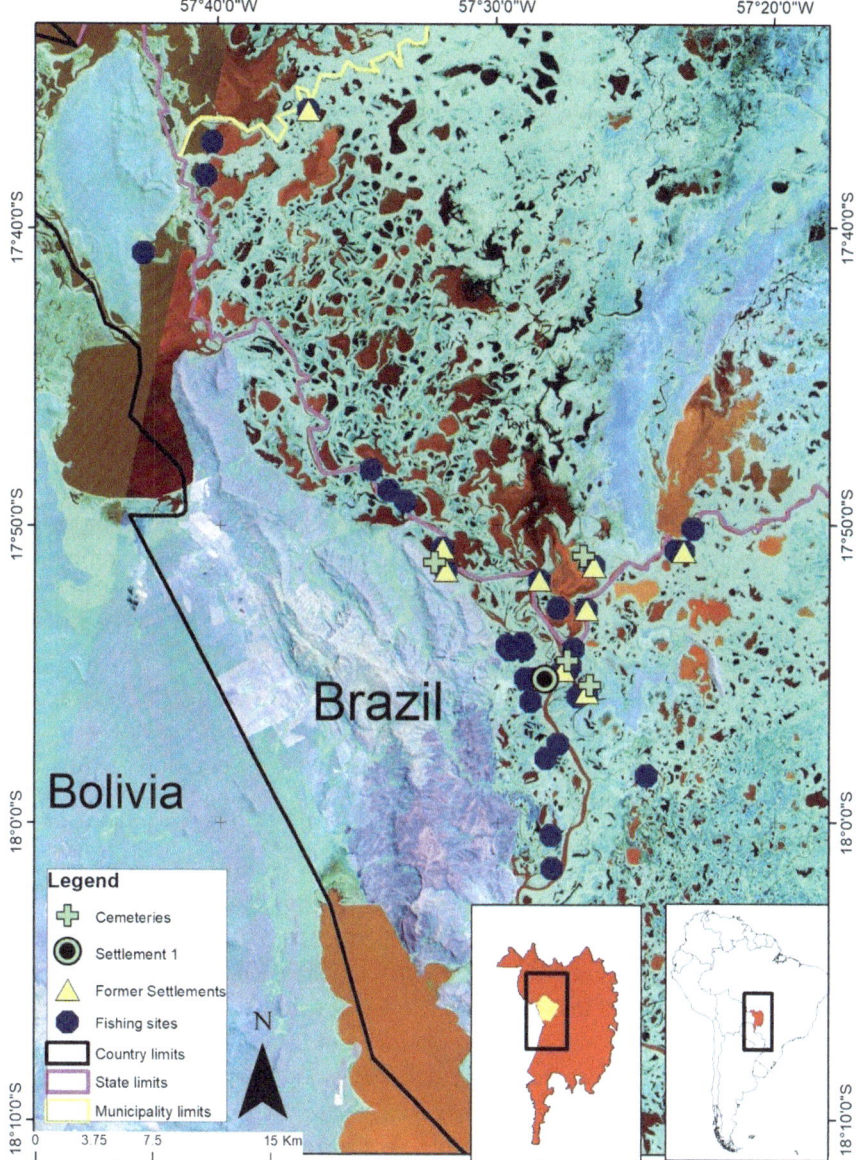

**Fig. 14.8** Territory defined by local people from Settlement 1. Inset: location of the Pantanal in South America (right) and location of the study area in the Pantanal (left). Credit: Map created by Rafael Chiaravalloti 2019.

rotational resource use, alongside the existence of hundreds of inaccessible fishing grounds in the region, Chiaravalloti and Dyble (2019) have shown that the local people's fishing activity in Settlement 1 is indeed sustainable. Other authors have shown similar findings, pointing out no signs of overfishing in the Pantanal (Mateus, Vaz and Catella 2011), and highlighting in the western border of the Pantanal an 'excellent degree of biological integrity' (Polaz, Ferreira and Petrere Júnior 2017). Therefore, the evidence that the project has gathered using geographic citizen science has played a fundamental role in deconstructing the picture that conservationists have previously drawn in terms of how local people make use of natural resources in the Pantanal (Franco et al. 2013).

## 5. Conclusion

Participatory mapping and geographic citizen science are important in allowing local people (regardless of their scientific knowledge) to represent their knowledge, customary habits and management strategies in a scientifically valid way. The case study discussed in this chapter clearly reflects this.

Local people in the Pantanal were accused of squatting in several strictly protected areas and for overfishing local fish stocks. However, Sapelli, in combination with SSIs and participant observation, helped uncover customary practices and a historical precedent, and demonstrate a sustainable use of natural resources. The results triggered a new political environment. Today the local people are recognised as a traditional group, and a new community reserve to protect their livelihoods is under discussion (Chiaravalloti 2019). Given the success of this project, ECOA decided to buy another 10 mobile phones and installed the latest version of Sapelli developed in Settlement 1. They distributed the phones to 10 different families from five different communities in the Pantanal. The main goal of this new geographic data-collection activity is to identify the boundaries of territory and customary use. The project is currently underway, and so there are no results yet from this new project to share with the reader.

Nevertheless, it should be noted that it required months of intensive fieldwork to gain the necessary rapport to convince local people about the importance of using Sapelli to collect this type of data and also to understand how best to design a software solution and interface design which would meet local needs and be able to be used successfully. Therefore, the positive impact this project had on their lives and for sustainability in general was due to a combination of different methods, and, most importantly, the researcher's ability to spend a relatively large amount of time in the field. Under these circumstances, geographic citizen science may have a huge potential to offer in terms of truly supporting a paradigm shift in the context of conservation.

Time, nonetheless, is scarce for the majority of conservation initiatives. Projects tend to be led by large NGOs who hire local organisations for a short period of time to use a framework developed somewhere far from the local reality (Rodríguez et al. 2007). Without investing the necessary time to build rapport, understand user issues and work around technical limitations, geographic citizen science is bound to fail. Given time and a participatory approach to engage with people fully, however, it offers a unique way to support local people and conservation organisations to meet their goals of sustainable development.

## 6. Lessons learned

- Geographic citizen science has great potential to support better participatory initiatives as part of conservation projects.
- Time spent with local people building rapport and trust plays fundamental roles in the success of geographic citizen science initiatives.
- Other qualitative methods, such as participatory observation and SSIs, need to work together with geographic citizen science to allow effective adaptions to represent local people's needs better.
- People with natural resource-based livelihoods may experience financial loss while collecting data. The length of time required to collect data has to be seriously considered, and compensation should be considered.
- Successful geographic citizen science depends on a change to how conservation initiatives approach local people as a whole. They should focus on real needs instead of silver-bullet solutions.

## Acknowledgements

This study was funded by Science Without Borders CNPq/Capes (GA 237737/2012-4), WWF Russell E. Train Fellowship (SW14), Rolex Award for Enterprise and Handsel Scholarship for Wildlife Conservation. I am grateful to Artemis Skarlatidou for the critical review and discussions, local people who hosted me during field trips and my beloved friend Gil Conquest, who introduced me to Sapelli, helped with the complicate coding and will always be remembered.

## Note

1   The real name of the community is anonymised in order to preserve local people's identities.

## References

Abbott, James G., and Lisa M. Campbell. 2009. 'Environmental histories and emerging fisheries management of the Upper Zambezi River floodplains', *Conservation and Society* 7: 83–99.

Alho, Cleber J.R., and Roberto Esser Reis. 2017. 'Exposure of fishery resources to environmental and socioeconomic threats within the Pantanal wetland of South America', *International Journal of Aquaculture and Fishery Sciences* 3: 22–9.

Assies, Willem. 1997. 'The extraction of non-timber forest products as a conservation strategy in Amazonia', *European Review of Latin American and Caribbean Studies* 62: 33–53.

Berkes, Fikret. 2006. 'From community-based resource management to complex systems: The scale issue and marine commons', *Ecology and Society* 11: 45.

Bertassoni, Alesandra, Nilson L. Xavier-Filho, Fernanda A. Rabelo, Stephanie P. S. Leal, Grasiela E. O. Porfírio, Viviane F. Moreira and Angelo P. C. Rabelo. 2012. 'Paraguay River environmental monitoring by Rede de Proteção E Conservação Da Serra Do Amolar, Pantanal, Brazil', *Pan-American Journal of Aquatic Sciences* 7: 77–84.

Catella, Agostinho C., Selene P. Albuquerque, Fânia Lopes de Ramires Campos and Darci Caetano dos Santos. 2014. 'Sistema de Controle Da Pesca de Mato Grosso Do Sul SCPESCA/MS – 20 – 2013', *Boletim de Pesquisa E Desenvolvimento* 127: 57.

Chiaravalloti, Rafael M. 2016. 'Is the Pantanal a pristine place? Conflicts related to the conservation of the Pantanal', *Ambiente e Sociedade* 19: 305–10.

Chiaravalloti, Rafael M. 2017. 'Overfishing or over reacting? Management of fisheries in the Pantanal wetland, Brazil', *Conservation and Society* 15: 111–22.

Chiaravalloti, Rafael M. 2019. 'The displacement of insufficiently "traditional" communities: Local fisheries in the Pantanal', *Conservation and Society* 17: 173.

Chiaravalloti, Rafael M., and Mark Dyble, 2019. 'Limited open access in socioecological systems: How do communities deal with environmental unpredictability?', *Conservation Letters* 12: e12616.

Chiaravalloti, Rafael M., Katherine Homewood and Kirsten Erikson. 2017. 'Sustainability and land tenure: Who owns the floodplain in the Pantanal, Brazil?', *Land Use Policy* 64: 511–24.

Cooke, Steven J., Edward H. Allison, T. Douglas Beard, Robert Arlinghaus, Angela H. Arthington, Devin M. Bartley, Ian G. Cowx et al. 2016. 'On the sustainability of inland fisheries: Finding a future for the forgotten', *Ambio* 45: 753–64. https://doi.org/10.1007/s13280-016-0787-4.

Da Silva, Carolina J., and Joana A. F. Silva. 1995. *No Ritmo Das Águas Do Pantanal*. Nupaub-Usp. São Paulo: NUPAUB/USP.

Franco, José L. A., José A. Drummond, Chiara Gentile and Aldemir I. Azevedo. 2013. *Biodiversidade E Ocupação Humana Do Pantanal Mato-Grossense: Conflitos E oportunidades*. Rio de Janeiro: Garamond.

Girard, Pierre, and Icléia Vargas. 2008. 'Turismo, desenvolvimento e saberes no Pantanal: Diálogos e parcerias possíveis', *Desenvolvimento e Meio Ambiente* 18: 61–76.

Hanski, Ilkka. 1998. 'Metapopulation dynamics', *Nature* 396: 41–9.

Kothari, Ashish, Philip Camill and Jessica Brown. 2013. 'Conservation as if people also mattered: Policy and practice of community-based conservation', *Conservation and Society* 11: 1–15.

Kümpel, Noelle F., Eleanor J. Milner-Gulland, Guy Cowlishaw and Marcus Rowcliffe. 2009. 'Assessing sustainability at multiple scales in a rotational bushmeat hunting system', *Conservation Biology* 24: 861–71.

Lewis, Jerome. 2012. 'Technological leap-frogging in the Congo Basin. Pygmies and Geographic Positioning Systems in Central Africa: What has happened and where is it going?', *African Study Monographs* 43: 15–44.

Mace, Georgina. M. 2014. 'Whose conservation?', *Science* 345: 1558–60.

Mateus, Luiz, Marcos Vaz and Agostinho Catella. 2011. 'Fishery and fishing resources in the Pantanal'. In *The Pantanal: Ecology and sustainable management of a large neotropical seasonal wetland*, edited by Wolfgang J. Junk, Carolina J. Da Silva, Cátia N. Cunha and Karl M. Wantzen, 621–47. Sofia, Bulgaria: Pensoft.

Moraes, André S., and Luizio W. Espinoza. 2001. 'A captura e a comercializacao de iscas vivas em Corumba, MS', *Boletim de Pesquisa* 21: 38.

Ostrom, Elinor. 1990. *Governing the Commons*. Cambridge: Cambridge University Press.

Polaz, Carla N. M., Fábio C Ferreira and Miguel Petrere Júnior. 2017. 'The protected areas system in Brazil as a baseline condition for wetlands management and fish conservancy: The example of the Pantanal National Park'. *Neotropical Ichthyology* 15.

Rodríguez, Jon P., Andrew B. Taber, Peter Daszak, Raman Sukumar, Claudio Valladares-Padua, Suzana Padua, Luis F. Aguirre et al. 2007. 'Environment: Globalization of conservation: A view from the south', *Science* 317: 755–6.

Sandbrook, Chris. 2015. 'What is conservation?', *Oryx* 49: 565–6.

Silva, Miguel Vieira. 1986. *Mitos e Verdades Sobre a Pesca No Pantanal Sul-Mato-Grossense*. Campo Grande, Brazil: FIPLAN.

Sunderlin, William D., Brian Belcher, Levania Santoso, Arild Angelsen, Paul Burgers, Robert Nasi and Sven Wunder. 2005. 'Livelihoods, forests, and conservation in developing countries: An overview', *World Development* 33: 1383–402.

Tocantins, Nely. 2011. 'Do Mar de Los Xarayes Ao Complexo de Áreas Protegidas Do Pantanal Mato-Grossense', *Revista Do Instituto Histórico E Geográfico de Mato Grosso* 1: 17–27.

Wilson, James A., James M. Acheson, Mark Metcalfe and Peter Kleban. 1994. 'Chaos, complexity and community management of fisheries', *Marine Policy* 18: 291–305.

# Chapter 15
# Digital technology in the jungle: a case study from the Brazilian Amazon

Carolina Comandulli

## Highlights

- Cultural aspects should be considered when designing technologies for specific segments of society. The social organisation and internal rules of communities should be respected and followed for success.
- Free, prior and informed consent is a continuous process that should be carried out as projects evolve.
- User feedback in an application should be quick and clear. The information captured by the user should be easily and rapidly made available. Otherwise, it may lead to a lack of interest and motivation for using the technology.
- Training on equipment maintenance can be as important as training on how to use a specific technology.
- Technologies should ideally be incorporated in daily use. Otherwise, they may be forgotten.

## 1. Introduction

This chapter is about the experience of developing and using the Sapelli software in an Ashaninka indigenous village in north-western Brazil, mediated by a process of extreme citizen science methodologies and tools. The period of research training, practice and observation described in this chapter took place between January 2015 and June 2017.

The interdisciplinary Extreme Citizen Science research group (ExCiteS) at University College London (UCL) was set up in 2012, with the aim of developing tools and methodologies to establish a collaborative, plural, participatory space, where multicultural interactions can be made explicit, struggles can be observed and recorded, and the conditions for the design of local solutions can emerge. One of the main tools developed to facilitate this process is the Sapelli software – a data-collection and -sharing platform designed for illiterate users with little or no previous communication technology experience (as described in detail in Chapter 11). Sapelli is intended to be adaptable to specific contexts to overcome accessibility issues to digital technology tools such as non-literacy and numeracy, the inability to read maps and a lack of electricity and network connectivity.

My engagement with ExCiteS led me to work with the Ashaninka people from Apiwtxa village, with the extreme citizen science approach making my research with them viable, as it offered a practical application to a concrete problem they were facing. As an anthropologist, I also investigated other themes while living with them for two and a half years.

In this chapter, first I present who the Ashaninka from Apiwtxa are, describe the environment where they live and explain why they were interested in engaging in a geographic citizen science project. Then, I detail the work of ExCiteS and the development of the Sapelli application in that specific context. Finally, I comment on the practical outcomes and challenges of the initiative.

## 2. The Ashaninka from Amônia River and their fight against invasions on their land

The Ashaninka are Arawak-speaking people who inhabit the Peruvian and Brazilian Amazon rainforest. They number more than 100,000[1] people and are probably the biggest indigenous population of lowland Amazonia. In Brazil, they live in Acre State and inhabit six indigenous lands (Ricardo and Ricardo 2017). This study took place with the Ashaninka from Kampa do Rio Amônia Indigenous Land (or the Ashaninka from Apiwtxa), situated on the border between Brazil and Peru, in Marechal Thaumaturgo town, Acre State. Their land was titled in 1992 with 87,205 ha after a great struggle against loggers. It is surrounded by protected areas of forest on both the Brazilian and Peruvian sides. There are nearly one thousand[2] Ashaninka dwellers in this land, and most of the population currently live in a single village called Apiwtxa. Apart from the main village, there are four other Ashaninka settlements in the indigenous land.

Apiwtxa village has limited communication and energy resources. When I was in the field, there were two public phones which were often broken and an Internet connection via satellite in a small house, which was also often in need of maintenance. The Internet via satellite was installed in 2003 through a digital inclusion project by the Brazilian Ministry of Culture. Despite the Internet access and the existence of two computers in the village, few people were allowed to access them, as the equipment was meant to be used exclusively for the community organisation. There was a solar panel that provided energy for the computer room only. A couple of other houses had gasoline generators, but in general, electricity was scarce.

Besides the computer room, there were two buildings for the cooperative. One was a house to stock the crafts produced in the village and to sell them to visitors, and the other was the cooperative market where there were goods from town which were mostly accessed in exchange for the crafts delivered by Ashaninka families. The range of goods available consisted of a list of items considered of prime necessity, such as salt, fishing hooks, lighters and soap.

The village had a primary education school. Secondary education was eventually offered, depending on demand. The primary school was mostly taught in Ashaninka, and all teachers were Ashaninka from Apiwtxa. There was no permanent medical assistance, but a team of doctors and nurses would come to the village approximately every two months. The village had five indigenous health agents hired by the government to help give guidance in case someone in the village was sick. There was no sewage system, and there were only a few toilets in the central area of the village.

Even though most families lived quite close to each other, the houses were scattered in the area. Population density was greater close to the school, the public phones and the cooperative buildings. In the busiest part of the village, the Ashaninka could no longer have gardens around their houses due to lack of space. So, their gardens were set in areas further from the main village, reachable by boat and/or by foot.

## 2.1 Protecting the land

The Ashaninka from Amônia River constantly fight illegal activities in their territory, but they often struggle to be heard by governmental enforcement agencies. They have monitored their land mostly by themselves, organising voluntary monitoring expeditions to the limits of their territory on a regular basis. They often invite relevant governmental authorities to participate in their actions, but they carry out their expeditions regardless of institutional support. When they identify invasions, they always report them to the responsible authorities. After several attempts with no practical results, they then take action themselves.

At the beginning of the 2000s, for instance, their land suffered a serious logging invasion from Peru, leading to issues of hunger. The loggers were both consuming animals and scaring them away, and community members feared meeting an invader if they went hunting and fishing. After many attempts to denounce the invasion to Brazilian and Peruvian authorities, the Ashaninka decided to act themselves and captured a number of loggers in the middle of the forest. Only then did the governments take measures to control the invasion. The measures were very powerful and gave considerable visibility to the Ashaninka, such as obliging the Brazilian government to install offices of the Federal Police, National Foundation for Indigenous Affairs, the Brazilian Institute of Environment and Renewable Natural Resources and the army in Marechal Thaumaturgo town, and to undertake regular helicopter flights to check the frontier line. At the time I was in the field, their main concern was invasions by illegal poachers coming from Marechal Thaumaturgo. In most cases, the poaching was not happening inside their land but in the Peruvian frontier with the Ashaninka native communities of Sawawo and Shawaya, just across the border, along the Amônia River. Notwithstanding the location, Apiwtxa community members were already feeling the impact of the illegal hunting and fishing on their food sources. Another entry point for poachers was the Arara River – a small river on the eastern limit of their land.

Apiwtxa's efforts to stop the poaching activity were wide ranging, from monitoring expeditions and denouncing to authorities, to awareness raising meetings with Arara settlement (inside their land) and the Sawawo and Shawaya communities. Stopping poachers by themselves was a last resort. They rightly understood that it was not their role to confront poachers but rather the role of the state, and feared that by confronting them directly, they would suffer death threats when going to Thaumaturgo town.

The Ashaninka were interested in collecting geographic data using extreme citizen science processes, as they thought that by improving their monitoring strategies, refining the quality of the evidence they collected and speeding up the communication with enforcement institutions with the use of digital technologies, they could get a more effective response from authorities. Appointed members of the community were already acquainted with the use of other digital technologies such as digital cameras, Global Positioning Systems (GPS) and audio recorders, and the Ashaninka were especially interested in quickly capturing geolocated photographs of the invasions accompanied by additional details.

Accordingly, a monitoring project was established using Sapelli to improve the evidence collection of invasions while quickly and accurately informing the relevant authorities about what was happening inside the Ashaninka land. The project was materialised in collaboration with Comissão Pro-Indio do Acre (CPI-AC) – a grass-roots Brazilian non-governmental organisation (NGO) that had been working with the community for decades, and in synergy with the additional activities of a project financed by the Amazon Fund – a REDD+ mechanism created to raise donations for non-reimbursable investments to prevent, monitor and combat deforestation, as well as to promote the sustainable use of the Brazilian Amazon.

## 3. Extreme citizen science and Sapelli software: the development process

To begin the process of building a Sapelli application with the Ashaninka from Apiwtxa, I followed the detailed participatory methodology proposed by extreme citizen science, which encompasses:

(1) A process of free, prior and informed consent (FPIC) 'to ensure that project activities and their potential consequences are fully understood by the majority of the community before monitoring activities begin' (Lewis and Nkuintchua 2012, 155). The FPIC process involves informing and discussing with participants project objectives, benefits and risks (and how to address them), and asking for the parties' consent. In this process, people's timing and processes of decision making must be respected.
(2) A period of iterative, participatory software design, during which local communities collaborate with ExCiteS' facilitators and other project stakeholders to identify the issues they wish to resolve in order to create tailored data collection tools based on an adaptable software platform (in the current case using Sapelli software) that will allow them to evidence the issues.

(3) Building community protocols (CPs) for engagement with (a) the project itself and (b) other stakeholders implicated in the problems local people have identified (e.g. NGOs, companies, the government, etc.). The idea is 'to strengthen the political organisation and participation of communities' (Lewis and Nkuintchua 2012, 155), with the definition of what resources each party will commit, their roles and responsibilities, access levels to the data, how it is going to be used and so on.

FPIC, software development and CPs have to be flexible enough to be changed or updated if necessary during the course of action. Consent and protocols should be registered in modern ways (either filmed and/or written) as well as in locally relevant ways.

My first field trip to Apiwtxa was in January 2015. At that time, I began the FPIC process with the community, which continued during my fieldwork. On that occasion, I presented the ExCiteS' work to them, and we talked about the possibility of collaboration. One of the exercises carried out was to discuss in groups what they wanted to monitor; the result was the invasions on their territory. Another exercise was focused on types of evidence they identified to characterise such invasions. With this material in hand, I went back to UCL, and after being trained how to work with Sapelli software, I built a prototype of a monitoring application to take back to Apiwtxa.

When I returned to the village, in the middle of April 2015, I scheduled a series of training days for the monitoring team, taking into account participants' availability and the village rhythm. The Ashaninka had decided to train a group of 10 monitors at first, which in the end became a group of 13. The group of monitors was composed of literate and illiterate people, mainly men aged between 17 and 42 years old, and a young woman.

During the training period, which lasted until September 2015, I evaluated with the monitors whether the pictograms used in the Sapelli interface design were easy to understand. I adapted the pictograms that were not very clear to them, incorporating Ashaninka drawings whenever possible, introduced them to smartphones (as most of them did not have contact with mobile phones) and trained them on the use of the app with hypothetical exercises around their land. In every training session (Figure 15.1), we assessed the difficulties and challenges they were having using the device and the app, and discussed the risks in doing this kind of activity.

**Fig. 15.1** Group discussion about their understanding of the pictograms. Credit: Photograph taken by Carolina Comandulli 2017.

At the same time, I began training a group of five Ashaninka on how to transfer data from the smartphone to the computer and how to visualise it. Only two people in the village had a computer, and neither of them was in the group of monitors. During the training, they quickly learnt how to extract data, while I noted down anything that could be improved on Sapelli's user interface design based on user feedback and reported it back to the team in London, who worked on improving the design whenever possible. The idea was to make the interface as user friendly as possible so that the community could then manage the whole data-collection process independently.

After several training sessions, we did a longer data-collection exercise which encompassed a three-day expedition on the Amônia River, up to the border with Peru, and checked the protection of their territory along the way. We also took the opportunity to test the use of alternative phone chargers (including a Japanese hot-pot phone charger – Figure 15.2), solar panels and power banks.

In September 2015, after Sapelli's interface had been fully developed to accommodate the end users' needs, we carried out a CP meeting. The meeting was held in Apiwtxa village, with the participation of the community and partner organisation CPI-AC. After presenting all the steps followed since the beginning of the project, there was a group session in

**Fig. 15.2** Testing the hot-pot phone charger. Credit: Photograph taken by Carolina Comandulli 2017.

which the community responded to two questions: (1) How can the project be improved? and (2) What are the risks of collecting this information, and how can we avoid or mitigate them? The monitors requested continuing training both in the use of the application and in the data transfer. As to the risks, some suggestions were made to protect the identity of the monitors, not to confront the invaders and to organise an educational activity for the schools in the surrounding areas to raise awareness about the prohibition of hunting, fishing and logging in indigenous lands. In the afternoon, the discussion was on information flows (e.g. Who can have access to the information? Who is allowed to send it to other institutions?), and about the contents of the final version of the CP.

The final protocol document had an opening section with questions and answers about the functioning of the project (e.g. Who collects the data? Who is responsible for the equipment? How to reduce the risks of the monitoring activity?), then a section on technical and methodological support, another about the logistics support and finally one on data-sharing protocols. With the project fully established in the village, I took part in several monitoring planning meetings and trips which provided opportunities to put the application into practice.

## 4. Sapelli in practice versus an assemblage of digital technology tools

The group of selected monitors did not face many difficulties in learning how to use the Sapelli land monitoring application in its final design, even though there were a number of difficulties for users 'before' and 'after' accessing the application which I will describe in this section.

### 4.1 Navigating a smartphone interface and using a computer

In my field site, most people had low literacy and numeracy and little or no familiarity with digital technologies. The ultimate objective of Sapelli is to be accessible to such users, enabling them to set up and design their applications and to collect and visualise the data.

First, Sapelli at the time did not offer an authoring tool that enabled Ashaninka users to build their own application. Interface development required XML coding knowledge, which the author had but the end users did not. In terms of data collection, the final version of the application was easy for users to handle overall, but there were some technological barriers before being able to access the application. Samsung Galaxy XCover smartphones, used in this case, must be switched on and off with quick and long presses, and their screens require unlocking. The backlight may turn off after several seconds/minutes, depending on the settings, and one must be able to switch it back on again. None of these required actions were obvious or previously known to the community, and they needed to be practised. Otherwise, they were forgotten. Unlock-

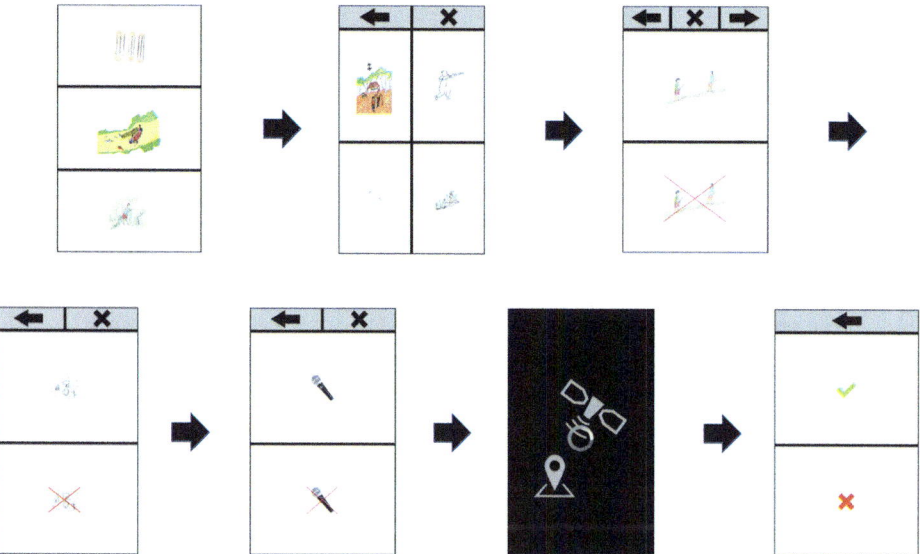

**Fig. 15.3** Example of a sequence of data collection. Credit: Sapelli platform UCL Extreme Citizen Science research group (ExCiteS).

ing the phone through swiping the screen proved to be particularly difficult in a humid environment and when one's hands were rough, which was commonly the case.

Also, with regards to the application itself, the 'decision tree' logic, which is the way Sapelli organises and structures the pictograms that are included in the interface design and which is used to navigate from abstract to more specific pictograms, was sometimes problematic. Especially towards the end of the data-collection chain of the decision tree, there was the option of selecting between the 'tick' and 'X' symbols (Figure 15.3) to either confirm or cancel the new data entry, but this metaphor was not obvious to the Ashaninka, and we found no adequate drawing to replace it in order to make it more understandable.

In relation to data visualisation, the challenges were even greater. During the time this case study took place, it was not possible for the user to see the data collected automatically. For the data to be visualised, it first had to be extracted following a series of complicated steps and then loaded into specialised software (such as an image editor, audio editor or a geographic information system tool) to see the collected data on a map. Even though the people trained to carry out this process were fast learners, they easily forgot the steps that they had to go through, as they didn't repeat the process often enough so that it could be memorised.

### 4.2 Managing equipment

Even though the app on the smartphone was quite straightforward to use when accessed, first we needed operating, charged, correctly set up and memory-free mobile phones. To begin with, in general, power is scarce in remote areas. Phones do not keep their battery charged for many days. So, it is necessary to constantly charge them so that they can be used when necessary. Internet connection is also scarce and unstable, which prevents the phone software from being properly updated and can cause the equipment to become very slow. Also, before the phones are put to use, the equipment must be adequately set up – especially when working with risky activities in the jungle. For instance, our phone set-up involved the following requirements: eliminating sound and flash, to be in an energy-saving mode and to free up the phone's memory after each use in order to make room for new data and to allow the previous data to be collected, organised and saved.

Besides being vulnerable to humidity, which is a clear challenge to equipment maintenance in forest areas, neither smartphones nor computers are 'one piece' devices. They have cables with plugs that may not match local sockets, the cables also detach themselves from the plugs to enable data transfer, the phones usually need a memory card to store extra information and so on. Also, when going on expeditions, it was necessary to be able to recharge phones, remembering the importance of switching phones off when not using them in order to save their battery. The Japanese hot-pot and the solar panels were not effective, while the power banks were light and provided a good number of complete charges, making them a preferable choice in the field. Still, they needed to be fully charged before the expeditions, and no cables could be lost!

### 4.3 Complying with agreements and the importance of considering cultural traits

The collective agreement was that monitors should take the phones with them on monitoring trips and during their individual traditional activities, such as when going hunting and fishing. In addition, there were people in charge of keeping the equipment ready for expeditions.

After the project was set up, there were many opportunities to use the phones, such as the monitoring trips and the situations in which there were invasions. Nonetheless, with time, I observed that when there was a monitoring trip, the community usually expected me to organise and set up the equipment. Also, the monitors would not take the phones with them on their daily activities. I tried to understand why this was happening, offered extra training sessions and encouraged them to manage the equipment by themselves, but this did not seem to work.

People maintained their roles as monitors, but there was not much reinforcement within the community or encouragement of others to use the technology and set up the equipment for the monitoring trips. In my assessment, this had to do with people being busy providing for their family every day (hunting, fishing and gardening) and also with the fact that monitors were involved in many community projects. Moreover, no one felt like taking the lead and encouraging others to use or improve the resources, given Ashaninka's high appreciation and respect for people's autonomy, examples of which can be found in the Ashaninka ethnography. For example, in his thesis, Killick explains: 'An adult Ashéninka man will seldom go to another for advice or help in any matter. I was repeatedly struck by how both men and women would deal with all facets of life without consulting others. All work can be done by single individuals' (Killick 2005, 70), and 'no individual is willing to tell another what to do. They might offer their opinion about something, but they never give a direct order' (Killick 2005, 100). Also, Pimenta, in his thesis, mentions one of Apiwtxa's leaders saying that people may feel inferior if they ask someone for guidance on how to do something (Pimenta 2002, 266).

During the whole project process, I sought to open space for criticism so that improvements could be made as they saw fit, but the Ashaninka users never openly criticised either the process in general or Sapelli specifically. This is also likely to be related to some Ashaninka characteristics which, on the one hand, avoid voicing direct criticism and, on the other, prevent them from asking for help when in doubt.

The fact that the Ashaninka did not use Sapelli to collect data did not mean they found digital technology unhelpful in protecting their lands. Below, I describe three situations to make this point clearer and to demonstrate further the Ashaninka's concrete actions in defending their territory.

## 4.4 Being selective with the available digital technologies

In December 2015, there was a community meeting to plan monitoring trips for the following year. As usual, during the meeting, the community discussed, with the use of maps, which areas of their land were more vulnerable at each time of the year and what was needed to carry out the expeditions. In January 2016, they did the first monitoring expedition of the year on the Arara River. Five trained monitors from Apiwtxa went to Arara settlement, held a meeting in the settlement and showed the new monitoring tool to the dwellers. On the following day, the five monitors and six members from Arara settlement started the expedition along the Arara River towards the Peruvian border.

Along the way, they encountered poachers' camps and met two illegal hunters. They stopped them and explained that they were not allowed to hunt on indigenous land. After that, the Ashaninka confiscated what the game poachers had on their boat and told them to leave the area. During this monitoring trip, they also visited neighbouring non-indigenous communities to inform people about the intensification of the monitoring activities and to remind them about the prohibition on extracting natural resources from indigenous land.

When the monitors came back, I asked to see the 'evidence' they had collected about the poachers. They had taken very good pictures and even produced a couple of videos of them confiscating game – but they did not use Sapelli to collect much evidence. The only Sapelli data were a couple of GPS coordinates close to the hunters' camps. One of the monitors told me the group that went to the poachers' camp had forgotten to take the phones with them. With the support of the London team, a map was produced with the coordinates, which was later used by Apiwtxa in a letter to authorities, alongside the pictures they had taken.

On another occasion, in November 2016, during a community meeting, Ashaninka leaders received information that representatives from a Peruvian logging company were going to cross the Amônia River to get to Sawawo native community and offer them a contract to extract timber. The leaders prepared some community members to monitor the river and to inform them when the invaders were passing by. They stopped the boats and told the company's representatives to come up to the middle of the community meeting and asked them to explain what they were doing. Not happy with their responses, the leaders soon decided to prohibit their passage to Peru.

Again, when under pressure, the Ashaninka did not resort to Sapelli. The situation was registered with cameras (pictures and videos), and they also asked me to voice record the meeting. Finally, on another occasion when poachers passed by their village at dawn, they chose to register and communicate the passage with mobile phones, taking pictures with the phones and sending them via WhatsApp, together with audio messages.

## 5. Conclusion

The lack of Sapelli utilisation is not to do with an Ashaninka lack of interest or capacity in engaging with modern digital tools to help in their monitoring strategies. Their preference for choosing to use other equipment seem to have been primarily due to the difficulties Sapelli presented in its general functioning – not so much the application per se, but the 'before' and 'after' processes.

In practical terms, the Ashaninka found WhatsApp more useful in one of the contexts, as it also allowed rapid voice recording and photograph sharing. The instant feedback provided by WhatsApp certainly was a significant factor in their preference for it over Sapelli. The community also used separate digital cameras and voice recorders, as they could capture better-quality information, while in Sapelli, these facilities were limited and were not available until several steps down the decision tree.

This research was carried out to support the Ashaninka to widen their community monitoring efforts more effectively and to combat illegal logging and poaching activities, and I have demonstrated through a process mediated by geographic citizen science technologies and the extreme citizen science methodology that this can be done successfully. It was evident that the Ashaninka monitors were already well-trained data collectors who could pick and choose the best available tools to collect the evidence they were after and which best suited their specific context and environmental conditions.

## 6. Lessons learned

- FPIC should be an ongoing process when implementing projects with local communities.
- Any project that involves the use of technology needs to consider cultural aspects and how these influence common individual and collaborative behaviours, such as ways of communicating, local organisational structures, forms of expressing criticism and others.
- Rapid feedback and media quality are important features of technologies to encourage use by local communities.
- In any project involving the use of digital technologies in remote areas, it is not enough to train people on how to use specific tools. Equipment maintenance is crucial and must be taken into account and included in any initiative that expects to succeed in such areas.

## Notes

1 In Peru, the estimate is 141,183 people (Peru, 2017, http://bdpi.cultura.gob.pe/pueblo/ashaninka. Accessed 22 March 2019) and in Brazil 1,645 (Siasi/Sesai, 2014, https://pib.socioambiental.org/pt/Povo:Ashaninka. Accessed 22 March 2019).
2 In 2014, the official number was 940 (Siasi/Sesai, https://terrasindigenas.org.br/es/terras-indigenas/3716#demografia. Accessed 22 March 2019).

## References

Killick, Evan. 2005. 'Living apart: Separation and sociality amongst the Ashéninka of Peruvian Amazonia'. PhD diss., London School of Economics and Political Science.

Lewis, Jerome, and Téodyl Nkuintchua. 2012. 'Accessible technologies and FPIC: Independent monitoring with forest communities in Cameroon', *Participatory Learning and Action* 65: 151–65.

Pimenta, José. 2002. '"Índio Não é Todo Igual": A Construção Ashaninka Da História e Da Política Interétnica'. PhD diss., Universidade de Brasília.

Ricardo, Beto, and Fany Ricardo, eds. 2017. *Povos Indígenas No Brasil 2011/2016*. Sao Paulo, Brazil: Instituto Socioambiental.

# Chapter 16
# Community mapping as a means and an end: how mapping helped Peruvian students to explore gender equality

Peter Ward and Rebecca Firth

## Highlights

- Many parts of the world remain unmapped. They are disproportionately located in low- and middle-income countries, with a higher risk of natural disasters, public health issues and socio-economic problems.
- The Humanitarian OpenStreetMap Team (HOT) supports a global volunteer community that contributes to geographic citizen science by mapping unmapped areas. As of 2019, more than two hundred thousand volunteer mappers had contributed to this work.
- The maps HOT creates help combat issues through providing better information for public, private and state organisations, but the act of creating, analysing, using and sharing data on gender issues in their local context is also a valuable tool for school children to learn through experience.
- 'Off-the shelf' open-source applications can provide effective tools for school children to use, enabling project funds to be spent on other areas, such as scaling and data analysis.

## 1. Introduction

Ever since John Snow used a hand-drawn map to identify the point source of cholera in London in 1854, map data have helped make decisions that have saved and improved people's lives in crises, including natural disasters or infectious disease outbreaks (Shiode et al. 2015). Disasters kill nearly a hundred thousand people and affect two hundred million people every year (IFRC 2018). Yet, many of the world's most vulnerable places – home to one billion people – are still 'missing' from any map. Governments and disaster responders lack information on where people live, the roads they use, the rivers they cross and the schools and doctors they rely on, and so struggle to make time-critical, data-driven decisions. In any humanitarian/development context, practitioners need to answer a set of basic questions to assess the challenge they face: (1) Where did the event occur? (2) What are its implications on local people? (3) What do they need? Up-to-date, detailed maps have always been one of the most critical resources in answering these questions (Koch 2005).

OpenStreetMap is an open-source mapping platform which anyone can contribute to, with a similar model to Wikipedia, where anyone can edit an article (Heron, Hanson and Ricketts 2013). The platform has had one million citizen 'volunteer' mappers globally. However, mappers from low- and middle-income countries have traditionally been under-represented. The Humanitarian OpenStreetMap Team (HOT) exists to close this gap through engaging communities and volunteers across the world who are not represented on the map to map their communities. To achieve the goal of closing the gap, the HOT mission has three guiding principles:

(1) Everyone is counted on the world map, ensuring a minimum standard of data are available globally on OpenStreetMap, such as buildings and roads.
(2) Data are accessible and used; OpenStreetMap data are openly accessible by anyone and used in humanitarian and development projects.
(3) Everyone can engage and contribute to OpenStreetMap.

Societal challenges vary across high-, middle- and low-income countries, and the availability of both maps and other data sources, such as an up-to-date and full-coverage census, also varies (SDSN 2015). In turn, this means that the needs for geographic data also vary. In scenarios where there is no available or up-to-date map, OpenStreetMap has increasingly been used by humanitarian responders to understand basic information (e.g. rapid population counts in the absence of up-to-date census data) when planning for or responding to crises. The information they need includes basic population data, such as where people live and what transport routes are available. In scenarios where base data are available, such as in middle-income countries where the majority of the world's extreme poor live, often the greatest challenge is understanding who the extreme poor are and where they live when interspersed among privileged populations (Pande, McIntyre and Page 2019). Example questions answered by mapping in HOT projects are: In the absence of a census or digital birth registration in a location, how can we estimate population size? When trying to spray houses with insecticide for malaria prevention, how can an organisation plan logistics for their campaign, or understand how it is progressing? When trying to provide water for one million refugees across 10 camps, how can we know how many people are in each one? HOT uses a broad toolset, ranging from computer-based mapping applications, which use high-resolution satellite imagery, to low-tech apps, which work offline on US$30 smartphones. The range in type and technical complexity enables a wide variety of people to map their communities. HOT volunteers range from semi-literate rural women to rehoused refugees to digital natives[1] volunteering around the world.

There is a clear need for up-to-date, accurate, reliable information to address these types of questions, and this need is currently not addressed by traditional mapmakers, such as the government and the military, who in many cases keep their maps private, or by mapping companies, who focus on mapping areas where they have a strong customer base, which do not commonly overlap with places experiencing crisis (Firth 2017). The impact of maps in improving capacity of decision makers is widely known, but what about the impact of engaging with OpenStreetMap on the mapper? This case study examines the experience of working with secondary school students in four districts of Cusco, Peru, to collect, analyse, use and share data on gender issues that impact them using open-source geospatial tools, including OpenStreetMap, the HOT Tasking Manager, OpenStreetMap iD Editor and KoBoToolbox.

## 2. Development, education and technology in Cusco: changes and challenges

According to many economic measures, Peru is a success. In the early 1990s, at the height of combating the Marxist Shining Path guerrilla uprising, 55 per cent of the population lived in poverty (Banco Central de Reserva del Perú 1997) and around 14 per cent had never enrolled in secondary school (UNESCO Institute for Statistics n.d.). Nearly 30 years later, Peru has changed considerably and has made huge advances in economic well-being (OECD/CAF/ECLAC 2018). Yet, despite this progress, the learning results achieved within schools remain poor. Nationally, only 12 of every 100 children aged 13 are reading and writing at the expected level – an average which drops to below 4 in every 100 in some of Cusco's 13 provinces (Ministerio de Educación 2018). Peru is one of the world's most unequal countries in terms of education results, comparable to South Africa with its enduring legacy of apartheid (Castro and Rolleston 2015). Cognitive development varies widely, especially regarding indigenous children in rural areas who consistently lose ground to non-indigenous children, even before formal schooling starts (Arteaga and Glewwe 2014). Research finds that adolescence is an important opportunity to correct this and to develop new skills and abilities, as it is the time when the brain goes through a process of pruning, reducing the synaptic connections previously developed to improve the efficiency of the brain (Balvin and Banati 2017). Furthermore, it highlights the social importance of adolescence as a period of time in which children come increasingly into contact with, and are impacted by, social constructs such as gender norms (Banati and Lansford 2018). The World Health Organization (WHO)'s definition of gender aligns with this concept, defining it as a social construct that defines the characteristics of both men and women, including roles, norms and relationships that vary across cultures and communities. The definition explains the negative health consequences that those who do not 'fit' gender norms in their communities face, highlighting the need to improve awareness of the issue in schooling (WHO 2011).

Strict social norms are especially powerful in Peru, where deeply rooted racism and sexism flourish, and self-proclaimed conservative groups of parents protest against gender being included in the national curriculum. The group 'Con Mis Hijos No Te Metas', or 'Don't Mess With My Children', has grown to be especially powerful, with marches proclaiming 'Gender ideology, never again', as well as more sexually explicit messages (Figure 16.1).

At all levels of government, from national ministries to the local district level, there are serious struggles with generating accurate data, let alone analysing and sharing it. In terms of maps, whilst medical and education facilities are mapped, gaps exist, and even decentralised local authorities do not know how many houses exist in the areas they administer, let alone their location. Without this basic information, it becomes harder to track data around behaviours and social norms, such as attitudes towards gender, as people cannot identify the exact locations where behaviours occur, meaning they are restricted in understanding what may drive them and so how to change them. Where official government data

**Fig. 16.1** The conservative parents' group 'Con Mis Hijos No Te Metas', or 'Don't Mess With My Children', protesting in Lima, Peru. Credit: Mayimbú. CC BY-SA.

do exist, even at a more general, non-spatial level, they are viewed with some scepticism: the estimate that only 17.8 per cent of women are employed in unpaid family domestic work is seen as very low, and does not take into account the extra unpaid hours they work compared to men (INEI 2018). With issues such as gender-based violence, data have existed only since the early 2000s. The data show that more than 6,300 people, or 35 every day, were victims of sexual violence in Cusco in the first six months of 2019 (Ministerio de la Mujer y Poblaciones Vulnerables 2019). This figure is generally assumed to be underestimated by those who live in the areas.

## 2.1 When new technology means more of the same

A series of education ministers have led impressive reforms, and the new national curriculum, with its goals of enabling students to develop cognitive and socio-emotional abilities, exercise their rights and respect others, contrasts with the education provided 40 years ago, which was focused on simply acquiring knowledge (Ministerio de Educación 2016). Yet, despite this, classroom practice remains largely content and theory focused, with teachers struggling to focus on the development of cognitive abilities such as logically structuring content or questioning or, in many cases, the practical application of knowledge. There is little space given for students to be the protagonists in their learning, and although alternative schools have sprung up, adjustment in state schools has been very slow. The reasons for this vary but include the belief held by some teachers that these changes are an imposition from above (despite the participative way in which the curriculum was created) and that there is a simple lack of understanding of how to implement the new focus and a lack of practical support for teachers (Guerrero 2018).

The national curriculum states that as part of their education in information technology, students should develop the competency of 'Designing in constructing technical solutions to solve issues in their context' and 'interpret, modify and optimise virtual environments during the development of learning activities and in social practices' (Ministerio de Educación 2016, 7). Yet, despite this aim, the reality of the implementation of technologies still focuses on providing hardware over other interventions, despite the evidence from similar initiatives that this simply does not work, such as the One Laptop Per Child Program (Cristia et al. 2017). As public authorities invest heavily in hardware and Internet connections for so-called innovation rooms, the way in which these new tools are used remains traditional. During a visit by the project team to a provincial school in Cusco, the head teacher proudly showed 30 brand-new desktop computers to the team, explaining how he had succeeded in getting funding for them from local authorities. Looking over his shoulder, it was clear that the students had their standard notebooks out – each one was copying pages of PDF text from their computer screen into the books, word for word.

## 3. Open geographic citizen science to address education challenges

The Global Active Learning Group, or GAL Group, a Cusco-based education organisation comprised of a low-cost private school and a consultancy, engages youth in relevant topics as protagonists, not simply as recipients of information and data, aiming to achieve the competencies outlined in the national curriculum. One of the pedagogical approaches used by the GAL Group is project-based learning, engaging youth in practical investigations based on their own interests, using technologies and other tools as required, rather than for their own sake. It aims to develop not only an understanding of the content that projects focus on (in this case technology and gender), but also other abilities, with the intention of forming a well-rounded child. All these are developed in a context where skills are used to create a real, tangible product, making learning more meaningful (Phillips, Burwood and Dunford 1999).

The project forms part of a Women Connect Challenge pilot funded by USAID and implemented in Peru by HOT and the GAL Group. The overarching programme objective was to empower women through improving access to and use of technology. HOT provided a specific focus for open maps as a tool for data generation, analysis and community building, and the GAL Group developed and adjusted pedagogical approaches to enable secondary school students and teachers to make the most of the technology.

As the project was a pilot, the team decided to work with diverse groups of male and female secondary school students across Cusco. These diverse groups, chosen to enable the team to understand a range of different realities and points of view, created investigative mapping projects using geospatial data in different ways in order to explore gender issues they were interested in and where government data were scarce. To provide some focus, these issues were broken down into topics: use of public space, violence, education, computer games, political representation, health and work. Within these broad topics, groups focused on a specific area of interest and created questions and hypotheses to guide their investigations. Thinking routines, developed by Harvard University's Project Zero, among others, to help to structure and visualise thinking skills such as analysing, justifying, empathising and questioning, were used to structure the analysis students performed. From a technical perspective, the project incorporated OpenStreetMap and related open-source applications, as listed in Table 16.1. The HOT Tasking Manager and OpenStreetMap iD Editor were used to organise group mapping activities with international and local volunteers, whilst KoBoToolbox was used for offline geolocated surveys and questionnaires, for example, to investigate opinions on the different uses of public spaces by men and women.

OSMTracker, an Android-based Global Positioning System (GPS) tracker application with offline functionality, was used to track routes and identify points of interest with text and voice notes, as well as photos and videos. Mapillary was used to create geolocated street-level imagery. Maps.me, which enables users to download maps to work offline as well as to map specific points of interest, was also used, although to a lesser extent. In terms of analysis of data, additional software such as QGIS, an open-source and free geographic information system application, was

**Table 16.1** Technologies used as part of the project

| Technology | Description |
|---|---|
| HOT Tasking Manager | A mapping tool designed and built for the collaborative mapping process in OpenStreetMap available at tasks.hotosm.org. It divides mapping projects into separate tasks, which can be locked by individual users whilst they edit them, preventing duplication of mapping efforts. |
| OpenStreetMap iD Editor | An Internet browser-based editor which does not require plugins and is available for use at openstreetmap.org. |
| KoBoToolbox | An open-source suite of tools developed to enable data and analysis collection using text, photos, videos and Global Positioning System data in resource-constrained environments. |
| OSMTracker | A free mobile app for Android mobile phones. It enables users to track journeys, mark waypoints with tags, record voice notes and take photos. |
| Mapillary | A platform focused on street-level imagery that aims to automate mapping. It includes a free mobile app for capturing imagery. |
| Maps.me | An open-source mobile app that provides downloadable, editable OpenStreetMap maps and guides. |
| QGIS | An open-source geographic information system desktop application that enables users to view, edit and analyse geospatial data. |
| MapHub.net | An open platform based on OpenStreetMap enabling users to personalise and add data to maps, as well as share maps with others. |

Source: KoboToolbox.org GNU AfferoGeneral Public License v3.0.

used for basic analysis, such as creating heat maps of higher incidences of certain attitudes towards domestic violence. The much simpler free website MapHub.net, which also allows basic visualisation and editing, was used to present analysis and images of issues such as sexist publicity, which students collected using OSMTracker. Across the project, HOT Tasking Manager, iD Editor and KoBoToolbox were the most popular with students, who easily understood their value to their projects. For this reason, we will focus on them specifically in this case study.

## 3.1 The HOT Tasking Manager and OpenStreetMap iD Editor: divide and conquer

The first part of the process for students was to identify if the areas they wished to investigate were already mapped in OpenStreetMap, which invariably they were not. For this reason, the HOT Tasking Manager and OpenStreetMap iD Editor, which are the easiest tools with which to edit OpenStreetMap from a computer or laptop, were the most commonly used tools.

The HOT Tasking Manager is a mapping tool designed and built for the collaborative mapping process in OpenStreetMap. As Figure 16.2 shows, the user interface is simple and uses colour-coding for the mapping tasks. It was quickly picked up by even younger secondary school students.

The tool uses a specific area needing to be mapped as an input, and divides the mapping project area into smaller micro-tasks each representing less than 10 km² of mapping. The division of the total project area in the tool into these small tasks enables many people to map the same overall area at the same time, without duplicating effort or overlapping work. The selection of where to map can be made by the user, or by selecting a task at random. People then remotely map the area using satellite imagery

**Fig. 16.2** HOT Tasking Manager screen once a project is selected. Source: Humanitarian OpenStreetMap team (tasks.hotosm.org/project /5807?task=35). Basemap © Humanitarian OpenStreetMap. CC BY-SA.

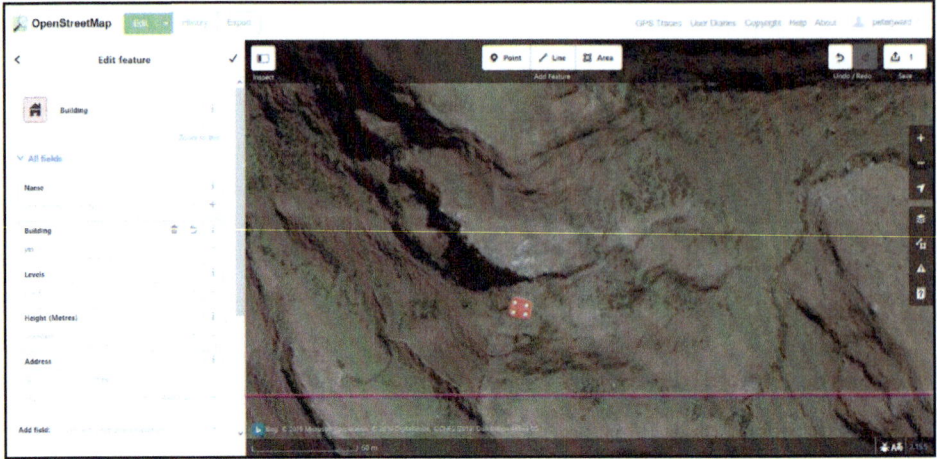

**Fig. 16.3** OpenStreetMap iD Editor, with a building mapped. Basemap © OpenStreetMap contributors. CC BY-SA.

and an editor, such as the OpenStreetMap iD Editor, pictured in Figure 16.3, and, once completed, save the task for validation. Students found the simple layout and obvious colour coding very easy to use, meaning that training was easy to facilitate.

Although other options exist, the easiest way to map remotely is through the OpenStreetMap iD Editor, an open-source, web-based tool which enables users to make edits in OpenStreetMap. The user can select from a variety of satellite images and map within an area assigned specifically to them by the HOT Tasking Manager. They trace objects of interest, such as buildings, roads or rivers, using the point, line or area functionality. Once traced, the user tags them according to the OpenStreetMap taxonomy, which is regularly updated on the OpenStreetMap Wiki. The OpenStreetMap iD Editor only surfaces a limited number of tags available with OpenStreetMap, but the simplified nature of the editor is usually sufficient for the first stages of a mapping project. The HOT Tasking Manager reports in real time the progress of mapping and validation, as well as tracking statistics on mappers' activities in a gamified manner. Mappers can count the number of edits made, and buildings or roads tagged, as they advance through different 'levels'.

The validation of the remote mapping is done initially by remote volunteers, who have more mapping experience and have received extra online training to become validators. As with remote mappers, they select specific areas to validate in the HOT Tasking Manager, and either approve the tasks for final validation in the field or reject them, providing mappers with feedback comments on what to update or how to improve their mapping. Finally, the maps are validated in the field through physically visiting the locations mapped and are updated.

### 3.2 KoBoToolbox: open-source, offline, georeferenced surveys

Given the cultural diversity in Cusco, most students wished to use surveys to compare opinions on gender issues in different parts of the region as a key part of their investigation, and so chose to use KoBoToolbox (Figure 16.4).

KoBoToolbox is divided into three sections. The first section is an online form builder, which enables users to create surveys or questionnaires using a variety of types of questions, data validation, skip logic, geolocation and share responses. The second section is a free Android application, Kobo-Collect, which works on basic smartphones. It enables users to collect both text and multimedia content in areas with no mobile or Internet signal. Once the user has an Internet signal, they can send responses from the phone in bulk to the third component: the reporting dashboard. The reporting functionality automatically creates charts and graphs of information, presenting them in a visually friendly way, and enables the downloading of data in Microsoft Excel Spreadsheet (.xls) format. Importantly, when geolocation is included in the survey, results can be disaggregated and visualised on OpenStreetMap, providing valuable geographic context to the results. Students generally used KoBo for creating questionnaires that were used in and around their communities in order to understand different views on gender and how locations people lived in or worked in affected these.

## 4. Learning through projects: the art of the possible

The first stage of the project involved training teachers and introducing them to the concepts of gender, project-based learning, thinking routines and the technologies. In the space of an afternoon, the GAL Group team trained teachers in using HOT Tasking Manager, OpenStreetMap iD Editor, as well as familiarising them with the tools for data generation, described in Section 3. The idea of this session was not to train them fully, but rather to provide them with enough experience to overcome their hesitation of using the technology and to understand what they and their students could achieve once trained to use it. This was an important step, as

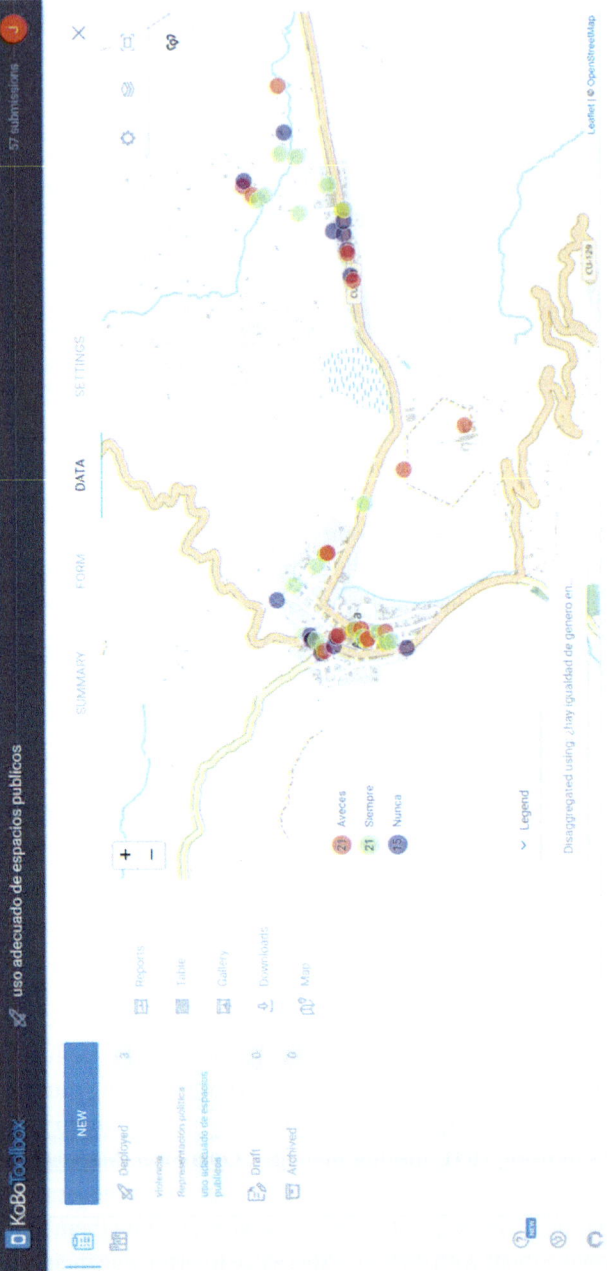

**Fig. 16.4** KoBoToolbox data-analysis screen. In this example, responses have been disaggregated by responses to the question 'Is there gender in the use of public space in your area?' Credit: Work developed by a group of secondary school students from Jose Maria Arguedas, District of Accha, Province of Paruro. Source: Kobotoolbox.org.

anecdotally many teachers are afraid of being humiliated by students who learn technologies more quickly than them, and so prefer not to use them at all. To the extent of removing this fear and stimulating the use of geospatial technologies, the training – and later in-person support and remote support through WhatsApp and phone calls – was a success.

### 4.1 Volunteers and students: a global mapping model

It was clear from the beginning of the project that the areas where students lived were not mapped in OpenStreetMap. Rather than encourage the students to engage in the time-consuming act of mapping their community using satellite imagery, the GAL Group and HOT engaged the international volunteer community to support this, with more than eight hundred volunteers supporting the mapping. This meant that students saw their communities appearing on the map for the first time, and felt that they were part of a much bigger project.

Despite the initial excitement, the project ran into challenges, principally around data quality. The students were able to grasp the basics of the HOT Tasking Manager and OpenStreetMap iD Editor quickly, mapping buildings, roads and rivers, but the quality of their mapping was often poor. There were multiple reasons for this. First, students were easily distracted, and would often abandon their 'task' to look for something more interesting, creating conflicts with other mappers. Whilst the HOT Tasking Manager and OpenStreetMap iD Editor provide clear instructions not to abandon an assigned task, the tools do not prevent users from doing so. Additional nuances, such as ensuring that buildings do not overlap, or overlap roads, were not always prioritised by mappers. Both local and international mappers also encountered issues in assuming buildings to be houses and tagging them as such, tagging multiple buildings as one in order to save time and misclassifying roads. Students were also guilty of rather less honest mistakes, such as tagging non-existent buildings or inventing names (Figure 16.5). The validation aspect of the process proved invaluable, as did the feedback function, where validators could message mappers individually, commenting on their edits. The realisation that volunteers around the world were checking their work improved the accuracy of the students' mapping, both reducing intentional errors and improving concentration.

Whilst the GAL Group team had checked Internet coverage and speed before the project, during group mapping sessions, even when using only 10 computers, Internet bandwidth availability was often overwhelmed, slowing down the tools. This created frustration for the students

**Fig. 16.5** An example of poor-quality community mapping. Note the misalignment of buildings and duplication of mapping. This work would be checked and corrected at the validation stage. Basemap © OpenStreetMap contributors. CC BY-SA.

and made it harder to convince them that the tools were effective. Their sense of frustration also contributed to carelessness, meaning more errors in their mapping. Workaround solutions were found, through 'hotspotting' mobile phones and purchasing extra portable routers for workshops. Once these measures were introduced, students mapped effectively and enjoyed the gamified aspect, enabling them to compare how much they had mapped relative to their peers.

## 4.2 Students using citizen-generated geospatial data in social analysis

As previously mentioned, the project provided students with a range of technologies for their investigative projects. KoBoToolbox was selected by students who were attracted by the functionality of comparing survey results across different areas, for example enabling them to analyse gender norms in urban, peri-urban and rural areas, as well as in specific districts with reputations for gender inequality and gender-based violence. Each group of students defined their own survey questions based around their interests, the overarching project questions and hypotheses they developed. For example, a group which chose political representation as their topic created the overarching question: 'Does a culture of machismo make it harder for women to be political leaders?' Their hypothesis was that the culture of machismo did make it harder, and they aimed to prove this through closed questions such as 'Who do you think are better leaders, men or women?' or 'When you attend a community meeting, who speaks more, men or women?' which were always followed up with the open question, 'Why?' Then, they compared answers to these questions across different geographic areas.

KoBoToobox's online form builder required an Internet connection to upload these questions, but it did not create the same issues as the HOT Tasking Manager in terms of slowing down the already poor Internet connection. In terms of usability, even those with less technical ability, such as younger students, aged 12, and those living in the most rural areas, found it easy to use the basic functionality with minimal instruction. Issues were encountered in the user interface in two areas. First, the 'Add Question' button, as depicted in Figure 16.6, was simply not big enough, and even when it was explained by the text in the middle of the page, many users simply did not notice it. Second, the descriptions of the types of questions, and the icons used to represent them, created confusion, as they did not correlate to the students' logic. This led to students creating questionnaires with inappropriate types of questions for the answers they hoped to receive, for example asking for a free-text response where a Boolean response would be more effective. Across all schools, the iterative process of creating and editing the questions so that students could generate useful data took more time and effort than learning to use KoBo itself.

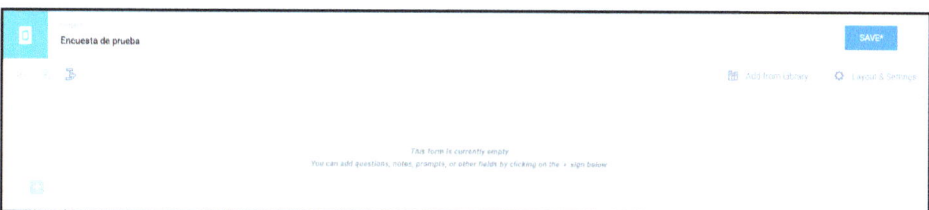

**Fig. 16.6** KoboToolbox screen for starting a questionnaire from scratch. Source: Kobotoolbox.org.

As part of the pilot, the GAL Group lent Android smartphones to the four schools and 160 students for their field investigations. Additionally, where teachers and students had their own Android smartphone, they were able to use the application to load the surveys onto their phones. It quickly became clear that data management was an alien concept to most teachers, and that many did not have the space to download the application or store the responses, which often included videos or photos.

During data collection, students used the geolocation functionality in two ways. First, they used it to identify the location where the questionnaire took place, using the phone's GPS. In this case, the application worked well, usually getting an accuracy of 14–15 m within a matter of seconds – more than enough for students to compare results. Second, the geolocation functionality was used to ask people who they questioned to identify a location that was not necessarily their physical location at the time of surveying. This use case was more challenging, as the lack of a culture of physical or digital map reading in rural areas meant many respondents were unable to identify places using the map on the phone used for the questionnaires. In both cases, the results enabled students to see issues through a geographic lens. The results dashboard proved to be easy to use, and the automatic presentation of results in graphs and on a map was extremely helpful in a context where creating graphs in Microsoft Excel is considered an advanced skill. Students saw the value in mapping as providing the base layer for their analysis and were able to focus their analysis around the issues in the data they found. One clear example came from a group studying violence in Accha, Paruro, who found that the women they interviewed in peri-urban areas were much more likely than those in urban areas to admit that their spouse had attacked them under the influence of alcohol or drugs. Another group analysed views on gender-based violence to challenge the hypothesis that the further away they were from the district capital, the more likely it was that they would encounter sexist attitudes. Yet another group focused on gender-based violence in the city of Cusco, and found that people were less willing to admit to having seen violence where they were currently being interviewed but more willing to provide geolocations of where they believed violence to be high in other areas. The value in automatically geolocating data and presenting the results in a simple dashboard should not be underestimated, as without this functionality, teachers and students would have had to dedicate considerable time and effort to achieve this visualisation manually. This time was instead invested in more valuable cognitive efforts, such as analysing their hypotheses, developing empathy for respondents and defining further questions, such as the differences in opinions in urban, peri-urban and rural areas.

The fact that youth were able to create and visualise spatial data easily as part of investigations, as opposed to simply being told the information in a one-way communication style, was one of the most important achievements of the project, going far beyond what most technology for gender equality programmes achieve. After the validation of hypotheses and final conclusions, the children were asked a simple question: 'Now what?' The idea was to stimulate children to respond to the problems uncovered by their investigations in whatever way they proposed. This led to groups developing videos on gender bias in politics, giving speeches in schools around gender-based violence and even expressing an interest in developing local radio programmes in Quechua, the indigenous language spoken more frequently than Spanish in many rural areas.

## 5. Conclusion

Whilst this experience represents a relatively small case study in terms of the impact of geographic citizen science, it shows the value of engaging high school students as volunteer geographic citizen scientists. Cusco is a region where there is little space for student-led investigation, let alone data generation, and where educational programmes to confront destructive gender norms rely on one-way transmission of information. IT education is still often limited to providing students with access to hardware, contributing to a challenging context for developing such a different project. Yet, despite these challenges, this project has shown that it is possible to develop cognitive ability through encouraging and enabling students to use open-source geospatial technologies. In the case of the pilot, this was measured informally through the changes seen in activities, such as logically structuring project plans and the ability to create questions and think critically using thinking routines, as opposed to a formal scientific study.

Students from different socio-economic groups have reported different positive outcomes. Those from more privileged backgrounds spoke of feeling empowered to help others through technology and advocacy, whilst others from poorer households shared how they felt that the project had helped them improve gender equality in their own schools and contexts. Teachers were also positive and shared their desire for students to go beyond simply creating maps in their investigations to using them actively to advocate for change at a broader level. The issues that were uncovered in the use of the tools, such as poor levels of concentration from the students leading to reduced data quality in OpenStreetMap, and more practical ones, such as teachers not being able to free enough memory on their own mobile phones to use KoBoToolbox, are minor, given the complexity of activities undertaken. The ease with which students in different contexts were able to adapt quickly to the range of open-source technologies used in the project and to learn how they are used is a testament to the design work of all those involved.

The open-source nature of the tools and information meant that the team was able to reduce technology costs significantly and invest in the actual generation of data and the development of cognitive skills in participating youth. Although only a small pilot, initial evidence shows that encouraging the next generations to participate in geographic citizen science by combining it with investigative projects and open-source technologies could provide a low-cost, sustainable path towards better data and better education results. We hope that this initial experience provides a useful model for implementing similar geographic citizen science projects in OpenStreetMap in secondary schools globally.

## 6. Lessons learned

- Building trust with gatekeepers – in this case, school teachers – is vital in the successful implementation and adoption of tools to enable geographic citizen science. Only once they are sure that the tools are not a threat to them will they embrace them and encourage others to do the same.
- Training in the use of the tools is necessary, but the value of accompaniment over a longer time cannot be underestimated, as it improves both uptake and data quality.
- Using open-source, 'off-the-shelf' technologies can work well with resource limited constraints. This enables significant reductions in investment in developing technology solutions.
- Users will always demonstrate behaviours that tools cannot – and should not – prevent. Building complex rules to prevent certain actions can hinder the usability of tools.
- There is value not only in the outputs of geographic citizen science but also in the process as an opportunity to create meaningful learning.

## Acknowledgements

With many thanks to USAID Women Connect Challenge for the inspiration and opportunity to make this project happen, and the Nethope Device Challenge for providing the devices to enable the students to get online.

## Note

1  By digital natives, we mean here a person born or brought up during the age of digital technology and therefore familiar with computers and the Internet from an early age.

## References

Arteaga, Irma, and Paul Glewwe. 2014. *Achievement Gap Between Indigenous and Non-Indigenous Children in Peru: An analysis of Young Lives survey data. Young Lives Working Paper 130*. Oxford: Young Lives.

Balvin, Nikola, and Prena Banati, eds. 2017. *The Adolescent Brain: A second window of opportunity – A compendium*. Florence, Italy: UNICEF Office of Research – Innocenti.

Banati, Prerna, and Jennifer E. Lansford, eds. 2018. *Handbook of Adolescent Development Research and Its Impact on Global Policy*. Oxford: Oxford University Press.

Banco Central de Reserva del Perú. 1997. Pobreza y Bienestar Social: Evolución en los Últimos Años. Accessed 2 February 2020. http://www.bcrp.gob.pe/docs/Publicaciones/Revista-Estudios-Economicos/02/Estudios-Economicos-2-1.pdf.

Castro, Juan F., and Caine Rolleston. 2015. *Explaining the Urban–Rural Gap in Cognitive Achievement in Peru: The role of early childhood environments and school influences. Young Lives Working Paper 139*. Oxford: Young Lives.

Cristia, Julian, Pablo Ibarrarán, Santiago Cueto, Ana Santiago and Eugenio Severín. 2017. 'Technology and child development: Evidence from the One Laptop per Child program', *American Economic Journal: Applied Economics* 9: 295–320.

Firth, Rebecca. 2017. 'OpenStreetMap and the Sustainable Development Goals', *Environmental SCIENTIST* 26: 9–13.

Guerrero, Gabriela. 2018. *Estudio Sobre la Implementación del Currículo Nacional de la Educación Básica en Instituciones Educativas Públicas Focalizadas*. Lima, Peru: FORGE, Fortalecimiento de la Gestión de la Educación en el Perú.

Heron, Michael, Vicki Hanson and Ian Ricketts. 2013. 'Open source and accessibility: Advantages and limitations', *Journal of Interaction Science* 1: 1–10.

Instituto Nacional de Estadística e Informática (INEI). 2018. *Perú: Brechas de Género, Avances Hacia la Igualdad de Mujeres y Hombres*. Lima, Peru: INEI.

International Federation of the Red Cross (IFRC). 2018. *World Disasters Report: Leaving no one behind*. Geneva, Switzerland: International Federation of the Red Cross.

Koch, Tom. 2005. *Cartographies of Disease: Maps, mapping, and medicine*. Redlands, CA: ESRI Press.

Ministerio de Educación. 2016. Currículo Nacional de la Educación Básica. Accessed 1 March 2017. http://www.minedu.gob.pe/curriculo/pdf/curriculo-nacional-de-la-educacion-basica.pdf.

Ministerio de Educación. 2018. Qué Aprendizajes Logran Nuestros Estudiantes? Evaluación Censal de Estudiantes (ECE). Accessed 21 September 2020. http://umc.minedu.gob.pe/wp-content/uploads/2019/06/DRE-Cusco-2016-Marzo-2019.pdf.

Ministerio de la Mujer y Poblaciones Vulnerables. 2019. Programa Nacional Contra la Violencia Familiar y Sexual: Boletīn Estadístico, Junio 2019. Accessed 21 September 2020. https://www.mimp.gob.pe/files/programas_nacionales/pncvfs/estadistica/boletin_junio_2019/BV_Junio_2019.pdf.

OECD/CAF/ECLAC. 2018 *Latin American Economic Outlook 2018: Rethinking institutions for development*. Paris: OECD Publishing.

Pande, Rohini, Vestal McIntyre and Lucy Page. 2019. 'A new home for extreme poverty: Middle income countries', *New York Times*, 28 January 2019.

Phillips, Diane, Sarah Burwood and Helen Dunford. 1999. *Projects with Young Learners*. Oxford: Oxford University Press.

Shiode, Narushige, Shino Shiode, Elodia Rod-Thatcher, Sanjay Rana and Peter Vinten-Johansen. 2015. 'The mortality rates and the space-time patterns of John Snow's cholera epidemic map', *International Journal of Health Geographics* 14 (21). https://doi.org/10.1186/s12942-015-0011-y.

Sustainable Development Solutions Network (SDSN). 2015. Data for development: A needs assessment for SDG monitoring and statistical capacity development. Accessed 21 September 2020. https://sustainabledevelopment.un.org/content/documents/2017Data-for-Development-Full-Report.pdf.

UNESCO Institute for Statistics. n.d. Education: Total net enrolment rate by level of education. (Total Net Enrolment Rate, Lower Secondary, Both Sexes (%)). Accessed 21 October 2019. http://data.uis.unesco.org/#.

World Health Organization (WHO). 2011. *Gender Mainstreaming for Health Managers: A practical approach. facilitator's guide*. Geneva, Switzerland: World Health Organization.

# Synthesis and Epilogue

# Geographic citizen science design: no one left behind – an overview and synthesis of methodological, technological and interaction design recommendations

Artemis Skarlatidou and Muki Haklay

## 1. Capturing the current state of geographic citizen science

At the opening of this volume, Haklay positions geographic citizen science at the intersection of volunteered geographic information and citizen science, and defines it as an activity (or a set of activities) which involves the utilisation of geographic information technology to collect, analyse and disseminate data collected by non-professional participants in a systematic and objective way. Although hundreds of citizen science applications currently exist, to which many of the underlying design principles and lessons discussed in this volume may apply, geographic citizen science entails a distinct subcategory of citizen science that relies on the collection of locational information through the use of specialised equipment (i.e. mostly mobile devices equipped with a Global Positioning System receiver). This equipment enables the systematic and objective collection of geographic objects, and sets the data on the road to becoming useful scientific knowledge.

The geographic citizen science projects which currently exist, including those covered in this volume, share some general themes in the ways in which they are used. First, it is increasingly common for citizen science initiatives to involve the collection and analysis of geolocated scientific data using geographic information technology. Citizen science platforms, such as Scistarter.org, iNaturalist, the Atlas of Living Australia and many others, include projects which show how scientific research (e.g. marine science, conservation biology and biodiversity) produces geographic information for research purposes and for updating and enriching existing scientific databases (e.g. biodiversity databases) with new observations from all over the world or for specific geographic locations. Second, geographic citizen science applications, such as OpenStreetMap, rely on participants producing accurate and reliable geographic information which is used to extend digital geographic coverage, even to the most remote areas, and to create free and open data sets which can be used for further analysis in various disciplines and for various purposes (e.g. for planning a humanitarian response to a disaster). Finally, geographic citizen science applications are increasingly used to enable volunteer-initiated participatory action research projects to address issues of local concern and to influence policy arguments and outcomes. For example, Wynn (2017) describes in detail how, in the Pepys Estate project, collecting noise-related geographic data influences the formation of both expert and non-expert policy argumentation.

The 12 case studies in this volume capture many facets of the current state of geographic citizen science and the wide variety of purposes that these applications are addressing, as well as the audiences that they engage. Each case study aims to provide information about the design and development process, to communicate previously anecdotal evidence from the user interaction implications and to provide lessons to inform the process of designing more user-friendly geographic citizen science applications in the future. To support this, the case studies in this volume are divided into two parts.

The first of these parts (Part 2 of this volume) comprises five case studies of geographic citizen science applications which are mainly used for environmental and urban planning purposes. Their target users are located mainly in urban centres of the Global North, where access to electricity, network connectivity, previous experience in the use of technological artefacts, including geographic interfaces, literacy and other user characteristics which influence interaction, are all assumed in the design and promotion of rather 'Westernised' interface designs. Specifically, Part 2 discusses geographic citizen science applications for capturing: ice-skating conditions, grassland bird sightings, wildlife health morbidity and mortality events (Chapter 5); noise pollution and the use of quiet areas in urban centres (Chapter 6); the current state of forest conditions and their changes to influence their effective management (Chapter 7); urban cycling routes and cyclists' habits (Chapter 8); and the reporting of non-emergency issues by members of the public directly to their relevant local authorities (Chapter 9). These cases offer insights into different types of geographic citizen science initiatives from 'contributory' initiatives such as the Hush City app (Chapter 6) and the Cyclist Geo-C app (Chapter 8), which have the potential to generate larger spatio-temporal data sets to support scientific research and analysis further, to 'collaborative' (e.g. ImproveMyCity application in Chapter 9) and 'co-created' initiatives (e.g. RinkWatch application in Chapter 5 and those discussed in the third part of this volume), which have the potential to not only contribute to scientific endeavour, but also 'facilitate in-depth civic participation that may contribute to more direct management advice and policy-change and more indirectly support education and capacity building in research' (Hecker et al. 2018, 471).

There is no doubt that citizen science, in general, and geographic citizen science initiatives, in particular, are more widely implemented in Western societies, enabled by social trends such as access to education, exposure to science and the wide use of digital technology. Nevertheless, Hecker et al. (2018), who discuss citizen science and innovation, highlight that in opening up participation in science, 'it is equally important to include indigenous and local knowledge as an added benefit to science, for example, in framing questions, designing projects, analysing results and understanding their possible impacts upon decision-making processes'. The authors continue that 'Danielsen et al. [in Hecker et al. 2018] demonstrate that both ILK [indigenous local knowledge] and institutionally derived scientific understanding can be valuable in conservation planning activities. This knowledge inclusivity can bring specific expertise to citizen science projects and embed the results in the community affected' (Hecker et al. 2018, 468). We should not ignore the fact that geographic information technology is commonly used to enable these types of initiatives and to support knowledge production.

Therefore, the second part of the case studies (Part 3 of this volume) presents and discusses seven geographic citizen science initiatives with indigenous communities, and it reflects on how the sociocultural and environmental contexts as well as user characteristics influence how participants interact with them. It also considers user-centric design processes which can be followed to support these interactions better. In particular, this part covers cases where geographic citizen science is used to: consult and engage Canadian First Nation communities in reviewing resource extraction project applications (Chapter 10); support indigenous communities in the Congo Basin to record Traditional Ecological Knowledge and illegal logging and poaching in their areas (Chapter 11); support Baka communities in Cameroon to collect data about wildlife crime (Chapter 12); enable local communities in Cambodia to collect data on natural resources and monitor illegal logging and the illegal conversion of the Prey Lang forest (Chapter 13); support communities in the Pantanal wetland (Brazil) to collect data about their fishing practices and natural resource management to demonstrate their sustainable practices and counter existing scientific arguments and models which have resulted in their economic and physical displacement (Chapter 14); support the Ashaninka communities in Brazil to collect data about illegal invasions and activities in their territory, which are subsequently shared with enforcement agencies and other authorities (Chapter 15); and enable secondary school students in Peru to collect, analyse and share data on gender issues to explore and address gender inequalities and violence in their area (Chapter 16).

In the next sections, we summarise the main lessons learned from this volume's chapters, and conclude with a set of technological, methodological and interaction design recommendations to inform the development of future geographic citizen science applications.

## 2. Technological considerations and recommendations

The first set of lessons from the volume emerged from insights into the technological aspects of geographic citizen science projects. Technological developments created new opportunities for initiatives to involve participants and to facilitate the collection, analysis and dissemination of spatially enabled data. In Chapter 2 of this volume, Antoniou and Potsiou identify three key technological trends which have enabled these types of geographic citizen science initiatives: 'the combination of significant developments in open-source software, do-it-yourself (DIY) hardware proliferation and the equipping of mobile devices with multiple sensors'. We can envisage that this trend will continue, with more coverage of mobile broadband and an increase in the number of Global Navigation Satellite Systems.

Yet, we should also recognise that while there is ongoing development, there is still a long way to go. From a technical point of view, citizen science platforms – stand-alone projects which rely on sophisticated geographic information system (GIS) solutions, social media–enabled tools and collective intelligence platforms for collaborative problem solving and instant communication – are all complex socio-technological systems. They are not simple to assemble, update or maintain. This can be challenging for the current maintainers of citizen science applications, who are usually operating with limited budgets and access to technical skills. There is therefore a need to learn lessons from other activities in the field which can ensure better development of applications and tools. From the chapters in this volume, we can point to the following lessons.

First, there is a need to improve understanding and awareness of the technological landscape, including its opportunities and limitations. Chapters 1 and 2 noted the ongoing challenge of implementing the best possible technical solution due to the constraints that these projects face, and an awareness of the pace of technological change and the need to update and upgrade the digital infrastructure of any given project constantly. Therefore, the adoption of a proper technological approach should follow a generic but long-term plan. Linked to that is Antoniou and Potsiou's (Chapter 2) recommendation that citizen science project owners need to build up their skills and awareness – a prerequisite for enabling them to identify barriers and opportunities in choosing the right technological infrastructure in specific contexts of use characterised by their own unique conditions while being flexible enough to accommodate future needs and implications and to keep up with rapid technological changes and new trends. As a solution to the updating challenge, they suggest identifying opportunities for improvements whenever they emerge. One way to achieve this is to ensure that technological solutions to specific projects do not work in isolated silos but consider international standards from the beginning and through the project in order to ensure the interoperability of not only the data collected (so that these can be added to existing databases or combined with data from other sources when necessary and feasible) but also the technological solutions that are being implemented. In other chapters, we have seen that mapping activities should support the input of data from external sources to perform additional and deeper analyses, such as in the case of ImproveMyCity (Chapter 9) where the authors report that this can provide valuable information by linking the data to analytical dashboards and other useful tools.

Linked to this general technological challenge is the need to address application extensions and updates – together with the problem of limited Internet connectivity – even with given technical resources to keep the software up to date. Nevertheless, it should be noted that the act of updating is difficult in non-urban cases, and this can further determine the successful utilisation of citizen science initiatives using digital devices.

Another way to address the technological upkeep challenges, highlighted in Chapters 2 and 10, is by building on existing platforms such as well-established open-source projects, and adding extensions to them to support the need for a specific activity. For example, social media networks can be used as a feedback mechanism to inform, motivate and keep participants engaged, as well as other technologies such as DIY sensors or platforms that record and store data from them.

Next, throughout the case studies, we see the need to consider the choice of the right technological infrastructure. This includes taking into account the mobile devices that the participants already have access to or, when they do not have access to any device, introducing a device which fits local conditions (i.e. literacy levels of potential users, the durability of devices, lack of access to electricity, etc.). Within that context, there is also a need to consider the visualisation and display of maps and geographic information on the technological device itself. While smartphone displays have improved significantly, there are advantages of using larger displays and capacities afforded by a tablet, laptop or a desktop PC. Of particular importance are additional technologies – from drones to DIY electronics – which can be relevant not only for prototyping but also for deployment. Finally, data storage remains an issue, as well as the cost of communication, and therefore consideration of the type of storage device (e.g. microSD) and the characteristics of the contracts offered on SIM cards need to be taken into account.

Next is the ability to consider the aspects of sustainability of the body that maintains and runs these technological apparatuses. As Corbett and Derrickson noted (Chapter 10), we need to consider the business models throughout the development of open-source geographic citizen science tools. For a very small organisation, sharing a product as free and open-source software might undermine the business model. A similar challenge is in the maintenance of software such as CyberTracker, which has evolved slowly over the past 25 years. The emergence of generalist software is also relevant for different geographic citizen science applications, for example the use of tools such as Gather or the ArcGIS mobile data-collection tools. As seen in Chapter 10, there is a need to balance the development tools required, and for the project coordinator to be open minded about what is needed; that is, just because it worked in the past or in another community does not mean that it will work in other contexts.

Last, we need to remember that the reason for the range of tools and approaches is the age-old lesson that there is no one-size-fits-all solution that will be suitable for all cases. This was noted in the early chapters and throughout this volume. The case studies clarify that there is a need to examine closely the context in which the application is being used and to choose the right technological solution to address important problems such as battery life, signal coverage, Internet connection, type of networks that are available locally and so on. Of equal importance is thinking about the adaptability of tools to fit specific contexts of use, as many examples in this volume have shown.

## 3. Methodological considerations and recommendations

All case studies in this volume use various methods to involve users in the development, implementation or evaluation of geographic citizen science applications. They all conclude with suggestions emphasising the importance of using human–computer interaction and anthropological approaches to work directly with end users. Yet, there are different implications, and the design approaches which are utilised to achieve this in specific contexts of use vary significantly.

The majority of the cases in the second part of this volume provide an insight into how end users are mainly engaged towards the final stages of product development and implementation, using mainly simple user feedback mechanisms (e.g. Chapters 6, 8 and 9). This enables improvement of the interface design by further incorporating additional user needs and addressing user issues and their suggestions when possible. Chapter 7, in Part 2, further provides a useful overview of a mixed methods approach to understand user needs, behaviour patterns and usability issues and to incorporate user suggestions in the design of the Global Forest Watch application. Feick and Robertson (Chapter 5) suggest that regardless of the contextual characteristics, geographic citizen science applications which require significant place-based knowledge benefit from higher levels of co-design engagement.

Noteworthy is that the cases in this part provide anecdotal evidence of user interactions to the wider geographic citizen science community. Yet, the theoretical and methodological frameworks which are used to inform user involvement are less clearly defined, and for that reason, they are perhaps harder to replicate. This is a limitation which is more broadly observed in the wider citizen science field. It is also what inspired the publication of this volume – to inform future development of geographic citizen science design by uncovering interaction issues and providing theoretical and methodological frameworks (Chapters 3 and 4) in order to improve user experiences with geographic as well as generic citizen science applications.

The case studies included in Part 3 of this volume all have the similarity of working with mainly indigenous communities in contexts where cultural, environmental and technological particularities pose various challenges to the successful adoption and utilisation of technological solutions. These are driven by the realisation that 'technologies should ideally be incorporated in daily use. Otherwise, they may be forgotten' and that 'cultural aspects should be considered when designing technologies for specific segments of society. The social organisation and internal rules of communities should be respected and followed for success' (Comandulli, Chapter 15). In most the cases, this requires an in-depth understanding of the local cultural context, knowledge systems, values, customs, everyday activities and so on, which is mostly achieved using participatory design methods to engage with users at all stages of project execution.

In Chapters 11, 12, 14 and 15, the authors employ participatory design with anthropological influences. Anthropological methods are not the only key to informing participatory design frameworks, but since most of these cases focus on conservation issues and environmental sustainability, they also support 'translating Traditional Ecological Knowledge (TEK) into data sets that can be placed in dialogue with current scientific conservation and policy models' (Fryer-Moreira and Lewis, Chapter 4). There are two main methods which are used to initiate and support a participatory design process and which are extensively discussed in Chapter 4: obtaining free, prior and informed consent (FPIC) and establishing a community protocol, which is an iterative process that requires the continuous involvement of the local community at all stages of project execution. Participatory design here also entails working closely with the potential users to understand interface design icons, prompts and interaction modes which are locally understood and which are therefore unique for each specific context of use, as well as ethnographic observation to observe how the technology is being utilised.

Building trust with end users and addressing their security, privacy, legal and ethical considerations are highlighted as major concerns by different case studies. These influence the success of an initiative and should therefore inform the design and development of geographic citizen science applications. For example, Pajarito Grajales et al. (Chapter 8), in the implementation of the Cyclist Geo-C app, encounter different ethical and privacy user concerns in different cultural contexts (e.g. users in Münster were much more concerned with data privacy and sharing settings compared to those in Castelló and Valletta, which influenced the ways they were willing to use the app in the future). Similarly, Hoyte (Chapter 12) explains that user concerns about data privacy required the use of the application SureLock, which was 'an important addition not only for security, but also to ease interaction with the devices', as well as implementing an anonymous ID system. Feick and Robertson (Chapter 5) explain that in the RinkWatch application, participants requested increasing privacy so that they could view solely their own data, which only became evident through participatory design sessions. Therefore, we suggest that to uncover and address these issues which are highly localised and context specific in nature, user input should be incorporated from the early stages of the application design and development. Involving users early and during all stages and incorporating their needs into the application further helps to build a trusting relationship and mutual respect.

Finally, an important consideration is the appropriateness and relevance of implementing specific methods in specific contexts of use. For example, while usability testing can be helpful in evaluating geographic citizen science applications and identifying interaction barriers, outside Western contexts the method should be used with caution and should be modified to address cultural specificities. For example, in Chapter 11, Vitos discusses how, in a usability texting experiment in the Congo Basin, tasks have to be designed in such a way that they are integrated into people's actual daily activities rather than introducing them via hypothetical situations. The author also reports that individual preliminary usability testing sessions had to be replaced by group sessions, as working in groups was most culturally appropriate. In other cases (e.g. Chapter 15), it is reported that specific communities avoid criticism – even when its subject is some technological interface – and therefore evaluation requires research teams being particularly creative in identifying the best ways to obtain honest feedback without making participants feel intimidated.

## 4. Common interaction barriers and suggestions for improvement

Throughout the case studies in the volume, a range of interaction barriers were identified, and ways to overcome them were proposed. Some of the most critical are summarised in this section.

One of the most important requirements that several of the authors in this volume (in both Parts 2 and 3) identify when it comes to participants collecting geographic data is the importance of the geographic interface and the ability to view data instantly or soon after the data-collection task has been completed. For example, in Chapter 15, Ashaninka trackers lost interest in using the app for collecting data about illegal land invasions and their activities, as it did not provide an interface to view and share data instantly (i.e. its spatial location, together with the photographs and videos that they captured as evidence). In a very different context, that of cycling in urban centres, Chapter 8 reports a similar issue. Users felt frustrated by not being able to view the data that they collected, and they expressed their need to view these data, the cycling frictions the app identifies, as well as share their cycling routes with other cyclists. The need to view the data is also noted in Chapters 12 and 13, which further demonstrate the effectiveness of utilising mapping interfaces for participants to see the broader geographic representation (through aerial imagery) of the area in question.

Providing a mapping interface and the ability to view or share the data with others, depending on the privacy and security characteristics of the project, may increase the impact of the citizen science initiative in different ways. First, being able to view the data helps participants decide whether they want to act based on the information provided, which instantly sets the foundations for two-way communication and which therefore places the initiative at the higher levels of the four-level engagement typology which was introduced in Chapter 1. Second, provision of instant access to the collected data can help keep participants motivated and interested in the project, especially if they are able to see how their data look in relation to the rest of the community's efforts, which according to Chapter 6, also favours knowledge dissemination and data democratisation. Third, being able to view the data that participants collect instantly serves as a verification mechanism which eventually improves data accuracy and reliability.

When a geographic interface is provided to view the collected data and support further analysis, a careful decision needs to be made as to whether this should be the central or a peripheral component of the provided application; that is, using the map to link to all major tasks and the application's functionalities, or using the map to complete tasks which are concerned with the geographic aspect of the information collected. The significant majority of geographic citizen science projects utilise the first approach, despite the fact that some major user tasks that they provide may not always be spatially relevant. In Chapter 10, Corbett and Derrickson discuss how the interface was redesigned from being map-centric towards a more text-centric visualisation due to space constraints, as well as to enable the easier completion of management-related tasks which did not require immediate access to the geographic component.

Feick and Robertson (Chapter 5) suggest designing and implementing geographic citizen science tools in ways which harness their power in terms of leveraging and developing geographic expertise and local knowledge. Moreover, the success in designing and implementing user-friendly geographic interfaces which can be operated by their end users requires an appreciation of their complexity and identifying ways to support interaction better. The inherent complexity of geographic data and geographic interfaces is frequently overlooked by citizen science practitioners. It may refer to simple tasks, such as a user – who has never used a map before – using a geographic citizen science interface to view the data on a map. It may also refer to more complex tasks, such as those described in Chapter 7, where the interface is used to support complex analyses for decision-making purposes which require advanced functionality, and which is overwhelming for the average and less technically minded user.

Scholars from the fields of geography and geographic information science have a long-standing research interest in understanding how non-experts interact with spatial interfaces. Yet, the increased availability and utilisation of open-access mapping software and web-mapping solutions supported by application programming interfaces (such as Google Maps) has resulted in the use of geographic interfaces without considering their suitability for their intended user audiences. To avoid this, we should consider Antoniou and Potsiou's (Chapter 2) and Feick and Robertson's (Chapter 5) suggestions that geographic skills and knowledge, as well as users' prior experiences in interacting with geographic interfaces, should be all considered in the design of suitable geographic visualisations and functionality to support the purposes of the project. In cases where potential users have not used a map or other types of geographic interfaces, these types of projects may significantly benefit from participatory design approaches to shape with end users the technologies and the tasks of data collection, analysis and dissemination.

A different issue from that of the complexity of geographic interfaces is that of connectivity. In the second and third parts of this volume, the authors highlight the effects of bandwidth and Internet connectivity in terms of how users interact with geographic citizen science applications, and they suggest that problems influence user experience in different ways. The issue of wireless connectivity, mentioned in Section 2 of this chapter, is a major concern in terms of keeping the device and software updated for optimum performance as an essential requirement for the successful utilisation of geographic citizen science applications. Also, as explained in Section 3 of this chapter, the majority of end users who use geographic citizen science applications, in both urban contexts and in developing countries with indigenous communities, have an expectation of viewing the data they collect instantly (or soon afterwards) which also requires a wireless signal or Internet coverage. When this is not possible, the authors suggest that the applications should provide an 'offline mode' option where tasks are executed normally and data are uploaded and viewed later when a connection is established. An interesting observation, highlighted by the authors of Chapters 7 and 16, is that slow bandwidth or intermittent connectivity generates a negative user experience (perhaps even more so than the cases where there is no connectivity at all), and this has an effect in the quality of the data collected, as users may become, due to their frustration, less careful with the accuracy of their observations. In these cases, it would be perhaps more effective if the application suggests to the user (e.g. through a pop-up) to use the 'offline mode'.

Geographic citizen science relies on technological apparatuses in addition to the needs of ensuring that the data are fit for purpose. So, there is an important role in training participants in using and managing the technological infrastructure, especially in cases where they have never interacted with similar technologies before. For example, in Chapters 11 and 12, where none of the participants had used a mobile device before, a period of training on basic use of the phone was required; that is, training participants how to hold the device, how to use the touch screen, the location of the camera and so on. In Chapter 15, the authors also encountered the need to train participants in basic operations, which are usually taken for granted, such as activating the phone and swiping the screen. Training also provided the opportunity to uncover problems with devices, such as the need to moisten the fingers to enable touch-screen operations.

In Part 3 of this volume, the authors also report that beyond training, there is a need for the ongoing management of the technological infrastructure in the field, for example training participants to reduce the sounds and lights that the phones produce, charging and updating the phones, as well as managing the infrastructure such as cables, solar panels and power banks which are all needed to operate the devices. This becomes an integral part of the geographic citizen science initiative implementation, and therefore such activities should not be overlooked – they are fundamental in ensuring the smooth operation of devices and therefore how participants interact with the applications to collect the required data and participate in analysis.

Provision of training and support is also needed to ensure the effective use of the geographic citizen science application itself. When these are used with indigenous communities and people who have less experience in interacting with similar applications, it is essential to ensure that participants fully understand the navigation flows, interaction modes (e.g. how long participants should press on an icon for their tap to be registered), the pictograms used in interface design for data collection, as well as the visual design of icons which are used for other inputs such as collecting audio files or for marking the completion of a task. Chapter 11 suggests that co-designing these with participants, evaluating their usability and improving their design when required may ensure that participants can use the actual application with as little help as possible from the very early stages.

While the case studies in Part 3 of this volume highlight the importance of providing help and training in the use of the actual hardware and software that are being utilised, the cases in Part 2 mostly highlight the need for training provision in terms of improving participants' scientific content knowledge to 'ensure that they collect data similar in quality to that collected by experts' (Feick and Robertson, Chapter 5). The cases in Part 2 do not have the same barriers and challenges to overcome, which in Part 3 are mostly associated with the limited textual and technological literacy of participants. Their end users are less clearly defined, since these applications can be used by anyone with an interest in the topic and who has access to the application and relevant technological infrastructure. Participation is more opportunistic, which makes it harder to provide the same training and support, even when these might be beneficial for the initiative. Yet, it is surprising that most citizen science projects provide extensive scientific training kits but overlook the importance of help provision (e.g. through tutorials) to ensure the effective and efficient use of their complex geographic interfaces and the overall use of their applications, which are critical components for success. Another way to deal with the increased complexity, even in cases where users are experts or have an advanced set of skills, is using the progressive disclosure principle to guide users gradually from simple to more complex tasks, as discussed in Chapter 7.

Additional interaction design lessons for geographic citizen science applications used with users with low textual and technological literacy skills include the use of pictograms and the avoidance of hierarchical structures to support navigation. ICT for development research proposes the use of icon-based interfaces with limited or no text when end users have low textual literacy skills (i.e. limited or no skills in reading or writing). Most of the case studies in Part 3 use hand-drawn pictograms, which are co-designed with participants in the field (Chapters 11, 12, 13 and 15), or pictograms replaced by photorealistic images, as proposed by the participants themselves (Chapter 14). The authors report that participants find interaction easier with pictograms that visualise data items to be collected, while they find it harder to understand categorical, top-level pictograms, which are used in support of hierarchical navigational structures. Ideally, these should be avoided. It should be mentioned that there are geographic citizen science applications based on complex hierarchical structures (e.g. CyberTracker) which are successful in their utilisation and in terms of achieving their aims in similar contexts. These usually satisfy two important conditions. They are used repeatedly and over long periods of time by a specific set of trackers who have first received specialised training. In encouraging the use of geographic citizen science applications by everyone in the community, it is important to minimise these complexities, and for that purpose, in Chapter 11, Vitos suggests that 'decision trees should be avoided in cases where low-literacy or non-literacy prevails'. Alternative technological solutions, such as tangible interfaces (e.g. see Tap&Map in Chapter 11), have the potential to improve usability and, as a result, 'participants' performance and satisfaction' (Chapter 11).

For geographic citizen science applications, the choice of the right technological equipment is of utmost importance, although this aspect is usually overlooked. Power capacity, durability, brightness and the size of the screen are just a few of the criteria that should be carefully considered in the choice of the most appropriate technology or perhaps a combination of different technologies to support different interaction tasks. For example, while mobile devices are generally preferred and are perhaps more effective, in supporting data-collection tasks, especially for the implementation of more opportunistic geographic citizen science projects, a desktop computer is much more effective in supporting complex geographic tasks which may benefit from larger screen displays. For example, in Chapter 5, Feick and Robertson demonstrate that digitisation tasks are more effective when they are carried out on a desktop computer environment with a larger screen size. The tasks should therefore be carefully considered to inform the right choice of technological equipment which will be subsequently used to enable them.

Finally, there is evidence that although usability is critical in all contexts and for all user interactions with geographic citizen science applications, designers and development teams have to make the occasional choice of sacrificing usability over the 'scientificness' of data quality. In Chapter 6, Radicchi explain that users of the Hush City app found the questionnaire, which is used to guide data collection, too long, and they were annoyed at not being able to skip specific questions. Nevertheless, the importance of collecting robust scientific data, which could be used for research purposes, was more highly rated, and they made the decision to keep the long and complex questionnaire without changes. There are several examples where scientific teams create scientific protocols which are tailored to the specific needs of citizen science projects so that participants are not overburdened with data collection and other tasks. The importance of usability requires perhaps similar trade-offs (when this is possible) to inform the design and development of digital technologies which can be effectively used in citizen science.

## 5. Why does geographic citizen science design matter?

Digital technologies have boosted the growth and implementation of citizen science initiatives, which have attracted the attention of scientific, mainstream media and policymaking sources, which in turn highlight its strengths, benefits and positive impacts to society. An increasing number of citizen science projects rely heavily on the use of geographic information science to enable the objective and systematic collection, analysis and dissemination of spatially enabled data. There is already – mostly anecdotal – evidence to suggest that citizen science initiatives are most likely to fail if the technologies they utilise cannot be effectively used by their intended audiences. Despite being literally flooded with such applications and discussions around data quality and coverage issues, until recently, little attention has been paid to user interaction – the theoretical, technological and methodological aspects which should be taken into account to inform and improve interaction and subsequently maximise the potential impact of these initiatives (Skarlatidou et al. 2019).

This volume provides a unique insight into the way geographic citizen science initiatives are currently being implemented in urban areas of the Global North and in more remote areas where they are used to assist indigenous communities in addressing major issues they currently face and their local needs. The case studies which are presented in this volume provide a state-of-the-art description of current practices and developments in the field. The aim of their analysis is to emphasise how participants interact with the applications used in their initiatives; participants' needs, goals and the most common interaction barriers; and the most important lessons for the design and development of future geographic citizen science applications which are used in similar contexts.

This volume was created to uncover anecdotal evidence from user interactions with geographic citizen science and further highlight the need for a holistic, participant-centred design, which takes into account the local context, cultural considerations, needs and purposes of the citizen science initiative. It provides insights into the theoretical, technological and methodological frameworks which can be used to assist user involvement in the development and utilisation of geographic citizen science applications, and which are user-friendly, trusted and suited to the purpose for which they were built. In this form, geographic citizen science can build on the extensive experience that has emerged from the practice of participatory mapping and the work of Chambers and others in the 1980s (Chambers 2006), and the rapid development of participatory GIS and Public Participation GIS (PPGIS) in the past quarter of a century (see Sieber 2006). Attention to usability issues can be linked to the discussions that started within the PPGIS literature (Haklay and Tobon 2003).

The current imperatives of ecological and climate emergencies, and the need to reach sustainability while ensuring that everyone – regardless of gender, ethnicity, social position or level of education – can be an active producer and user of information that can help them in protecting their environment and themselves, make the call for inclusive geographic citizen science application a more urgent one.

This volume demonstrates how geographic citizen science can contribute to the urban planning, environmental and conservation contexts. We hope that it will assist others to ensure that no one is left behind in the process of finding the path for sustainability and participation in scientific debate, and in achieving responsible research and innovation and open science for all.

## References

Chambers, Robert. 2006. 'Participatory mapping and geographic information systems: Whose map? Who is empowered and who disempowered? Who gains and who loses?', *The Electronic Journal of Information Systems in Developing Countries* 25: 1–11.

Haklay, Muki, and Carolina Tobón. 2003. 'Usability evaluation and PPGIS: Towards a user-centred design approach', *International Journal of Geographical Information Science* 17: 577–92.

Hecker, Susanne, Muki Haklay, Anne Bowser, Zen Makuch, Johannes Vogel and Aletta Bonn. 2018. *Citizen Science. Innovation in open science, society and policy*. London: UCL Press.

Sieber, Renée E. 2006. 'Public participation geographic information systems: A literature review and framework', *Annals of the Association of American Geographers* 96: 491–507.

Skarlatidou, Artemis, Marisa Ponti, James Sprinks, Christian Nold, Muki Haklay and Eiman Kanjo. 2019. 'User experience of digital technologies in citizen science', *Journal of Science Communication* 18: 1–8.

Wynn, James. 2017. *Citizen Science in the Digital Age: Rhetoric, science and public engagement*. Tuscaloosa: The University of Alabama Press, 2017.

# Index

page numbers in *italics* refer to Figures, Tables, Illustrations

Aboriginal Mapping Network, 212
Amazon Fund, 306
anthropological methods, 6, 7, 8, 64, 87–102, 345, 346
   ethnographic observation in, 87, 89, 94, 99, 101
   in 'extreme' GSC projects, 88, 89, 93, 99, 101
   free, prior and informed consent (FPIC), use of. *See* FPIC
   in-depth interviews, use of, 91–92
   and participative evaluation, 98–99, 100
   prototype pictogram decision trees, use of, 94–96, 98
   and technical practices, 94, 101
   translation, notion of in, 101
   *See also* community protocols (CPs), development of; ethnography; ExCiteS; indigenous communities
ArcGIS Online and Server, 117, 118, 123, 344
Ashaninka communities (Amazon) project, 12, 101, 302–315, 342
   Comissão Pro-Indio do Acre (CPI-AC), collaboration in, 306, 308
   data collection/visualisation in, 311, 347
   ethnographic considerations in, 101, 313
   FPIC (consent/protocols) in, 302, 306, 310, 315
   and government monitoring support, 304–305, 342
   illegal logging/poaching, effect on, 303, 305
   Internet and computer access in, 304, 312
   land monitoring in, 12, 304–306, 312–313, 314, 347
   literacy and numeracy levels in, 310
   monitor training period in, 307–308, 310, 313
   participatory design methodology in, 306
   pictogram adaption in, 307, *308*, 311
   Sapelli app development/testing in, 302, 306, 307, 308, 310, 314, 315
   and technological challenges/barriers, 304, 310–312, 314, 315, 347
   technological training/maintenance in, 302, 307–308, 313
   user feedback in, 302, 308–310, 313, 315
   *See also* Brazil, Pantanal fisheries project
Association of Geographic Information Laboratories, 171
Atlas of Living Australia, 339

Bee Lab project, 75
   design workshops in, 75
Berlin Plan of Quiet Areas 2019–23, 133
Bike Citizens, 167
BikeMaps.org, 167
Biko, 167
biodiversity and conservation, 35, 154, 230, 247, 251, 340
   biocultural approach to, 251, 262
   community-centred approaches to, 248
   community empowerment in, 247, 248, 251, 262
   extreme citizen science approach to, 251, 262
   indigenous local knowledge, use in, 247–248, 249–250, 262, 278, 279, 298
   noise pollution, effect on, 130, 131, 132
   *See also* Cameroon, Congo Basin projects; ExCiteS
birdwatching/monitoring, 21, 65, 110, *116*, 118, 121, *123*, 124, 125, 156, 340
   in Grasslander app, 118, *120*, *121*

355

Brazil
   deforestation in, 160
   GWF protected forests map, *155*
   indigenous displacement in, 282
   NGO sustainability programmes in, 283–284, 306
   traditional peoples, national policy on, 286
   *See also* Ashaninka communities (Amazon) project; Brazil, Pantanal fisheries project
Brazil, Pantanal fisheries project, 11, 61–62, 282–299, 342
   conservation biology approach in, 282, 283
   data confidentiality and sharing, 292
   decision tree/pictogram, use in, 288, *289*, 292–294
   displacement of local people in, 282, 284, 285, 342
   Ecologia e Ação (ECOA) support/organising in, 286, 290, 298
   ethical consent process in, 290–291, 292
   ethnographic methodology, use of, 61–62
   fishing tourist business, local effect of, 284–285
   and government policy/regulations, 283, 285
   government protected areas expansion in, 283–284, *285*
   local fishers, blame and restrictions on, 283, 285–287, 298
   local territory definition in, 296–298
   The Nature Conservancy (NGO) support in, 283–284
   participant observation in, 288, 296, 298
   participatory mapping approach in, 283, 289–290, 296, 298
   Protection and Conservation Network for the Amolar Region partnership, 284
   Sapelli app, test and use in, 61, 287–288, *290*, 291–298
   sustainable resource use, data results, 298–299
   user challenges and feedback in, 62, 292
   *See also* Ashaninka communities (Amazon) project
British Columbia, University of, 214, 216
   SpICE lab at, 210, 214

Cambodia, Prey Lang communities project, 11, 266–280, 341–342
   data ownership/sharing decisions in, 271, 275, 280
   decision trees, use in, 271, 272, *273*, 276
   development costs in, 277, 280
   forest dependence and safety in, 269–270, 279
   forest patrols in, 269, 274–276, 277–278
   government interaction/regulations in, 269, 271, 274, 277–278, 279, 280
   illegal logging/forest monitoring in, 11, 266, 268, 269, 274–276, 278, 279, 342
   'It's Our Forest Too' project, 268, 269
   literacy/technological difficulties in, 272, 274, 276–277
   offender conflict resolution strategies in, 269
   participatory design in, 270–274, 278
   Prey Lang Community Network (PLCN) in, 268, 269, 270–280
   Sapelli platform use in, 271
   smartphone app design, use in, 268, 270–274, 275, 276–277, 278
   user testing and feedback, 272, 275–277, 280
Cameroon, Baka communities project, 11, 248, 250, 251–263, 341
   Baka exclusion/discrimination in, 250, 254, 262
   Baka extensive knowledge in, 250
   challenges in, 252, 257, 260, 262
   community-led design/management in, 248–249, 254–255, 256, 262, 263
   Community Maps interface use in, 259, *261*
   cultural diversity in, 149, 256
   data utilisation and sharing decisions in, 255, 259, 260–261, 263
   extractive industries in, 249, 250
   illegal wildlife trade in (IWT), 149, 150, 251, 254, 341
   non-literate design in, 249, 256, 261, 263
   participants in, 252, 262
   report verification in, 259–260
   Sapelli icon-based tool co-design in, 256–259, 260, 263
   technology design and interaction in, 255–259, *260*
   traditional knowledge (TEK) use in, 100, 101, 102, 252, 261–262
   trust and consent in, 252–255, 262
   wildlife monitoring in, 11, 248, 252, 341
   Zoological Society of London (ZSL) partnership in, 248
Canada, 10–11, *116*, *119*, 120–122
   Grasslander project in, *116*, 118, 120–121
   RinkWatch project, *116*, 117, *119*
   Wildlife Health Tracker project, *116*, 121–122
   *See also* Canada, First Nations communities project; GrassLander; RinkWatch project
Canada, First Nations communities project, 10–11, 209–226, 341
   development consultation process in, 210, 211–212, 213, 216–217, 222–223, 224
   Firelight Group consultation in, 212, 213–214, 222
   Gather tool development for. *See* Gather First Nations tool
   land rights and management issues in, 210, 211, 213, 216, 225
   resource development, impact on, 10, 209–210, 211, 212, 213

Royal Proclamation (1763) land recognition in, 211
Saulteau First Nations (SFN) in, 212, 213, *214*, 215, 223, 224
software/data management challenges in, 212, 216–217, 223, 224
SpICE lab tool co-design in, 210, 212, 214, *217*
stewardship land focus in, 211, 213
and Treaty 8 territory, 213
Wabun Tribal Council (WTC) in, 212, 213, *214*, 215, 222–223
*See also* Gather First Nations tool; SpICE lab
citizen, use of term, 20
citizen-government communication, 186–204
and app utilisation by authorities, 200
collaboration tools in, 186, 187
and community-driven engagement, 193
community sense, strengthening of in, 186, 188
non-emergency issue reporting in, 186, 187, 340
transparency promotion in, 186, 188, 204
*See also* ImproveMyCity (IMC) app; Greece, Thessaloniki IMC platform
citizen science
coordination and supervision in, 45, 46
definition of, 3, 18
education, role of in, 25, 26, 56, 81, 341
experts/expertise, role of in, 23–24, 108
extreme citizen science approach in. *See* ExCiteS
'the Five Cs' typology (Shirk et al.), 22, 24–26, 27, 31, *33*
and geographic interfaces, 3–4, 16, 18
levels of engagement (Haklay), 21–24, 26, 27, 28, 31, *33*, 45, 109–110, *123*, 182, 187, 348
models of participation, 22, *23*, 24–28, *33*
motivations for participation in, 20, 23, 34, 35, 67–68, 80, 109, 348
terminology analysis, 19–20, 21, 29
types of activities in, 26–29
*See also* crowdsourcing; ExCiteS; geographic citizen science
Citizen Science Global Partnership, 251
Citsci.org, 64
Climate change, 117–118, 354
community-led monitoring, role in, 280
deforestation, effect on, 267
Intergovernmental Panel (IPCC) guidelines, 267
RinkWatch project, *116*, 117–118
Climate Prediction network, 27
cloud computing, 43, 194, 216
collaborative intelligence, 251
community/civic science, 28, *30*, *33*
Community Peace-Building Network (CPN), 268

community protocols (CPs), development of, 9, 73, 82, 87–88, 96–98, 181, 223, 231, 240, 254, 259, 263, 307, 310, 346
in indigenous communities, 9, 87–88, 90, 92–94, 231–232, 240, 302
*See also* FPIC (free, prior and informed consent)
Congo Basin forest management projects, 11, 228, 230–245, 248, 249, 341, 347
action research approach in, 231
Baka hunter-gatherers in, 11, 237, 250
community consent/protocols (FPIC) in, 231–232, 240, 253
cultural interpretation challenges in, 229, 231, 237, 238–240, 241–242, 244–245, 347
EU Voluntary Partnership Agreement, 230–231
ExCiteS aims and design initiatives in. *See* ExCiteS (Extreme Citizen Science), UCL
forest population in, 230
Forest Stewardship Council (FSC) accreditation, 231
illegal logging/poaching monitoring in, 11, 232, 341
logging/resource extraction in, 230–231, 234–235
NGO involvement in, 231, 232, 249
participatory/co-design in, 232, 238, 241, 247
and private-sector investment, 230
Sapelli interface, use in. *See* Sapelli interface
Traditional Ecological Knowledge (TEK) collection in, 11, 101, 229, 230, 237–242, 341
UCD approach in, 231
user barriers/feedback in, 233–234, 235–245, 347
*See also* Cameroon project; Sapelli interface; Traditional Ecological Knowledge (TEK); rainforest management
conservation biology, 282–283, 340
conservation projects, 61, 88, 149–151, 282–283, 299
biocultural conservation, success of, 150–151
as 'bioimperial', 249
and 'conservation refugees', 150
Eurocentric models of, 249–250
'green militarisation' in, 150
local needs inclusion in, 282–283
NGO domination of, 149, 150, 299
participatory approach to, 299, 346
use of indigenous knowledge in, 88–89, 100, 102, 247, 341
*See also* projects in Brazil, Cambodia, Cameroon, Congo Basin; biodiversity and conservation; environmental monitoring
contextual inquiry, 65–66, 69, 76

Creek Watch, 66
crowdsourcing, 15, 16, 19–20, *22*
   in citizen science, 22, 27, 31, *33*, 44, 45, 110, 166
   origin and use of term, 19–20
   US guidance on, 44
   *See also* Cyclist Geo-C mobile app
CyberTracker, 16, 114, 344
   use with indigenous expertise, 114, 233
cycling, urban, 165–183, 340, 348
   benefits of, 165–166, 167
   crowdsourcing information, use of, 166, 167
   data platforms, use of, 166, 167–168, 180, 182
   mobile devices and LBS use in, 166, 167, 180, 181
   and route mapping, 166
   *See also* Cyclist Geo-C mobile app
Cyclist Geo-C mobile app, 4, 9–10, 165, 168–183, 341, 346
   barriers reported in app use, 179–180, 182, 347
   collaboration- and competition-based rewards in, 175, *176*, 181, 182–183
   crowdsourcing data collection in, 168, 175, 181, 182
   cultural and social context in, 168–170, 171, 181
   data collection and analysis in, 171, 175–177, 178, 179
   friction areas, identification of, 165, 168, 172, *174*, 175, *177*, 178, 179, 182, 348
   gamification strategy, use of in, 165, 168, 179, 180, 181, 182
   geo-spatial route data collection in, 168
   semantic tags, use in, 165, 168, 171, 172, 178
   upscaling and improving of, 182–183
   user privacy and data sharing issues in, 180–181, 182, 346
   volunteer feedback/improvement suggestions, 178–179, 180
   volunteer testing of, 168–169, 171, 178–179, 182
   *See also* under Germany, Malta, Spain; Open City Toolkit

Danmission, University of Copenhagen, 268
data collection and management, 4, 8, 15, 22, 46–47, 48–50, 67, 81, 108, 109, 348
   community protocols for sharing, 98
   in contributory and collaborative projects, 109, 110, 341
   data accuracy/reliability methods, 348
   data-analysis tools, 49–50
   expert-amateur divide in, 109
   extreme citizen science approach to, 251
   local knowledge contribution in, 108, 109, 229, 247–248, 278, 298, 299, 341, 347
   mobile apps, use of, 4, 39, 47, 140, 143–145
   open access in, 143, *145*, 348
   positional accuracy in, 47
   privacy and sharing issues in, 182, 346, 348
   and spatial analysis, 45–46, 50
   and spatial databases (SDBs), 49
   storage and security in, 47, 49, 344
   and technological innovation, 43, 48, 343
   UCD approach to, 60, 63
   volunteer computing, use in, 49
   *See also* FPIC; Gather software; human-computer interaction (HCI); questionnaires and surveys
decision trees, 94–96, 98, 311, 315
   cultural difficulties in adaption of, 311
   flow diagram, *95*
   literacy levels, effect on use of, 243, 245, 352
   *See also* Brazil, Cambodia, Congo Basin projects; Sapelli interface
deforestation, 44–45, 149–151, 162, 267, 269
   and climate change effects, 267
   indigenous communities, effect on, 269
   model of main drivers of (Curtis), 150–151
   *See also* Cambodia, Prey Lang communities project; forests, protection of; Global Forest Watch platform/app
DIYBio, 28, 30
DIY science, 26, 28, 30, *33*
DreamLab (Vodaphone app), 28
drones, 39, 48, 52, 344

Ecologia e Ação (ECOA), 286, 290, 298
Ecotrust Canada, 212
environmental and ecological observations, 26, 27–28, 108, 116–117, 118, 120–122
   linked web dialogs, use of, 116
   map-centric design, use in, 116
   use of VGI in, 18, 29, *30*, *33*
environmental monitoring, 88, 90–91, *116*, 248, 266
   ethnographic approach to, 100
   indigenous community engagement in, 266, 279
   use of participant observation in, 90–91
   *See also* biodiversity and conservation; forests, protection/management of; rainforest management
environmental noise projects, 130–146, 340
   Berlin Plan of Quiet Areas 2019–23, 133
   Hush City mobile app, development of. *See* Hush City app
   people-centred app recommendations, 131, 142–145
   public quiet areas, identification of, 130, 131, 132, 133, 340
   soundscape approach in, 131, 132, 143, *144*
   *See also* noise pollution
environmental projects, 11, 24–25, 61, *116*

relationship models in, 24–25
*See also* Ashaninka; Brazil; Cambodia; Cameroon; Congo Basin; Peru projects; conservation projects
EpiCollect, 233
EsriLeaflet app, 118, 122, 123
ethnography, *57*, 61–62, 65–66, 88, 89, 99–100, 346
   community engagement practices in, 89, 90–91
   use in interactive system designs, 62–63, 89
   *See also* anthropological methods; participant observation; participatory design
European Citizen Science Association, 132
European Commission
   Environmental Noise Directive (END), 131, 133
   Horizon 2020 Framework Programme, 44
European Joint Research Institute, 279
ExCiteS (Extreme Citizen Science), UCL, 230, 232–245, 247, 248, 251, 254, *257*, 261, 303, 307
   collaborative intelligence in, 251
   community self-management/protocols in, 254, 259, 263, 307
   and data sovereignty, 254
   design mission/objectives of, 232, 233, 237, 251, 252, 303
   Mapping for Change in, 259, *261*
   non-literate data usability focus in, 233, 236, 239, 240, 243, 245, 248, 249, 261, 303
   participant empowerment in, 251, 252, 254, 306
   Sapelli interface, development of, 230, 232, 233, 235–242, 248, 256–259, 287, *290*, 303, 306
   Tap&Map tool, creation/use of, 236–237, 238–240, 241–242, 243, 352
   usability/barriers analysis by, 237–245
   *See also* projects in Brazil, Cambodia, Cameroon, Congo Basin; Sapelli interface
experts and expertise, 23–24, 108–110, 111–112, 126
   conceptualisations of, 108–109
   contributory and interactional expertise, 111–113
   and experience-based knowledge, 111
   expertise, concepts of, 111
   Expertise Space Diagram (ESD) model (Collins), 112–113
   *See also* geographic expertise

farming/farmers, *120*, *123*, 252, 268
   and conservation projects, *116*, 118, 120–121
   *See also* GrassLander
fisheries, freshwater, 282–283
   and conservation biologist practices, 283
   and local people's needs, 282–283
   *See also* Brazil, Pantanal fisheries project
Floracaching app and game, 75–76
   use of prototyping in, 75–76
focus groups, *57*, 68–69, 74, 76
   advantages of, 68–69
   citizen science design and use of, 69, 76
Foldit, 48
Forest Atlases, 156, 159
forests, protection/management of, 149–162, 231, 248, 266, 267–268, 340
   awareness and data access, 150, 151–152, 266
   and biocultural conservation, 150–151
   and carbon capture capacities, 267
   community-based monitoring in, 267, 268, 279
   environmental costs of, 150, 267
   global picture of, 150
   and illegal logging/ forest conversion, 267, 268
   IPCC data guidelines for, 267
   obstacles/risks for local involvement in, 267–268, 279
   *See also* Cambodia, Prey Lang communities project; deforestation; Global Forest Watch platform/app
Forests Monitor, *234*
Forest Stewardship Council (FSC), 231
Forest Watcher app, 156, 158, 161, 163
FPIC (free, prior and informed consent), 92–94, 97, 231–232, 240, 253, 302, 306, 307, 315, 346
   key elements in, 92–93
   steps for citizen science, 93–94
   use with indigenous communities, 231–232, 240, 253–254, 302, 306, 307, 315
   *See also* community protocols (CPs), development of

Galaxy Zoo, 67
   participant motivation survey use in, 67, 68
gamification, 43, 48, 63, 181, 201
   collaboration-based strategies in, 181
   use for data collection, 181
   *See also* Cyclist Geo-C mobile app
Gather First Nations tool, 10, 209, 210, 212–226, 344
   basic aims of, 217–218, 224–255
   co-design approach in, 209, 210, 215–216
   commercial providers, tensions with, 222–223
   community-based research/workshops in, 215, 216, 217, 221
   community-government tensions, impact on, 223
   data gathering and management in, 209, 213, 216, 219, 221, 222

Gather First Nations tool *(cont.)*
  implementation challenges for, 209, 219, 221–222, 224
  Indigenous Mapping Workshop in, 215, 221
  'just-in-time' programming in, 215–216
  mobile app development in, 219
  open-source basis of, 212, 222, 224, 225
  project partners in, 212, 213–214, 215, 221, 224
  referrals management tool in, 218–219, *220*, 223, 224, 225
  usability issues in, 225
  user interface (UI) design in, 216–217, 218–219, 224
  *See also* Canada, First Nations communities project; SpICE lab
gender issues, 35, 342
  awareness/education about, 12, 320
  gender, definition of (WHO), 320
  as social norms construct, 320
  *See also* Peru, Cusco community mapping project
General Data Protection Regulation (GDPR), 171, 182
GeoCities, 16
geographic citizen science
  anthropological approaches in. *See* anthropological methods
  cultural context in, 12, 55, 56, 58, 89, 90, 92, 93, 94, 98–99, 102, 229, 302, 315, 341, 346, 347
  data in. *See* data collection and management
  definition of, 16, 18, 339
  design challenges in, 6, 8, 17, 18, 31–32, *33*, 34–35, 47, 56, 63, 109, 110, 345, 351, 353
  'digital divide' challenges in, 17, 18, 56
  as an educational pathway, 107, 127
  Global North/Western influence in, 340, 341
  and government-citizen interaction. *See* citizen-government communication
  inclusivity in, 15, 18, 35, 45, 97, 354
  and literacy skills, 3, 5, 6, 7, 10, 17–18, 32, 34, 93, 228, 243, 344, 351
  mass mobile computing in, 16–17
  participant-centred design, importance of in, 351, 353
  participant demographics in, 6, 7, 40, 56, 81
  participant expertise differences in. *See* geographic expertise
  participant recruitment/engagement in, 32–33, 35, 45, 47, 50, 81, 97, 107, 145
  power differences and relations in, 34, 35
  progressive disclosure principle, use in, 161, 163, 351
  project management in, 46, 50–51, 52–53, 80
  social networking, use in, 48
  technology, use of. *See* technological development and innovation
  and VGI intersection. *See* volunteered geographic information (VGI)
  and volunteer training, 110, 334, 351–352
  *See also* citizen science; ExCiteS; human-computer interaction (HCI)
geographic expertise, 9, 107–128, 348–349, 350
  and binary expert-amateur divide, 109
  and co-design engagement, 108, 109, 113, 126–127, 209, 345, 351
  dimensions of expertise in, *125*
  educational pathways, creation of, 107, 127, 128
  effects on participant engagement, 107–108
  geographic ESD model (GESD) use of, 107, 112–114
  and geographic knowledge types framework, 109
  and local knowledge, 9, 108, 109, 111, 113, 126, 127, 298, 341, 348
  and participation levels, 109–110, *123*
  place-based expertise, types of, 113–114, 115, 127
  social and cultural context in, 115–116
  technological training and support in, 350–351
  tool and activity design interaction in, 9, 107, 108, 109, 110, 111, 114–115, 122–126, 127, *144–145*, 350
  and volunteer's knowledge variation, 110, 350
  within 'cultures of practice', 115, 127
  *See also* experts and expertise
geographic information systems (GIS), 46, 49–50, 78, 80, 154, 171, 175, 216, 260, *261*, 339, 342, 354
  Public Participation GIS (PPGIS), 354
  QGIS app, 323–324
GeoTag-X, 27
Geotec research, Jaume I University, *173*, *177*
Germany, Münster, *169*, 170, 172, *173*
  Cyclist Geo-City app test, 168, *169*, 170, 171, 178, 180, 181
  promotion of cycling in, 170
  road network in, 172, *173*, *174*
  *See also* Cyclist Geo-C mobile app
Global Active Learning Group (GAL Group), 322–323, 327, 329, 332
  project-based learning in, 322–323
  *See also* Peru, Cusco community mapping project
Global Forest Watch platform/app, 9, 149–162, 279, 345
  barriers encountered in, 157, 160–161
  core and contextual data sets in, 154, 156, 160
  data access and sharing in, 150, 154, 158, 159–160, 161

Forest Watcher mobile app in, 156, 158, 161, 163
Global Land Analysis and Discovery (GLAD) layers in, 154, 155, 158, 279
global platform database key users, 151–152, 154
Google Analytics reviews/interviews in, 153, 154, 159–160
key lessons in, 162–163
mapping component in, 154, *155*, 156–157, 159–160
mixed-methods approach in, 149, 153, 154, 162, 345
prototypes use in, 153, 154
usability testing in, 153, 154, 160, 161, 162
user-centred design (UCD) in, 152–153
user needs research, 149, 150, 153, 157–160, 161–162, 163
user personas method in, 151–152, 154, 157
*See also* deforestation; forests, protection/management of
Global Navigation Satellite Systems, 342
Global Positioning System (GPS), 5, 15, 16–17, 29, 46, 168, 191, 237, 271, 287, 306, 314, 323, 339
mass use of, 16–17
Google Analytics, 153, 154, 159–160, 202
Google Maps, 117, 123, 193–194, 216, *217*, *218*
GrassLander, 115, *116*, 117, 118, 120–121, 124–125
farmer expertise/design engagement in, 118, 120, *123*, 124–125, 126
OSCIA as intermediary in, 120, 124
technology platform in, 124
Greece, Thessaloniki IMC platform, 188–193, 198, *199*, 200, 201, 203
community-driven user engagement in, 193
goal of, 188
IMC launch and technical management of, 191
impact of, 203
metropolitan area/population in, 188
municipality organisation of, 188–189
participation rates in, 191, *192*, 193, 201
pre IMC issue management methods, 189–191
reported issues and resolution numbers in, 191, *192*
*See also* ImproveMyCity (IMC) app

Harvard University, Project Zero, 323
HOT Tasking Manager, 319, 323, 324, 325, 326–327, 329, 331
object tracing in, 326
statistics tracking in, 326
validation process in, 326–327, 329
*See also* Humanitarian OpenStreetMap Team (HOT)

human-computer interaction (HCI), 6, 8, 32, 41, 55–82, 230, 345, 350
cultural context in, 56–57, 58, 65
design approaches/methods model, *57*
ethnographic approaches in, 89–90
heuristic evaluation in, 79–80
impact on citizen science, 56, 59, 61, 63
inspection methods in, 79–80
interaction design and user experience (UX), *57*, 62–63, 78, 79, 80
participatory design in. *See* participatory design
progressive disclosure principle in, 161, 163, 351
usability, ISO definition, 58
usability testing process in, 59–60, 63, 64, 69, 71–74, 75, 78, 79, 153, 154
user-centred design (UCD), 57–60, 63, 64, 81, 152–153, 231
user empowerment/encouragement in, 61, 63, 249–250
and user interface (UI) evaluation, 8
*See also* interviews; prototyping; technological development and innovation; storytelling
Humanitarian OpenStreetMap Team (HOT), 317, 323, 329
base data provision by, 319
and crisis response, 318–319
Cusco, Peru case study. *See* Peru, Cusco community mapping project
gender issues data sharing in, 317, 319
guiding principles in, 318
HOT Tasking Manager in. *See* HOT Tasking Manager
mapping toolset used in, 319
'off-the shelf' open source apps in, 317, 334
school children's involvement in, 317, 319
unmapped/underpriviledged areas, focus on, 318–319
volunteer global mappers in, 317, 319, 329
*See also* OpenStreetMap (OSM)
Hush City app, 9, 28, 130, 131–146, 341, 352
benefits and barriers in, 131, 140–143
data crowdsourcing process in, 133–134, 137–138, 142
education/civic awareness aim in, 132
and Hush City Ambassadors, 134, 143
interface design in, 134–135, *137*, 139–140
participatory methodology use in, 132, 140
quiet areas open access map in, 130, 132, 133, 134, *138*, 140
rationale for development of, 131, 132, 142
and user feedback analysis, 140–143
user questionnaire in, 133–134, *135–136*, 139, 140, 141, 352
users of, identification analysis, 133, 141
*See also* environmental noise projects; noise pollution

INDEX **361**

ICT for development (ICT4D), 230, 243
ImproveMyCity (IMC) app, 4, 10, 30,
    186–204, 341, 343
  administration's utilisation of, 195, 197,
      198, 202, 203
  aim of, 187
  analytics dashboards, use in, 197–198,
      *199*, 343
  citizen volunteer role of in, 187, 193
  design recommendations from, 202,
      203–204
  duplication problem in, 200–201
  evidence-based decision making support
      of, 188
  heat maps/geohash grids, use in, 191, *192*
  interactivity and responsiveness in, 186,
      188, 193
  issue reporting and discussion in, 187, 194,
      195, 200–201, 202
  and local knowledge improvement, 187
  as map-centric application, 193–194
  political concerns with, 188
  sentiment analysis in, 203
  service model of, 195
  'smart everything' paradigm in, 187
  user motivation in, 201
  *See also* citizen-government communication;
      Greece, Thessaloniki IMC platform
iNaturalist, 339
indigenous communities, 8, 10, 71, 73–74,
    87–88, 92–94, 99–101, 150, 266, 341
  consent and protocols, 9, 87–88, 90, 92–94,
      231–232, 240, 302
  ecological knowledge contribution of,
      88–89, 100, 102, 114, 247
  forced eviction and abuse of, 150
  participant observation in. *See* participant
      observation
  and project design/methodology, 8–9, 10,
      73, 87–89, 215–216, 345–346
  storytelling significance in, 71
  technological design and adoption in,
      115, 215, 221, 228, 302, 345–346,
      350–351
  Traditional Ecological Knowledge (TEK)
      data, 88, 101, 228
  trust and engagement with, 90, 99–100,
      253–255, 299, 346, 347
  usability testing in, 73–74, 81, 347
  *See also* projects in Brazil, Cambodia,
      Cameroon, Canada, Congo Basin;
      ethnography
interaction design, 62–63
  *See also* human-computer interaction (HCI)
Internet, 5, 16, 51, 141
  and access/speed difficulties, 17, 141, 156,
      161, 163, 221, 304, 343, 349–350
  app offline mode, use of, 349–350
  interactive maps, use of, 16, 50

user-generated content (UGC), 16, 18
interviews, *57*, 59, 63, 64–65, 76, 91–92
  citizen science use of, 64–65, 70
  and contextual inquiry, 65–66
  in ethnographic research, 91–92
  preparation strategies for, 64
  qualitative method, use of, 64
  storytelling/scenarios, use in, 65
  and user app research, 153, 162
  *See also* focus groups; storytelling
iSpot platform, 59, 64
Italian Public System for Digital Identity, 195

KoBo Toolbox, 319, 323, 324, 327, 330–331
  data analysis screen in, *328*
  survey/questionnaire form builder in, 327,
      330–331
  usability of, 331, 334

Malta, Valletta, *169*, 170, 172, *173*
  Cyclist Geo-City app test, 168, *169*, 170,
      171, 178
  National Cycling Strategy, 170
  road network in, 172, *173*, *174*
  *See also* Cyclist Geo-C mobile app
MapHub.net, 324
Mapillary, 323, *324*
maps and mapping, 16, 50, 117, 317–334,
    343, 348–349
  and disaster response, 317–318
  geographic interface provision, complexities
      of, 348–349
  and open-source mapping platforms, 318
  and unmapped world areas, 317, 318
  and user considerations, 348–349
  *See also* Cyclist Geo-C mobile app; Global
      Forest Watch platform/app;
      Humanitarian OpenStreetMap Team
      (HOT); OpenStreetMap (OSM); Peru,
      Cusco community mapping project
Maps.me, 323, *324*
Microsoft, Usability Guidelines, 80
mobile apps, 47, 130, 140, 142–146, *324*
  design recommendations, 9, 32, 143–146
  data sharing and open access in, 143,
      *144*, *145*
  motivational factors, use in, 31, 143,
      *144*, 145
  readability in, 141
  user centred design in, 143, *144*
  *See also* Cyclist Geo-C mobile app; Hush City
      app; Internet
MonkeySurvey tool, 140

The Nature Conservancy, 283–284
noise pollution, 4, 130, 131, 340
  biodiversity/environment effects of, 130,
      131, 132
  health effects of, 130, 131, 132

*See also* environmental noise projects; Hush City app
North Carolina Coastal Atlas, 78–79
  use of personas in, 78–79

Open City Toolkit, 171
Open Data Kit (ODK), 221, 233
open-source software, 39, 49–50, 194, 221, 323, 334, 342, 344
OpenStreetMap (OSM), 4, 12, 18, 27, 29, 34, 40, 43, 47, *122*, *155*, *173–174*, 193, *199*, 318, 319, 323, 325, 326, 334, 340
  citizen volunteer numbers in, 318
  humanitarian action of, 12, 318, 340
  iD Editor in, 319, 323, 324, 325–326, 327, 329
  as open-source platform, 318, 340
  and OpenCycleMap.org, 167
  Sustainable Development Goals, action on, 12
  *See also* Humanitarian OpenStreetMap Team (HOT); Peru, Cusco community mapping project
OSMTracker, 323, 324

participant observation, 9, 66, 90–91, 92, 288, 296, 298
  in environment monitoring, 90–91
  and researcher active participation, 91, 288
participatory design, 55, *57*, 58, 60–62, 97, 102, 232, 346–347
  citizen science use of, 61, 63, 67–68, 81, 346–347, 349
  ethnographic methodology in, *57*, 65–66
  FPIC and community protocols use in, 346
  HCI use, *57*, 58, 60–62, 65
  storytelling, use of, *57*, 58, 69
  and user-led design workshops, 74, 75
  *See also* anthropological methods; Cambodia, Prey Lang communities project; FPIC; prototyping; Sapelli interface
participatory sensing, 26, 28, 29, *30*, *33*
partner/partnership, use of term, 20
passive sensing, 26–27, 29, *30*, 31, *33*
  design implications in, 31
Peacebridges Organisation (PBO), 268
Pepys Estate project, 340
personas tool, 76–79
  common design mistakes in, 76
  in Global Forest Watch app, 151–152
  scenarios, use in, 76
  user needs evaluation in, 76, 78, 151
  *See also* North Carolina Coastal Atlas
Peru, 160, 305, 314, 319–335
  economic well-being in, 319–320
  education levels in, 320
  educational curriculum in, 321–322
  gender-based violence in, 321
  gender conservativism in, 320, *321*
  indigenous population in, 320
  social norms/behaviour data availability in, 320–321
  *See also* Peru, Cusco community mapping project
Peru, Cusco community mapping project, 12, 319–334, 342
  data quality/management challenges in, 329–330, 332, 334
  GAL Group project-based learning in, 322–323, 327, 329, 332
  gender issues analysis in, 323, 327, 330–331, 332–333, 342
  HOT mapping tools, use in, 323, *324*, 325–327
  Internet coverage and speed difficulties in, 329–330, 331
  KoBo Toolbox survey/questionnaire data in, 327, 330–331, 334
  open-source applications use in, 323, 334
  OpenStreetMap editing process in, 325–326, 334
  secondary school students in, 323
  student stimulation and learning in, 333–334
  technical usability in, 331–332
  technologies used in, 323–327
  training sessions in, 327, 329, 350
  and unmapped areas support, 329
  Women Connect Challenge pilot in, 323
  women's empowerment objective in, 323
  *See also* HOT Tasking Manager; Humanitarian OpenStreetMap Team (HOT); KoBo Toolbox; maps and mapping; OpenStreetMap (OSM)
pictograms, 93, 95, 98, 351
  and accuracy challenges, 294
  use on decision trees, 95
  and user co-design, 350–351
  *See also* Ashaninka, Brazil projects; Sapelli interface
PostgreSQL database, 117, 118, 123
prototyping, 69, 74–76, 153
  digital prototyping, 75
  prototypes, definition of, 74–75
  various forms of, 153
Public Laboratory for Open Technology and Science, 28

QGIS, 323–324
questionnaires and surveys, 66–68, 327, 352
  citizen science use of, 67–68
  design considerations in, 67, 352
  and user motivation, 67–68
  *See also* Galaxy Zoo; interviews

rainforest management, 228, 229, 249, 252, 255, 261
  and community literacy levels, 229

rainforest management *(cont.)*
  logging and resource extraction in, 229
  and resource mapping, 229
  and use of Traditional Ecological Knowledge (TEK), 229
  *See also* Ashaninka, Cameroon, Congo Basin projects; forests, protection/management of; Traditional Ecological Knowledge (TEK)
RapidEye, 289
Rechenkraft, 27
REDD+, 306
Representational State Transfer (REST) interface, 197
RinkWatch project, *116*, 117–118, *119*, 122–124, 341, 347
  participant expertise in, *123*
  rink visualisation in, *119*
  user design interaction in, 117–118, 122–124, 126

Sapelli Designer, 95
Sapelli interface, 11, 61, 95, 96, 221, 230, 232–242, 256, 259, 271, 287–288, 302, 303, 306–315
  audio feedback feature in, 239–241, 243–244, 245
  cultural interpretation challenges in, 229, 231, 237, 238, 241, 244–245, 302
  decision tree format in, 95, 96, 233, 234–235, 236, 237, 238, 240–241, 243, 245, 256–257, 288
  non-literate use of, 232, 233, 235–236, 239, 240, 242, 243, 245, 303, 310
  participatory design in, 61, 238, 241, 310
  pictogram-driven UIs in, 233–235, 236–237, 240, 241, 242, 243, 244, 245, 287
  survey function in, 233, 234
  user barriers/feedback in, 233–234, 235–245, 302
  *See also* projects in Brazil, Cambodia, Cameroon, Congo Basin
Scistarter.org, 339
sensors, 4, 5, 26, 38, 39, 48, 51, 342, 344
  and wearable technology, 51
social networking, 43, 48
Software-as-a-Service (SaaS), 194
South Africa, 320
Spain, Castelló, *169*, 170, 172, *173*
  Cyclist Geo-City app test, 168, *169*, 170, 171, 178
  road network in, 172, *173*, *174*
  sustainable transport strategy in, 170
  *See also* Cyclist Geo-C mobile app
species monitoring, 39, 52, 60, 88, 108, 110, *116*, 118, 121, 122, 238, 272, 276, 294
  *See also* birdwatching/monitoring; wildlife conservation

SpICE lab (Spatial Information for Community Engagement), 210, 212, 214, *217*, 218, *220*, 223
  digital mapping tools development of, 214–215, 221
  indigenous community involvement of, 214–215
  *See also* Gather First Nations tool
Stardust@home, 67–68
storytelling, 69–71, 76, 152
  in citizen science, 70–71
  cultural significance of, 71
  HCI use and design, *57*, 58, 65, 69–70, 71
  storyboarding process, use in, 70–71
  use of scenarios in, 70
  *See also* participatory design; personas tool
Strava, 167, 179
StreetBump app, 29
SureLock, 256, 346

technological development and innovation, 5, 8, 15, 16–17, 38–44, 51–53, 115, 342–345
  adoption considerations, 39–40, 41, 42, 43–44, 52–53, 343, 344, 352
  AI and machine learning (ML) advances, 51–52
  anthropological approaches to, 94–96
  and context-specific designs, 88, 115, 343, 344–345, 346–347
  and cultural variations, 101, 315
  and cultures of practice, 115, 127
  digital advances, use of, 30–34, 51–52, 63
  drawbacks and restrictions in, 42, 342–343
  and engagement opportunities, 39, *57*, 63
  ethical issues and rights, 42, 346
  interoperability principles in, 44, *144*, 343
  participant involvement in, 94, *144*, 349–354
  security/privacy considerations, 42, 346, 348
  and 'selfish' technology, 44
  and user experience design, 41, 56–57, *144*, 349
  virtual/augmented reality, use of, 51
  and wearable technology, 51
  *See also* human-computer interaction (HCI); indigenous communities; mobile apps
Traditional Ecological Knowledge (TEK), 11, 88, 100, 101, 228, 346
  data collection tools, use in, 233
  digital form, capture of, 229
  mapping user interface (UI) use of, 229, 348
  policymaking, use in, 229
  and rainforest management, 228, 229
  *See also* Cameroon, Congo Basin projects; Sapelli interface

United Nations
  Convention on Biodiversity, 253
  Declaration of the Rights of Indigenous Peoples, 253

Framework Convention on Climate Change, 267
GWF use by, 152
Millennium Development Goals, 17, 230
Sustainable Development Goals, 12, 149
2030 Agenda for Sustainable Development, 6
*See also* OpenStreetMap (OSM)
United States, *116*, 117–118
　climate-change projects, 117
　environmental monitoring projects, *116*, 118
University College London (UCL), 260, 261, 288, 303, 307
　*See also* ExCiteS
University of Copenhagen, 268, 275
unmanned aerial vehicles (UAVs), 39, 42
user, term of, 20
user experience (UX) design, 62–63
　*See also* human-computer interaction (HCI)

Vizzuality app, 156, 157, 161
volunteered geographic information (VGI), 4, 7, 16, 18, 20, 29–35
　definition of, 18
　intersection with citizen science, 18–19, 20, 29, *30*, *33*, 339
　quality and management of data in, 34
　*See also* geographic citizen science
volunteers/volunteering, 19, 20
　computer resources sharing, 26, 27, *30*, *33*, 49
　explicit and implicit contributions in, 21
　SETI@home project, 49
　use of term, 19, 20
　volunteer thinking, 26, 27, *30*, *33*

Weather Underground network, 26–27
Western, Educated, Industrialised and Developed (WEIRD) participants, 6, 56
WhatsApp, 314, 315, 329
Wikipedia, 29, 318
wildlife conservation, 250, 252–253, 340
　community initiatives in, 252–253, 255
　conservation agency disciplinary action in, 252–253
　and illegal poaching, 250, 252
　and local corruption, 252, 255
　*See also* birdwatching/monitoring; Cameroon, Baka communities project; species monitoring
Wildlife Health Tracker, 115, *116*, 121–122, *123*, 125–126
　participant expertise/design engagement in, *123*, 125–126
WordPress, 194
World Health Organisation, 131
　gender definition of, 320

Zoological Society of London (ZSL), 248, 255
Zooniverse projects, 80

The manufacturer's authorised representative in the EU for product safety is Easy Access System Europe, Mustamäe tee 50, 10621 Tallinn, Estonia, (gpsr.requests@easproject.com).

Printed and bound by CPI Group (UK) Ltd, Croydon, CR0 4YY

19/03/2026

02074820-0001